T0202986

Lecture Notes in Computer Science 12213

More information about this series at http://www.springer.com/series/7409

Heidi Krömker (Ed.)

HCI in Mobility, Transport, and Automotive Systems

Driving Behavior, Urban and Smart Mobility

Second International Conference, MobiTAS 2020
Held as Part of the 22nd HCI International Conference, HCII 2020
Copenhagen, Denmark, July 19–24, 2020
Proceedings, Part II

 Springer

Editor
Heidi Krömker
Institute of Media Technology
Ilmenau University of Technology
Ilmenau, Germany

ISSN 0302-9743   ISSN 1611-3349 (electronic)
Lecture Notes in Computer Science
ISBN 978-3-030-50536-3   ISBN 978-3-030-50537-0 (eBook)
https://doi.org/10.1007/978-3-030-50537-0

LNCS Sublibrary: SL3 – Information Systems and Applications, incl. Internet/Web, and HCI

This Springer imprint is published by the registered company Springer Nature Switzerland AG
The registered company address is: Gewerbestrasse 11, 6330 Cham, Switzerland

Foreword

The 22nd International Conference on Human-Computer Interaction, HCI International 2020 (HCII 2020), was planned to be held at the AC Bella Sky Hotel and Bella Center, Copenhagen, Denmark, during July 19–24, 2020. Due to the COVID-19 coronavirus pandemic and the resolution of the Danish government not to allow events larger than 500 people to be hosted until September 1, 2020, HCII 2020 had to be held virtually. It incorporated the 21 thematic areas and affiliated conferences listed on the following page.

A total of 6,326 individuals from academia, research institutes, industry, and governmental agencies from 97 countries submitted contributions, and 1,439 papers and 238 posters were included in the conference proceedings. These contributions address the latest research and development efforts and highlight the human aspects of design and use of computing systems. The contributions thoroughly cover the entire field of human-computer interaction, addressing major advances in knowledge and effective use of computers in a variety of application areas. The volumes constituting the full set of the conference proceedings are listed in the following pages.

The HCI International (HCII) conference also offers the option of "late-breaking work" which applies both for papers and posters and the corresponding volume(s) of the proceedings will be published just after the conference. Full papers will be included in the "HCII 2020 - Late Breaking Papers" volume of the proceedings to be published in the Springer LNCS series, while poster extended abstracts will be included as short papers in the "HCII 2020 - Late Breaking Posters" volume to be published in the Springer CCIS series.

I would like to thank the program board chairs and the members of the program boards of all thematic areas and affiliated conferences for their contribution to the highest scientific quality and the overall success of the HCI International 2020 conference.

This conference would not have been possible without the continuous and unwavering support and advice of the founder, Conference General Chair Emeritus and Conference Scientific Advisor Prof. Gavriel Salvendy. For his outstanding efforts, I would like to express my appreciation to the communications chair and editor of HCI International News, Dr. Abbas Moallem.

July 2020   Constantine Stephanidis

HCI International 2020 Thematic Areas and Affiliated Conferences

Thematic areas:

- HCI 2020: Human-Computer Interaction
- HIMI 2020: Human Interface and the Management of Information

Affiliated conferences:

- EPCE: 17th International Conference on Engineering Psychology and Cognitive Ergonomics
- UAHCI: 14th International Conference on Universal Access in Human-Computer Interaction
- VAMR: 12th International Conference on Virtual, Augmented and Mixed Reality
- CCD: 12th International Conference on Cross-Cultural Design
- SCSM: 12th International Conference on Social Computing and Social Media
- AC: 14th International Conference on Augmented Cognition
- DHM: 11th International Conference on Digital Human Modeling and Applications in Health, Safety, Ergonomics and Risk Management
- DUXU: 9th International Conference on Design, User Experience and Usability
- DAPI: 8th International Conference on Distributed, Ambient and Pervasive Interactions
- HCIBGO: 7th International Conference on HCI in Business, Government and Organizations
- LCT: 7th International Conference on Learning and Collaboration Technologies
- ITAP: 6th International Conference on Human Aspects of IT for the Aged Population
- HCI-CPT: Second International Conference on HCI for Cybersecurity, Privacy and Trust
- HCI-Games: Second International Conference on HCI in Games
- MobiTAS: Second International Conference on HCI in Mobility, Transport and Automotive Systems
- AIS: Second International Conference on Adaptive Instructional Systems
- C&C: 8th International Conference on Culture and Computing
- MOBILE: First International Conference on Design, Operation and Evaluation of Mobile Communications
- AI-HCI: First International Conference on Artificial Intelligence in HCI

Conference Proceedings Volumes Full List

http://2020.hci.international/proceedings

Second International Conference on HCI in Mobility, Transport and Automotive Systems (MobiTAS 2020)

Program Board Chair: **Heidi Krömker, TU Ilmenau, Germany**

- Angelika C. Bullinger, Germany
- Bertrand David, France
- Marco Diana, Italy
- Christophe Kolski, France
- Lutz Krauss, Germany
- Josef F. Krems, Germany
- Lena Levin, Sweden
- Peter Mörtl, Austria
- Gerrit Meixner, Germany
- Lionel Robert, USA
- Philipp Rode, Germany
- Matthias Roetting, Germany
- Thomas Schlegel, Germany
- Ulrike Stopka, Germany
- Alejandro Tirachini, Chile
- Xiaowei Yuan, China

The full list with the Program Board Chairs and the members of the Program Boards of all thematic areas and affiliated conferences is available online at:

http://www.hci.international/board-members-2020.php

HCI International 2021

The 23rd International Conference on Human-Computer Interaction, HCI International 2021 (HCII 2021), will be held jointly with the affiliated conferences in Washington DC, USA, at the Washington Hilton Hotel, July 24–29, 2021. It will cover a broad spectrum of themes related to Human-Computer Interaction (HCI), including theoretical issues, methods, tools, processes, and case studies in HCI design, as well as novel interaction techniques, interfaces, and applications. The proceedings will be published by Springer. More information will be available on the conference website: http://2021.hci.international/.

General Chair
Prof. Constantine Stephanidis
University of Crete and ICS-FORTH
Heraklion, Crete, Greece
Email: general_chair@hcii2021.org

http://2021.hci.international/

Contents – Part II

Contents – Part I

Studies on Driving Behavior

Hand-Skin Temperature Response to Driving Fatigue: An Exploratory Study

Leandro L. Di Stasi[1,2](✉) (iD), Evelyn Gianfranchi[1] (iD),
and Carolina Diaz-Piedra[1,3] (iD)

[1] Mind, Brain and Behavior Research Center-CIMCYC, University of Granada,
Campus de Cartuja s/n, 18071 Granada, Spain
distasi@ugr.es
[2] Joint Center University of Granada – Spanish Army Training and Doctrine
Command, C/ Gran via de Colon, 48, 18071 Granada, Spain
[3] College of Nursing and Health Innovation, Arizona State University,
550 N. 3rd Street, Phoenix, AZ 85004, USA

Abstract. Driving fatigue detection is core to road safety. Infrared thermography has gained increasing attention due to the chance to non-invasively monitor real-time fatigue-related variations. Here, we present an exploratory study regarding the use of the driver's hand-skin temperature as a fatigue index. Eleven participants drove along a monotonous circuit on a simulator for 2 h, while their hand-skin temperatures were recorded. The results showed a quadratic trajectory for the hand-skin temperature, with an increase in the first 20 min, followed by a decrease of about 2 °C, over the rest of the time. The initial increase might be due to the gradual lowering in griping force on the steering wheel, because of growing fatigue. The final decrease might reflect the driver's attempt to keep an acceptable level of performance while fatigued, increasing his/her griping force. The present study indicates that infrared thermography might represent a complementary method for unbiasedly track the driver's fatigue levels, without interfering with his/her comfort or task performance, and without compromising safety.

Keywords: Time on task · Body-skin temperature · Infrared thermography

1 Introduction

A key concept for road safety is mental fatigue consequent to extensive driving. It is defined as the "state of reduced mental alertness that impairs performance" [1] and it is usually occasioned by prolonged periods of a cognitive, high-demanding, and sustained activity requiring mental efficiency [2]. Driving fatigue is the cause of up to 24% of road crashes in European countries, with a mean of 15% *per* country [3, 4].

Nowadays, the theoretical models underpinning mental fatigue (for a broad discussion see Ref. [5]), as well as its practical implications (*e.g.*, impairs in cognitive and physical performance, as well as in emotion regulation [6–8]) are well known. However, the measurement of mental fatigue in complex and dynamic tasks, such as driving, is still challenging due to the difficulties in establishing an objective, reliable,

© Springer Nature Switzerland AG 2020
H. Krömker (Ed.): HCII 2020, LNCS 12213, pp. 3–14, 2020.
https://doi.org/10.1007/978-3-030-50537-0_1

non-intrusive and real-time tracking method of mental fatigue [9]. Indeed, fatigue, especially when assessed in a clinical context, has frequently been assessed by means of self-report measures (see Ref. [10] for a review). However, because of intrinsic limits of subjective measures, which cannot be employed to track fatigue variations in real-time, the use of neurobehavioral objective indices for fatigue measurement is gaining increasing attention. Although neurobehavioral measures have proved to be useful for fatigue detection and measurement, some limitations need to be considered when it comes to assess their suitability to real field settings, such as the driving context. A summary of limitations present in these measures is reported below.

Neurobehavioral Measures for Driving Fatigue

Among the most-employed neurobehavioral measures for fatigue tracking in driving contexts, eye movements play surely a key role [11–15]. Eye movements consists of frequent, quick movements, called saccades, interspersed with periods of steady gaze, called fixations. For instance, saccade-based metrics, particularly the saccadic peak velocity (i.e., the highest velocity reached within a saccade [11, 12, 15]) emerged as crucial means for tracking mental fatigue during driving [11]. Thus, eye-movement recording is for sure a widely used and reliable technique, able to track real-time fatigue variations. However, in real field settings some problems may emerge due to troubles that most eye-tracking systems have in data recording when people wear eyeglasses [16].

Another central psychophysiological index, frequently employed to assess driving fatigue is the EEG power spectra, which is the relative prevalence of the sinusoidal waves composing the EEG signal. In fatigue conditions, it is usually found a decrease of frontal beta and gamma power and an increase of frontal theta, alpha and delta [17–20]. For instance, during a 2-h driving along a simulated scenario, an inverted U-shaped quadratic trend in frontal delta power has been reported [18, 19]. Event-related potentials (i.e., changes in the brain activity occurring in response to a stimulus; ERPs) are sometimes employed to track driving fatigue [21]. Among them, ERPs linked to early and mid-stages of stimulus processing (such as the P3 component) have been investigated for their relation to automatic attentional processes that are crucial to driving. All the mentioned EEG indices have proved to accurately monitor mental fatigue. However, EEG measurement, even when conducted by means of wearable systems, is quite uncomfortable for the participant.

Concerning peripheral indices, cardiac activity (especially heart rate) is frequently employed to measure fatigue [9], also in driving contexts [9, 22, 23]. A recent review [9] identified at least fourteen studies reporting a linear increase in heart rate with the increase in the time spent driving, both in real and simulated contexts. Thus, heart rate has proved not only to be an easily measurable index, but also directly correlated with fatigue [9, 22, 23]. Another reliable and easy measurable index, electrodermal activity, has proved to be related to fatigue while driving [24–26]. Both skin conductance level and frequency of skin conductance responses show a general increasing trend with increasing stress and fatigue [27]. These indices have been recently included as predictors in algorithms aimed at classifying different drivers' state, along with other fatigue indices such as heart rate and respiration rate [28]. Some recent evidence pointed out a relation between superficial electromyography (EMG) [29] (i.e., the measurement of the muscular electrical activity) and fatigue, with an increase in the signal frequency with increasing fatigue. For instance, EMG signal recorded from Brachioradialis muscle located at the forearm during a 2-h simulated driving showed good degree of sensitivity to increasing levels of fatigue [29].

The measurement of heart rate, electromyography and electrodermal activity, although more comfortable than EEG, is generally motion sensitive (e.g., Ref. [30]): complex tasks, such as driving a vehicle, often require the person to move, thus movement artifacts could affect the reliability of the data. Finally, respiratory activity, in terms of respiration rate, although quite responsive to changes in driving fatigue [31, 32], is an index that should be always coupled with other measures (e.g., EEG), since breath variations can occur for a variety of reasons [see Ref. 33].

Another measure for assessing driving fatigue is actigraphy [9, 34], which involves the measurement of movements frequency over time, usually employing a wearable device, typically a wristband. The amount of movements within a given time is used to derive the typical metrics of this method, that, in comparison to a threshold, can discriminate between "sleep" and "awake" states. Thus, actigraphy is a well-known and widely employed method to produce reliable estimates of sleep/wake timing and duration [35] that are crucial inputs for fatigue-prediction models. However, the devices usually employed with this aim are scarcely suitable for monitoring alertness directly [35]; as a matter of fact, wrist movement activity has only been proved to consistently vary between wake and sleep, being unable to systematically track different levels of alertness during awake states [35].

Facial Infrared Thermography

Considering the above-mentioned shortcomings, an emerging and challenging research trend (e.g., HADRIAN project [36]) focuses on the identification of a non-invasive, yet reliable method, suitable to track real-time driver's state variations in real field settings. The ideal future scenario [36], as already pointed out by various researchers [32, 36], should involve several neurobehavioral measures, chosen among the most comfortable and precise, so as to obtain the most accurate and comprehensive estimate of the driver's state. An example of successful integration of several fatigue-sensitive measures is a recent technology developed by Panasonic (Japan) [37] that uses contactless sensing devices of neurobehavioral indices (body temperature, blink features, and facial expressions) to track the driver's states and in-vehicle environment (e.g., temperature, light conditions, air velocity) in real-time. This technology integrates all this information to assess drivers' drowsiness level, to adjust the inside-vehicle conditions and, in case of high levels of fatigue, to suggest resting. This application is just an example of how body temperature can be useful to detect driving fatigue in combination with other indices. Indeed, facial infrared thermography, might represent a good candidate to reach this aim [38]. In a recent study from our laboratory [20], we used infrared thermography to remotely measure nasal-skin temperature during a 2-h driving task. As the driving session progressed, arousal levels decreased. It was confirmed by the increased frontal delta EEG activity (inverted U-shaped quadratic trend), as well as subjective ratings of alertness and fatigue. The behavior of the nasal-skin temperature was coherent: it increased over the first 45 min of session, tending to gradually decrease in the last part of the task (for more details see Ref. [20]). Nevertheless, the presence of make-up and sweat, which cannot be controlled outside the experimental environment, can alter the reliability of this technique. A body region that is commonly uninfluenced by make-up or sweat is the dorsal part of the hands. Thus, hand-skin temperature measurement might represent a valid alternative to overcome these limitations. Studies monitoring hand-skin temperature to study driver fatigue, however, are just anecdotic (*e.g.*, Ref. [39]).

Here, we investigated the effects of a monotonous 2-h simulated driving session, a common method to induce fatigue at the wheel (*e.g.*, Ref. [11]), while we continuously monitored the drivers' hand-skin temperature. As the experimental session progressed, we expected the driver to reduce his/her grip force on the steering wheel [40]; and consequently we expected an increase in the driver's hand-skin temperature, due to changes in hand peripheral circulation and to the gradual mechanical decompression of the blood vessels (see, among others, Ref. [41]).

2   Materials and Methods

2.1   Participants

Eleven active drivers (mean age ± standard deviation = 25.36 ± 1.70 years; 6 women) were voluntary enrolled to take part to the present study (University of Granada's Institutional Review Board approval #484/CEIH/2018). All of them were non-smokers, held a valid driving license, had normal or corrected-to-normal vision, and were naïve to the hypotheses of the experiment.

2.2   Experimental Design

The experiment consisted of a 2-h simulated driving session along a monotonous course [11, 18] in a virtual environment. We chose this temporal window to be close to the maximum driving time that professional drivers are allowed before a mandatory break [42]. The study followed a within-subjects design with the driving time as the independent variable. As dependent variables, we considered right-hand-skin temperature over the 2-h driving session.

2.3   Thermographic Recordings and Analyses

We used the ThermoVision A320G Researcher Infrared Camera (FLIR Systems, USA). The camera (resolution of 320 × 240 pixels) was placed on a tripod 110 cm above the floor and ~140 cm from the driver. The camera has automatic focus that was always employed to focus the image recording. We stored the recorded signal using the program Researcher TermaCAMP 2.9 (FLIR Systems, USA).

The main relevant point (pixel) of interest (POI) was the skin surface of the dorsal proximal phalangeal joint of the third finger of the right hand (approximately, 2 cm below the knuckle). To control for possible room temperature changes, we selected another POI on the wall behind the driver seat. The coordinates for this POI were kept constant within the recording session. Two independent researchers manually performed the collection of the temperature for each POI and participant. The ICC estimate for the driver's right-hand-skin temperature was 0.98 [95% C.I. 0.97–0.98], which indicates excellent reliability. For the temperature analysis (see Sect. 3.1 Right-hand-skin temperature), we used the mean value of two researchers.

2.4 Procedure

The experimental protocol was designed following the recommendations of the consensus statement on thermographic imaging studies by Moreira and colleagues [43]. The experiment took place in a simulation laboratory (two adjacent windowless rooms of about 8.5 m² each: the test/preparation unit and the simulator unit) free from thermal noise sources, located at the Mind, Brain, and Behavior Research Center (Granada, Spain) (see Fig. 1). Once the participant signed the informed consent form, we recorded sociodemographic and health data in the test/preparation unit. All participants wore the same upper body clothing (a clean cotton t-shirt). Approximately 15 min after his/her arrival (i.e., adaption period to room temperature), the participant positioned him/herself on the car seat (in the simulator unit), he/she filled several questionnaires (see Ref. [20] for more details) and the thermographic camera was turned on to allow for its sensor to be stabilized. Participants drove a middle-sized car for two hours without breaks, around the same road without any other traffic present.

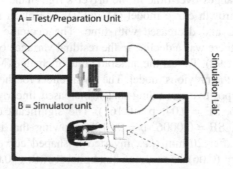

Fig. 1. Experimental setting. Two adjacent rooms (A and B), separated by a small corridor, constitutes the driving simulation laboratory

We developed a virtual two-lane, rounded rectangle monotonous grassy meadow road scenario using the OpenDS 2.5 software (OpenDS, Saarbrücken, Germany). A speed limit of 60 km/h was set. The driving simulator recorded the car speed at 20 Hz. The simulator is described in Ref. [20]. After a five-minute driving familiarization session, we calibrated the thermographic camera, and the driving simulation started. We instructed participants to hold the steering wheel at two marked points (10 o'clock and 2 o'clock positions). Furthermore, in order to avoid diurnal fluctuations that affect arousal levels [44], the experimental protocol always started at 8:30 am and the driving session, at 9:00 am.

2.5 Statistical Analyses

We categorized the full driving period into sixty 2-min bins, discarding the data from the first and the last bin to remove possible transient effects caused by the starting and ending of the driving session. To analyze changes over time (2-h driving session, 58 time points) in the right-hand-skin temperature, we tested different individual growth curve models (linear, quadratic, and cubic) using a maximum likelihood estimation to study intraindividual differences in the patterns of change of the dependent variables [45] (see Ref. [20] for more details). To select the best model, we used the Akaike's Information Criterion (AIC, a lower AIC value indicates a better fit). The results of a given growth model were shown, only if its AIC was lower than AIC for the precedent model. For each model, we present parameter estimates and standard errors (SE). Significance levels were set at $\alpha < 0.05$. This is a complementary report to reference [20], thus data concerning the effects of driving time on nasal skin temperature, EEG power activity, and speeding behavior have been already reported (see Ref. [20]).

3 Results

3.1 Right-Hand-Skin Temperature

To study trajectory changes over time in the driver's right-hand-skin temperature, we first tested the linear growth curve model (Model 2). The mean right-hand-skin temperature was 30.57 °C and decreased with time. The decrease was not significant (p = 0.142), although there was a decline in the residual variance of 1.18 from Model 1 to Model 2 (13.10 to 11.92). To test the quadratic rate of change (Model 3), we added a quadratic parameter in the previous model. The linear effect for the driver's right-hand-skin temperature was positive, revealing that it increased linearly over time, but not significantly ($\beta = 0.06$, SE = 0.04, p = 0.161). The significant quadratic effect was negative ($\beta = -0.001$, SE = 0.0006, p = 0.012), showing that the increasing effect gradually diminished after 20 min (*i.e.*, inverted U-shaped curve). Compared to the linear change trajectory (0.06), the rate of quadratic growth (−0.001) was small. We also tested any cubic changes in individual trajectories over time (Model 4), but they were non-significant. Thus, the hand-skin temperature showed a quadratic trajectory, $y \cdot ij = 29.59 + (0.059 \times time) + (-0.0015 \times time \cdot 2)$ (see Fig. 2). The patterns of change of the right-hand-skin temperature did not vary when we considered the room temperature. The ratio between the right-hand-skin and the room temperatures presented a similar significant cubic trajectory: $y_{ij} = 1.42 + (0.014 \times time) + (-0.0004 \ time^2) + (0.000003 \times time^3)$. From this, we can affirm that small variations in the environmental temperature (less than 1 °C) did not affect the sensitivity of the hand-skin temperature for detecting arousal changes.

Table 1 presents the results of the fitting the unconditional means model and the growth curve models.

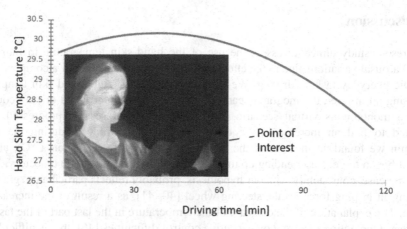

Fig. 2. Effects of driving time (time on task) on the drivers' hand-skin temperature. The blue curve represents the quadratic fit trajectory of the drivers' hand-skin temperature over the 2-h driving session. Insert: A driver during the experimental session with her two hands on the steering wheel. The point of interest is marked with a grey arrow.

Table 1. Results of fitting growth curve models for the trajectory of the driver's right-hand-skin temperature over a 2-h simulated driving session (n = 11). It presents parameter estimates and, in brackets, standard errors.

	Model 1	Model 2	Model 3
Fixed effects			
Intercept	29.59 (0.73)**	30.56 (1.23)**	29.59 (1.29)**
Rate of change			
Linear	–	−0.03 (0.02)	0.06 (0.04)
Quadratic	–	–	−0.001 (<0.001)*
Cubic	–	–	–
Variance components			
Level 1			
Within-subjects	13.10 (0.82)**	0.88 (0.05)**	0.59 (0.03)**
Level 2			
In initial status	4.63 (2.29)*	7.59 (3.42)*	7.60 (3.42)*
In rate of change	–	0.002 (0.001)*	0.002 (0.001)*
Covariance	–	−0.09 (0.05)	−0.09 (0.05)
AIC	2830.76	2798.84	2794.57

*Note. Model 1 = Unconditional means model; Model 2 = Linear growth model; Model 3 = Quadratic growth model. AIC = Akaike's Information Criterion. The intraclass correlation coefficient for Model 1 (magnitude of the residuals linkage) was 0.26. *p < 0.05 **p < 0.001.*

4   Discussion

The present study aimed to assess the use of the hand-skin temperature to monitor overall arousal variations due to the effect of fatigue, when engaging in a complex and dynamic everyday task as driving. We tracked participants' right-hand-skin temperature using an infrared thermography camera during a 2-h uninterrupted driving course along a monotonous virtual scenario. As the experimental session progressed, we expected to find an increase in the drivers' temperature. Indeed, during the first ≈20 min we found an increase of the hand-skin temperature (less than 1 °C); afterward, it began to decrease, ending up to 27.8 °C, following a quadratic trajectory. The initial increase, consistently with our hypothesis, probably reflects participants' gradual lowering in griping force on the steering wheel [40, 41], as a result of the increasing fatigue. The explanation of the decrease of the temperature in the last part of the task is less clear. One option may rely on executive control difficulties [46], that is difficulties in the ability to regulate perceptual and motor processes for goal-directed behavior. Fatigued drivers might have difficulties in regulating their perceptual and motor processes while driving [46, 47]; as a result, they would need to put more effort in regulating the vehicle trajectory. This, in turn, might lead them to increase their grip force on the steering wheel. This explanation is coherent with the results reported by independent investigations in non-driving scenarios, which showed a decrease in hand-skin temperature when the experimental subjects exerted constant pression (see for instance Ref. [41]). Future studies should control for the griping force on the steering wheel while monitoring hand-skin temperature variations.

Another explanation may rely on the role of stress [48]. Tasks requiring sustained attention (*e.g.*, vigilance tasks or even prolonged driving tasks) can be simultaneously (and paradoxically) de-arousing and stressful, due to the multidimensionality of stress responses [48]. Actually, the driving task employed in the present study, although monotonous, requested a constant level of attention that may have caused an increase in participants' stress. A well-known effect of stress is the decrease in the peripheral temperature [49]: stress seems to induce peripheral vasoconstriction, causing a rapid, short-term drop in skin temperature [49]. Thus, the sustained attention required by the monotonous driving task here employed may have cause stress responses, leading to vasoconstriction and to the decrease in the hand-skin temperature in the second part of the session. Further studies are needed to disentangle the effects caused by stress from those caused by fatigue.

Nevertheless, the present results are in favor of the validity of thermography to track hand-skin temperature variations as an index of drivers' fatigue, without interfering with task performance or compromising drivers' comfort and safety. A limitation of this method relies on the need to cope with the expectable variations in environmental temperature in a real vehicle. This might be overcome by using a ratio between skin and environmental temperatures (see Ref. [20]). Thermal variations could also be used to reduce driver distractions. For instance, Jaguar-Land Rover [50, 51] recently proposed a sensory steering wheel that uses haptic (mainly thermic) feedback technologies.

By slightly increasing or decreasing the temperature in various zones of the steering wheel, the technology would signal to the driver that he/she should change roadway or keep the eyes on the road because of a risky situation, such as a difficult crossroad.

In conclusion, the present results may represent a first step toward the development of a complex system with a huge impact on driver's safety, potentially able to detect (and even notify) safety-critical increases in fatigue, such as situations where the drivers should take a break, monitoring their drowsiness in a non-invasive, yet accurate and safe way.

Acknowledgments. This study was funded by the Ramon y Cajal fellowship program from the Spanish State Research Agency (AEI) (RYC-2015-17483 to LLDS). The sponsor had no role in the design or conduct of this research.

CDP and LLDS are supported by a Santander Bank - CEMIX UGRMADOC grant (Project PINS 2018-15 to CDP & LLDS). We thank Eduardo Bailon and Luis Henrique Alves de Melo for their help during data collection and pre-processing. We thank Dr. Francisco Tornay and Dr. Emilio Gomez-Milan (Department of Experimental Psychology, University of Granada, Spain) for their comments. Finally, we thank Federico Marafin and Fulvio Russo for their assistance in language edition.

References

1. Grandjean, E.: Fitting the Task to the Man: An Ergonomic Approach, 3rd edn. Taylor & Francis, London (1980)
2. Wang, C., Trongnetrpunya, A., Samuel, I.B.H., Ding, M., Kluger, B.M.: Compensatory neural activity in response to cognitive fatigue. J. Neurosci. **36**(14), 3919–3924 (2016)
3. ERSO – European Road Safety Observatory: Fatigue (2018). https://ec.europa.eu/transport/road_safety/sites/roadsafety/files/pdf/ersosynthesis2018-fatigue.pdf. Accessed 28 May 2020
4. National Transportation Safety Board. Report on the methodology of the NTSB most wanted list (2019). https://www.ntsb.gov/about/reports/Documents/NTSB-Most-Wanted-List-methodology-report-February-2019.pdf. Accessed 28 May 2020
5. Hockey, B.: The Psychology of Fatigue: Work, Effort and Control. Cambridge University Press, New York (2013)
6. Van Cutsem, J., Marcora, S., De Pauw, K., Bailey, S., Meeusen, R., Roelands, B.: The effects of mental fatigue on physical performance: A systematic review. Sports Med. **47**(8), 1569–1588 (2017)
7. Grillon, C., Quispe-Escudero, D., Mathur, A., Ernst, M.: Mental fatigue impairs emotion regulation. Emotion **15**(3), 383–389 (2015)
8. Strober, L.B., DeLuca, J.: Fatigue: Its influence on cognition and assessment. In: Arnett, P. A. (ed.) National Academy of Neuropsychology Series on Evidence-based Practices. Secondary Influences on Neuropsychological Test Performance: Research Findings and Practical Applications, pp. 117–141. Oxford University Press (2013)
9. Bier, L., Wolf, P., Hilsenbek, H., Abendroth, B.: How to measure monotony-related fatigue? A systematic review of fatigue measurement methods for use on driving tests. Theor. Issues Ergon. Sci. **21**(1), 22–55 (2018)
10. Whitehead, L.: The measurement of fatigue in chronic illness: A systematic review of unidimensional and multidimensional fatigue measures. J. Pain Symptom Manag. **37**(1), 107–128 (2009)

11. Di Stasi, L.L., Renner, R., Catena, A., Cañas, J.J., Velichkivsky, B.M., Pannasch, S.: Towards a driver fatigue test based on the saccadic main sequence: A partial validation by subjective report data. Transp. Res. Part C: Emerg. Technol. **21**(1), 122–133 (2012)
12. Di Stasi, L.L., Catena, A., Cañas, J.J., Macknik, S., Martinez-Conde, S.: Saccadic velocity as an arousal index in naturalistic tasks. Neurosci. Biobehav. Rev. **37**(5), 968–975 (2013)
13. Di Stasi, L.L., McCamy, M.B., Catena, A., Cañas, J.J., Macknik, S.L., Martinez-Conde, S.: Microsaccade and drift dynamics reflect mental fatigue. Eur. J. Neurosci. **38**, 2389–2398 (2013)
14. Schleicher, R., Galley, N., Briest, S., Galley, L.: Blinks and saccades as indicators of fatigue in sleepiness warnings: Looking tired? Ergonomics **51**, 982–1010 (2008)
15. Diaz-Piedra, C., Rieiro, H., Suarez, J., Rios-Tejada, F., Catena, A., Di Stasi, L.L.: Fatigue in the military: Towards a fatigue detection test based on the saccadic velocity. Physiol. Meas. **37**(9), N62–N75 (2016)
16. Fuhl, W., Tonsen, M., Bulling, A., Kasneci, E.: Pupil detection for head-mounted eye tracking in the wild: An evaluation of the state of the art. Mach. Vis. Appl. **27**(8), 1275–1288 (2016)
17. Wascher, E., et al.: Frontal theta activity reflects distinct aspects of mental fatigue. Biol. Psychol. **96**, 57–65 (2014)
18. Morales, J.M., et al.: Monitoring driver fatigue using a single-channel electroencephalographic device: A validation study by gaze-based, driving performance, and subjective data. Accid. Anal. Prev. **109**, 62–69 (2017)
19. Morales, J.M., Rabelo-Ruiz, J.F., Diaz-Piedra, C., Di Stasi, L.L.: Detecting mental workload in surgical teams using a wearable single-channel electroencephalographic device. J. Surg. Educ. **76**, 1107–1115 (2019)
20. Diaz-Piedra, C., Gomez-Milan, E., Di Stasi, L.L.: Nasal skin temperature reveals changes in arousal levels due to time on task: An experimental thermal infrared imaging study. Appl. Ergon. **81**, 102870 (2019)
21. Guoping, S., Kan, Z.: An ERP study of effects of driving fatigue on auditory attention. Psychol. Sci. (China) **32**(3), 517–520 (2009)
22. Vicente, J., Laguna, P., Bartra, A., Bailón, R.: Drowsiness detection using heart rate variability. Med. Biol. Eng. Comput. **54**(6), 927–937 (2016)
23. Wang, F., Chen, H., Zhu, C.H., Nan, S.R., Li, Y.: Estimating driving fatigue at a plateau area with frequent and rapid altitude change. Sensors **19**(22), 4982–4998 (2019)
24. Matthews, G., Wohleber, R., Lin, J., Funke, G., Neubauer, C.: Monitoring task fatigue in contemporary and future vehicles: A review. In: Cassenti, D.N. (ed.) AHFE 2018. AISC, vol. 780, pp. 101–112. Springer, Cham (2019). https://doi.org/10.1007/978-3-319-94223-0_10
25. Han, S.Y., Kwak, N.S., Oh, T., Lee, S.W.: Classification of pilots' mental states using a multimodal deep learning network. Biocybern. Biomed. Eng. **40**, 324–336 (2019)
26. Sharma, M.K., Bundele, M.: Cognitive fatigue detection in vehicular drivers using k-means algorithm. In: Khanna, A., Gupta, D., Bhattacharyya, S., Snasel, V., Platos, J., Hassanien, A. E. (eds.) International Conference on Innovative Computing and Communications. AISC, vol. 1059, pp. 517–532. Springer, Singapore (2020). https://doi.org/10.1007/978-981-15-0324-5_44
27. Boucsein, W., et al.: Society for psychophysiological research ad hoc committee on electrodermal measures. Publication recommendations electrodermal measurements. Psychophysiology **49**(8), 1017–1034 (2012)
28. Kumar, V.V., Grimm, D.K., Kiefer, R.J.: U.S. Patent No. 9,956,963. U.S. Patent and Trademark Office, Washington, DC (2018)

29. Mohd Azli, M.A.S., Mustafa, M., Abdubrani, R., Abdul Hadi, A., Syed Ahmad, S.N.A., Zahari, Z.L.: Electromyograph (EMG) signal analysis to predict muscle fatigue during driving. In: Md Zain, Z., et al. (eds.) Proceedings of the 10th National Technical Seminar on Underwater System Technology 2018. LNEE, vol. 538, pp. 405–420. Springer, Singapore (2019). https://doi.org/10.1007/978-981-13-3708-6_35
30. Menghini, L., Gianfranchi, E., Cellini, N., Patron, E., Tagliabue, M., Sarlo, M.: Stressing the accuracy: Wrist-worn wearable sensor validation over different conditions. Psychophysiology 56(11), e13441 (2019)
31. Lee, B.G., Lee, B.L., Chung, W.Y.: Mobile healthcare for automatic driving sleep-onset detection using wavelet-based EEG and respiration signals. Sensors 14(10), 17915–17936 (2014)
32. Pernice, R., Nollo, G., Zanetti, M., De Cecco, M., Busacca, A., Faes, L.: Minimally invasive assessment of mental stress based on wearable wireless physiological sensors and multivariate biosignal processing. In: IEEE EUROCON 2019, 18th International Conference on Smart Technologies, pp. 1–5. IEEE, July 2019
33. Tipton, M.J., Harper, A., Paton, J.F., Costello, J.T.: The human ventilatory response to stress: rate or depth? J. Physiol. 595(17), 5729–5752 (2017)
34. Fonseca, A., Kerick, S., King, J.T., Lin, C.T., Jung, T.P.: Brain network changes in fatigued drivers: a longitudinal study in a real-world environment based on the effective connectivity analysis and actigraphy data. Front. Hum. Neurosci. 12, 418 (2018)
35. Balkin, T.J., Horrey, W.J., Graeber, R.C., Czeisler, C.A., Dinges, D.F.: The challenges and opportunities of technological approaches to fatigue management. Accid. Anal. Prev. 43(2), 565–572 (2011)
36. HADRIAN - Holistic Approach for Driver Role Integration and Automation Allocation for European Mobility Needs - project (2019). Grant agreement ID: 875597. https://hadrianproject.eu/. Accessed 28 May 2020
37. Panasonic Corporation: Panasonic develops drowsiness-control technology by detecting and predicting driver's level of drowsiness. Press release (2017). https://news.panasonic.com/global/press/data/2017/07/en170727-3/en170727-3.html#010. Accessed 28 May 2020
38. Tashakori, M., Nahvi, A., Shahiidian, A., Ebrahimian-Hadikiashari, S., Bakhoda, H.: Estimation of driver drowsiness using blood perfusion analysis of facial thermal images in a driving simulator. J. Sleep Sci. 3(3–4), 45–52 (2018)
39. Watson, T., Krause, J., Le, J., Rao, M.: Vehicle integrated non-intrusive monitoring of driver biological signals. In: SAE International (eds.) Occupant Protection: Safety Test Methodology and Structural Crashworthiness and Occupant Safety (Paper No. 2011-01-1095). SAE International, Warrendale (2011)
40. Balasubramanian, V., Bhardwaj, R.: Grip and electrophysiological sensor-based estimation of muscle fatigue while holding steering wheel in different positions. IEEE Sens. J. 19, 1951–1960 (2019)
41. Silva, D.C., et al.: Evaluation of two PET bottles caps: An exploratory study. Procedia Manuf. 3, 6245–6252 (2015)
42. VOSA – Vehicle and Operator Service Agency. Rules on drivers' hours and tachographs: passenger-carrying vehicles in the UK and Europe. http://www.syta.org/downloads/UK%20and%20Europe%20Rules%20of%20the%20Road.pdf. Accessed 28 May 2020
43. Moreira, D.G., et al.: Thermographic imaging in sports and exercise medicine: A Delphi study and consensus statement on the measurement of human skin temperature. J. Therm. Biol. 69, 155–162 (2017)
44. Del Río-Bermudez, C., Díaz-Piedra, C., Catena, A., Buela-Casal, G., Di Stasi, L.L.: Chronotype-dependent circadian rhythmicity of driving safety. Chronobiol. Int. 31(4), 532–541 (2014)

45. Singer, J.D., Willett, J.B.: Applied Longitudinal Data Analysis. Oxford Press, New York (2003)
46. Van der Linden, D., Frese, M., Meijman, T.F.: Mental fatigue and the control of cognitive processes: effects on perseveration and planning. Acta Physiol. (Oxf) 113(1), 45–65 (2003)
47. Hockey, G.R.J.: Compensatory control in the regulation of human performance under stress and high workload: A cognitive-energetical framework. Biol. Psychol. 45, 73–93 (1997)
48. Warm, J.S., Matthews, G., Finomore Jr., V.S.: Vigilance, workload, and stress. In: Performance Under Stress, pp. 131–158. CRC Press (2008)
49. Vinkers, C.H., et al.: The effect of stress on core and peripheral body temperature in humans. Stress 16(5), 520–530 (2013)
50. Di Campli San Vito, P., Brewster, S., Pollick, F., White, S., Skrypchuk, L., Mouzakitis, A.: Investigation of thermal stimuli for lane changes. In: Proceedings of the 10th International Conference on Automotive User Interfaces and Interactive Vehicular Applications, pp. 43–52, September 2018
51. Jaguar-Land Rover Company. Sensory steering wheel keeps your eyes on the road. https://www.jaguarlandrover.com/news/2019/05/sensory-steering-wheel-keeps-your-eyes-road. Accessed 28 May 2020

Toward Driver State Models that Explain Interindividual Variability of Distraction for Adaptive Automation

Margit Höfler(iD) and Peter Moertl$^{(\boxtimes)}$ (iD)

Virtual Vehicle Research GmbH, Graz, Austria
Peter.Moertl@v2c2.at

Abstract. Although there exist many models of distracted driving, identifying a distracted driver is still challenging as distraction might appear differently for different drivers but also within an individual driver in different situations. Here we present a driver state model that focusses on safety-relevant driver-distraction by conceptualizing driving control as influenced by both environmental factors and individual preferences. Also, the model differentiates compensatory control from exploratory control movements to better diagnose driving distraction. We then test several predictions that are derived from this model in a driving-simulator study. In this study participants drove the same road with or without a secondary task while their eye movements and driving performance was recorded. Our results are consistent with previous findings that overall steering control actions increase in the distraction condition but also that exploratory steering movements are apparently more sensitive indicators for distraction than compensatory control actions.

Keywords: Driver distraction · Driver state models · Control-action rates

1 Introduction

Automated driving technologies are more and more introduced in vehicles in order to avoid critical driving situations and increase safety. However, as these technologies become increasingly complex, this often challenges the interactions with the human driver because drivers are required to monitor both the environment and automation, and to efficiently take over driving responsibilities if necessary. That is, although automated driving may relieve from most of the usual driving tasks, new tasks such as monitoring and (often immediate) responding to requests of the system arise. One way to address this issue of driver interaction are adaptive technologies that consider the state of the driver in order to offer appropriate functionalities or to initiate interventions. Technologies that sense the driver state such as eye tracking, tracking of facial expression, body movements, or other physiological and driving performance measurements (such as steering wheel parameters) are becoming widespread (e.g., [1]). The outcome of such measurements requires driver state models against which the observed driver behavior can be compared to. Nevertheless, inferring the correct individual driver state, and often most importantly, using observed data to correctly identify a

H. Krömker (Ed.): HCII 2020, LNCS 12213, pp. 15–28, 2020.
https://doi.org/10.1007/978-3-030-50537-0_2

distracted driver, is still challenging because human behavior exhibits large amounts of variability both in terms of within one individual over time and between individuals. For example, distracted driving can look very different between different drivers and the same behavior can sometimes indicate distraction and sometimes not. In this paper we report our efforts to address this challenge by investigating methods to segment the variability of driving performance of distracted driving and present a model on safety-relevant driver distraction.

1.1   Driver Distraction

Driver distraction is one of the most common factors reported when it comes to crashes or safety critical events: Previous research has shown that 5 to 25% of accidents can be attributed to the involvement of driver distraction [2]. Typically, visual distraction (when the gaze is off the road) and cognitive distraction (when the mind is off the road) are distinguished [2, 3]. It has been shown that, when drivers are engaged in complex visual and/or manual secondary tasks, they have a near-crash/crash risk three-times higher than attentive drivers [4].

There are several definitions of driver's distraction and its relation to inattention. For instance, in their taxonomy, Engström and colleagues [5] proposed different forms of inattention (namely insufficient attention and misdirected attention, in which only the latter is leading to distraction). They defined driver distraction as situations in which the driver's resources are allocated "*to a non-safety critical activity while the resources allocated to activities critical for safe driving do not match the demands of these activities*" (p. 35). That is, distraction occurs in situations in which attention is allocated away from activities that are important for safe driving to unimportant activities. Similarly, in the taxonomy of driver inattention of Regan et al. [6], driver distraction is proposed to be "*the diversion of attention away from activities critical for safe driving toward a competing activity, which may result in insufficient or no attention to activities critical for safe driving*" (p. 1776) and only one of several factors that are caused by a driver's inattention (see also [7]). Kircher and Ahlstrom reviewed several definitions of driving distraction and identified two different viewpoints as whether the results of distraction are included in the definition of distraction or not [8]. In this paper we take the perspective that the results of unsafe driving should be considered in the definition of distraction. This differentiates "safety relevant distraction" from "general distraction" and make the term more operationally relevant.

1.2   Modeling Safety Relevant Driver Distraction

In this paper we formulate a simple cognitive model for the mechanisms that underlie distracted driving as defined above. According to this model, safety relevant distraction is the result of a mismatch between a driver's apparent control processes (such as gaze orientation and steering) with the adaptive control demands of a situation. The model is explained in the following: The minimal driving information extraction rate for a situation can be theoretically determined and depends on the situation, the number of information elements, the driver, the complexity of the situation, and the vehicle speed (see Fig. 1). Thereby, the minimal driving information extraction rate is calculated from the number of critical road elements (road curvature, signs, traffic, etc.) that are encountered during a drive and need to be processed per time unit by the driver's cognitive system. The driver's cognitive system is based on a simple cognitive architecture such as [9–11].

Driver individual variations in driving and control action determination is reflected as individuation factor. This factor reflects individual differences of drivers and reflect their differences in strategies and skills to control a vehicle. In previous research we have found that drivers apparently adopt a certain control action rate that is relatively constant across different roads but shows differentiation from other drivers [12]. This could be based on a multitude of factors such as individual scanning behavior, driving expertise, driving style, or response time to external events. This second sub-model therefore calculates a driver specific adjustment factor with which the nominal minimum control action rate is corrected for the individual driver. This adjustment factor may also lead the driver to adjust the speed of the vehicle so that the nominal control action rate can be closer met.

The result of these two sub-models is a control action rate that consists of two components. First, *compensatory* control actions are intended to keep the vehicle safely on the road. Such control actions should only occur when the driver's gaze is on the street when information about the environment can be perceived as input for compensatory control. It is worth noting that issuing compensatory control actions in real time seems not a simple feat: because the control has to happen in real time, the right control action cannot just be "tried-out" but has to fit right away.

Second, and also as support the appropriate initiation of compensatory control actions, *exploratory* movements are seen to serve as a tactile orientation that put the driver in contact with the vehicle and road and can be generally observed in the tactile exploration of space [13] (p. 45 ff). The *exploratory* movements may also relate to the inherent physiological tremor of human motor systems (e.g. [14]) that are the side-effects of physiological processes that synchronize thousands of nerve cells necessary for the coordination of muscle movement toward appropriate motor control. We refer to them here as *exploratory* movements because they seem to provide useful sensory feedback for the motor system to quickly react to oncoming events: continuously moving the steering wheel without changing the direction of the vehicle gives information about the amount of steering that will be needed effect an appropriate change in the vehicle direction if required. In this way, a driver's exploratory steering wheel movements may serve as "stay-in-the-loop" behavior to facilitate sensory-motor integration where the magnitude and timing of the compensatory control actions can be prepared in real time. Also, to the extent to which such movements may be fueled by the physiological tremor of the driver these movements can be amplified by stress or fatigue [14]. Furthermore, and most importantly, such *exploratory* movements are indicators for distraction.

The results of these two processes is then a preferred control action rate (PCAR) for a given situation and driver. By comparing the PCAR with the driver's observed control action rate, a deviation can be calculated. A temporary deviation should be indicative for distraction. In fact, drivers may actually change the minimally needed adaptive control to meet their concurrently ongoing non-driving related activities. A driver speaking on the cell phone may slow down the vehicle speed to reduce the nominally needed control action rate to get closer to the preferred one that is reduced because attention is diverted toward a cell-phone conversation).

The essence of the model consists of the relation between distracted driving and the increased need for exploratory control movements under these conditions. As drivers get distracted, their rate of exploratory movements increases to improve their readiness to set the compensatory control actions if needed. This is the primary means of detecting safety relevant distraction according to this model.

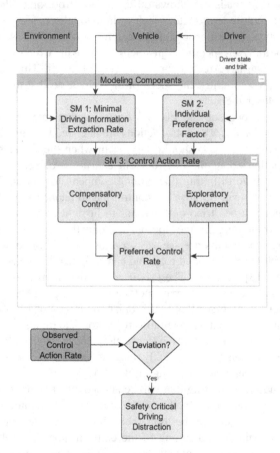

Fig. 1. A model for safety critical distraction based on preferred control action rate

The presented model has several main assumptions:

1. There is a minimally needed control rate for a given environment, vehicle, and driving situation that can be calculated.
2. Individual drivers have PCARs for a given situation that differ from other drivers.

3. Control actions are composed of compensatory control and exploratory movements to control the vehicle to meet driving goals under varying conditions. For manual driving they include steering, information extraction, accelerating, and decelerating.
4. *Exploratory* movements increase under distracted driving.

1.3 Measuring Driver's Distraction

There are many distracted driving detection algorithms that utilize driving performance measurements as well as physiological methods, especially gaze behavior, to categorize distracted driving (see e.g. [1, 15, 16] for overviews). A very simple assumption is that a driver is (visually) distracted once he or she is looking not on the street (or driving-related areas such as the mirrors or the speedometer) for some time. Previous findings showed that the time during which the focus is off the road is typically at a maximum of approximately 1.6 s ([17] cited after [18]) and that glances off-the-road longer than two seconds increase the risk of crashes significantly [4]. Many algorithms therefore use a "gaze-off-the-road" criteria to detect distraction (e.g. [4, 5]; see also [16]). For instance, the AttenD algorithm comprises a two-second time buffer that is filled up in real time while the driver is looking at the relevant driving field and degrades in case the driver looks away from it [19]. Other gaze parameters that have repeatedly been demonstrated to be related with both visual and cognitive distraction are mean fixation durations and the length as well as the speed of saccades, although the direction of change – longer or shorter for fixation durations and saccade lengths coming along with – is often less clear [2–4] Finally, gaze has also been shown to be more concentrated on the center of the road under high compared to low cognitive load [4].

Besides eye-movement behavior, driving-performance measurements can also provide valuable information about distracted driving. Related parameters often involve steering wheel positions/movements as well as lateral position of the vehicle. Regarding these latter performance measurements, distraction has been shown to cause changes in steering wheel operations [5] and in lane variability [6]. For instance, cognitive load resulted in less and visual load in more variation of lane keeping [20]. Similarly, [14] showed that, for instance, cognitive distraction increased steering wheel reversal rates as well as entropy compared to a baseline condition. However, these parameters are often closely related to gaze behavior [7]. For instance, the findings of [21] also showed that cognitive distraction increased gaze concentration towards the center of the road and [22] proposed that correlations of the horizontal eye position and steering wheel angle are affected by driver distraction.

2 The Current Study

In the current study we wanted to test all but the first assumption that were derived from the distraction model as stated above. That is, using steering wheel reversal rates (SWRs) as main variable of interest, we wanted to demonstrate that individual drivers have different PCAR for a given situation, that these control actions are either compensatory control actions or exploratory movement and that only exploratory movement increases under distracted driving. Note that we do yet report on our assumptions and investigation of minimal driving information extraction rates as this is work-in-progress.

From these assumptions the following predictions can be derived: If a PCAR is a characteristic of a driver, this rate should be similar within an individual driver across different conditions (and different between drivers). That is, we expected that the overall number of SWRs made across driving conditions differs between drivers but is highly correlated within a driver: The more SWRs are made by a driver in one driving condition the more he or she should also show in another driving condition. Compensatory control actions should become evident in an higher frequency of SWRs when the driver's gaze is on the street than when not (as compensatory control actions are the consequence of external requirements such as road shape, and should be less common/not occur when the gaze is off the street). This should be regardless of distraction. Finally, only exploratory movement should be affected by distraction; this should be seen in an overall increase of the frequency of SWRs during distraction compared to a non-distraction.

2.1   Method

Participants
Twenty-one participants took part in the study (17 male, 4 female; M_{age} = 36 years; SD_{age} = 6.5). Eleven participants owned a valid driving license for 11 to 20 years, eight for 21–30 years and two for less than 11 years. Written information consent was provided by all participants before the study. The design followed a 2×2 within-subject design with driving condition (with or without a leading car) and secondary task (with or without secondary task) as independent and various gaze and driving behavior outcomes (see below) as dependent variables. A driving scenario was created using an OpenDS driving simulator (see Fig. 2). In order to track the eye movements, a Dikablis eye tracking system was used. The road was a high-way scenario without traffic disregarding the leading car in two trials.

Procedure
At the beginning of the study, participants gave written informed consent and filled in a demographic questionnaire. Furthermore, a motion-sickness test was provided. After that he or she was introduced to the driving simulator and was asked to complete a training section in order to get familiar with the simulator. Before the actual study started, the eye tracker was mounted, and a calibration/validation procedure completed.

After that, the participants drove the same scenario four times under the four driving conditions that resulted from the two independent variables: with or without leading car and with or without a secondary task. The sequence of the conditions was counterbalanced across participants to avoid any effect of order. The route comprised a section of 5 km length and lasted about 3.5 min. Depending on the driving condition, participants were instructed to follow the leading car or to follow the speed limits as indicated by the traffic signs. The speed limits and the speed of the leading car (ranging between 50 km/h and 140 km/h, respectively) were matched such that the same speed was achieved at a specific section of the road in both conditions. Furthermore, in two trials, participants were required to work on the SuRT [23] which was presented on a tablet computer (Beneve; Android 7.0) mounted below the monitors to the right of the steering wheel. When working on the SuRT, participants were asked to indicate whether the biggest circle within an accumulation of smaller circles was in the left of right half of the tablet display by tapping on the respective half. A high score for the number of hits was provided but not reported back to the participants. After each drive, participants were required to fill in the NASA TLX [8] and a self-made questionnaire that addressed the perceived level of activity. After the participants had completed the four drives, they again filled in the motion-sickness scale and were then dismissed. Each session lasted approximately one hour.

Fig. 2. Setup of the driving simulator during the study.

2.2 Results

Gaze Data

Three different areas of interest (AOIs) were defined in order to analysis the gaze behavior: street, speedometer and display (in case of a secondary-task condition). Due to technical problems, eye movement data for one participant was not included in the following analysis (hence, data of 20 participants were analyzed in this section). Furthermore, for all following analysis, we included only 2.5 min of the individual drives (from Min 0:30 to Min 3:00).

Figure 3 shows example scanning patterns (i.e., x- and y-coordinates of the raw eye data as sampled by the eye tracker) of the of two drivers when driving with or without distraction (the data was merged across the two leading-car conditions for the sake of clarity). As can be nicely seen in the data, the gaze of both drivers is on the street area most of the time (as indicated by the cluster of fixations on the top of every graph) but there are also fixations on the speedometer (bottom left cluster of fixations in every graph) and on the tablet during in the distraction condition (bottom right cluster of fixations in the graphs of the bottom row). Although the overall distributions of the fixations look similar for the two drivers, the data clearly reflect individual driving and scanning behavior: Driver 11 seems to show larger horizontal variability than Driver 13 especially when looking on the street area.

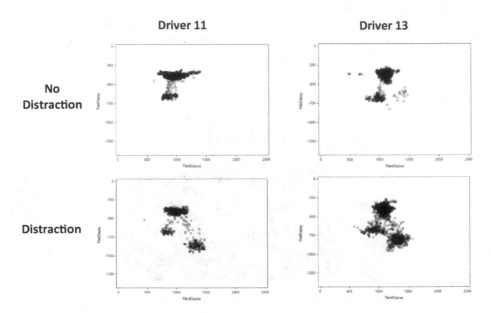

Fig. 3. Scanning patterns without distraction (top row) and with distraction (bottom row) for two individual drivers.

Table 1 provides the percentage of gaze time on the different AOIs separately for each driving condition. The data show that the drivers spent, on average, in the no-distraction condition about 88% of time on the road (with leading car even 96%, without leading car about 79%). This time decreased significantly under distraction to about 74%. Furthermore, the time spent on the speedometer decreased from 10% in the no-distraction conditions to about 7% in the distraction conditions. This effect seemed, however, to be driven mainly by the conditions without leading car: When there was a

Table 1. Average percentage of time spent on different AOIs in the four driving conditions. The individual range is presented in parentheses.

Experimental condition	AOI			
	Road	Speed	Display	noAOI
DistrLead	80.3 (63.4–91.1)	2.11 (0.2–7.4)	10.4 (0.1–28.0)	7.1 (0.8–36.5)
DistrnoLead	67.9 (47.1–82.6)	11.5 (3.7–25.5)	12.5 (0.4–34.1)	8.1 (0.8–39.5)
noDistrLead	95.9 (83.8–100.0)	2.5 (0.0–8.4)	–	1.6 (0.0–15.6)
noDistrnoLead	79.2 (58.2–91.2)	17.6 (8.5–26.1)	–	3.2 (0.3–17.2)
Mean Distr	74.1	6.8	11.4	7.6
Mean NoDistr	87.6	10.0	–	2.4

leading car, drivers spend only about 2% of the time on the speedometer, regardless of distraction condition. On the one hand, these findings confirm that our distraction manipulation was successful. On the other hand, the data again suggest that there are large inter-individual differences between the drivers.

Steering Wheel Reversal Rate

Driving-performance metrics collected during each drive included various measurements such as speed, lane position, steering wheel position/angle, brake pressed, but only the steering wheel reversal rate (SWR, as computed from the steering wheel position data) is of interest here. The SWR was defined as any change in direction of the steering wheel angle regardless of its size. As [24] stated, higher SWR rates indicate increased effort in maintaining the lane position due to the increased corrections of the steering wheel position.

To investigate the assumption that individual drivers have a PCAR for a given situation, we first correlated the total number of SWRs between driving conditions (see Fig. 4). There was a significant correlation across the conditions (Pearson's r from .78 to .89, respectively, all p's < 0.001), meaning that a driver with a higher/lower number of SWRs showed this behavior relatively constant in all driving situations. At the same time, the range of the number of SWRs between participants was large, supporting the assumption in our model that the PCAR differed between participants: Whereas some drivers made only about 60 SWRs during one drive, other showed more than 200.

To investigate effects of compensatory control actions and exploratory movements with regard to SWR rates (i.e., the frequency of SWRs per second), we merged data across driving conditions (with and without leading car) and also merged the "display" AOI and the "speed" AOI to a "no-street"-AOI. This allowed us to compare the SWR rates made during the driver's gaze was on the street vs. not. Note that gazes which could not be assigned to a valid area were excluded. As the drivers drove the same road four times, we expected that the compensatory SWRs should be stable across the drives and only be reflected in a difference in the SWR rates when the driver's gaze is on the street than when not. We found indeed a higher frequency of SWRs when the gaze was on the street area ($M = 1.1$, $SD = 0.3$, individual range from 0.6–1.6) than when not ($M = 0.9$, $SD = 0.4$, individual range from 0.3–1.8), $F(1, 20) = 26.80$, $p < 0.001$ (see Fig. 5). This was regardless of the distraction condition, $F < 1$. Furthermore, and according to the expected effect of distraction on exploratory movements, the SWR rates were reliably higher under distraction ($M = 1.1$, $SD = 0.3$, individual range from 0.5–1.8) than under no distraction ($M = 0.9$, $SD = 0.4$, individual range from 0.3–1.6), $F(1, 20) = 46.94$, $p < 0.001$.

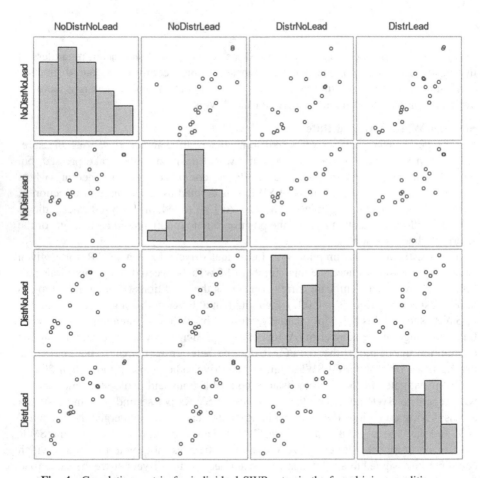

Fig. 4. Correlation matrix for individual SWR rates in the four driving conditions.

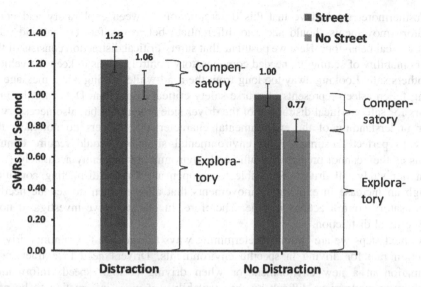

Fig. 5. Steering wheel reversal rate when gazes are on the street vs. no-street under the two distraction conditions. Error bars represent 95% CIs corrected for within-subject designs [25, 26].

3 Discussion

Our results are consistent with previous research that steering movements increase under distraction. Furthermore, we found evidence that exploratory steering movements are apparently more sensitive indicators for distraction than compensatory control actions. This was derived from our postulated model that complements prior research that did not provide such an explanation. We believe that this confirms the utility of our modelling approach to differentiate phenomena based on assumptions of the latent cognitive and perceptual processes that are behind them. Our model allows to explain why steering wheel reversals increase during distraction and that they do so even if drivers look away from the road. This is because exploratory movements should help to establish the controllability over the vehicle whereas drivers use compensatory control actions to actually control the vehicle in the intended direction.

We only investigated this notion for steering wheel reversals but plan to extend this also to other modalities. For example, we believe that gazing behaviour should resemble the differentiation established for steering movements. There as well, drivers use exploratory gaze behaviour to discover the environment in which they are driving to detect critical information elements on the road. These exploratory movements are then interspersed with information extraction gaze movements where information is actually processed and leads to other movements. For example, detecting a speed sign may therefore lead to check the speed of the own vehicle by glancing on the speedometer. Under distraction, the exploratory movements should increase as they represent a direct, probably highly automated behaviour when driving. We plan to investigate these predictions in further research.

Furthermore, we believe that this differentiation between exploratory and compensatory movements should allow to differentiate between safety critical and non-safety critical behaviour. Here we postulate that safety critical distraction consists of the drivers inability of setting the needed compensatory control actions to keep the vehicle and others safe. Looking away too long from the road while reading a text message or having fallen asleep represents of course safety-critical distraction. Detecting the precursors for safety-critical distraction of the driver side are critical but also necessary is better understanding of the environmental characteristics; drivers do not have the ability to perfectly estimate when environmental situations would require control actions as they cannot predict the traffic and environment sufficiently accurately.

In our study, all drivers were able to compensate for the distracting condition through an increase in exploratory movements that allowed them to set the needed compensatory control actions in time. Therefore, in our study we investigated non-safety critical distraction.

As next steps we are currently determining ways to calculate the minimal driving extraction rate for driving in specific environments. Drivers need to extract more information in a new environment or when driving at high speed. Information extraction is made more difficult by the availability of more information to be processed (multiple traffic signs that have to be decoded) or when driving in more complex environments (e.g. cities) where the critical information may be more difficult to locate. Driving in a familiar environment should allow to reduce the needed exploratory gaze movements to identify that information whereas more exploratory gaze movements are needed in a new driving environment. Therefore, the identification of safety critical distraction very much depends on the environment and the present conditions that need to be taken into consideration.

Another next step consists of testing the findings from the simulator studies to driving in a real vehicle. Furthermore, we will investigate the impact of non-visual distraction on the ways how people exert control during driving.

A common position in scientific research is that phenomenal descriptions are insufficient to explain the phenomenon and eventually predict it. For this, models and theories are needed that explain the data in appropriate concise ways and eventually allow predictions. Accordingly, we think it is important to differentiate between the observed behavior of distracted driving from the driver state that gave rise to distracted driving between models and measurements. Conceptually, the driver state causes various types of distracted driving behavior such as gaze orientation, motoric control adjustments, or multi-tasking. By differentiating the distracted driver state from the associated behavior, we think it should become possible to connect and interpret the joining of different observations across different modalities toward a meaningful interpretation of the driver state that then would even allow to predict future behavior. Also, it may allow us to differentiate between safety-critical and non-safety-critical distracted driving behavior. We think that this kind of modeling is needed to create adaptive automation that senses the driver state and can initiate appropriate safety interventions that are not perceived as annoyance while also increase driving safety.

Acknowledgment. The publication was written at VIRTUAL VEHICLE Research Center in Graz and partially funded by the COMET K2 – Competence Centers for Excellent Technologies Programme of the Federal Ministry for Transport, Innovation and Technology (bmvit), the Federal Ministry for Digital, Business and Enterprise (bmdw), the Austrian Research Promotion Agency (FFG), the Province of Styria and the Styrian Business Promotion Agency (SFG).

References

1. Papantoniou, P., Papadimitriou, E., Yannis, G.: Review of driving performance parameters critical for distracted driving research. Transp. Res. Procedia **25**, 1796–1805 (2017). https://doi.org/10.1016/j.trpro.2017.05.148
2. European Community: Driver Distraction (2018)
3. Liang, Y., Lee, J.: Driver Cognitive Distraction Detection Using Eye Movements, pp. 285–300. Springer, Berlin (2007). https://doi.org/10.1007/978-3-540-75412-1_13
4. Klauer, S., Dingus, T., Neale, T., Sudweeks, J., Ramsey, D.: The impact of driver inattention on near-crash/crash risk: an analysis using the 100-car naturalistic driving study data, vol. 594, January 2006
5. Engstroem, J., Monk, C.A.: A Conceptual Framework and Taxonomy for Understanding and Categorizing Driver Inattention, Brussels (2013)
6. Regan, M.A., Hallett, C., Gordon, C.P.: Driver distraction and driver inattention: definition, relationship and taxonomy. Accid. Anal. Prev. **43**(5), 1771–1781 (2011). https://doi.org/10.1016/j.aap.2011.04.008
7. Regan, M.A., Strayer, D.L.: Towards an understanding of driver inattention taxonomy and theory. Ann. Adv. Automot. Med. **58**, 5–14 (2014)
8. Kircher, K., Ahlstrom, C.: Minimum required attention a human-centered approach to driver inattention. Hum. Factors J. Hum. Factors Ergon. Soc. **59**(3), 471–484 (2017). https://doi.org/10.1177/0018720816672756
9. Card, S.K., Moran, T.P., Newell, A.: The model human processor an engineering model of human performance. Handb. Percept. Hum. Perform. **2**, 1–45 (1986)
10. Moertl, P., Wimmer, P., Rudigier, M., Rom, W., Watzenig, D.: The application of human mental models for engineering to improve acceptance and performance of driving automation. In: Proceedings of the 7th Transport Research Arena, TRA 2018, Vienna, Austria (2018)
11. Moertl, P., Wimmer, P., Rudigier, M.: Praktikable fahrermodelle mit psychologisch fundierten prozessannahmen. In: Conference Proceedings of the 9e VDI Tagung: Der Fahrer im 21en Jahrhundert, Braunschweig (2017)
12. Moertl, P., Festl, A., Wimmer, P., Kaiser, C., Stocker, A.: Modelling driver styles based on driving data. In: de Waard, D., et al. (eds.) Proceedings of the Human Factors and Ergonomics Society Chapter (2018)
13. Anan'ev, B.G., Lomov, B.F. (eds.): Problems of Spatial Perception and Spatial Concepts. National Aeronautics and Space Administration, Washington, D.C (1964)
14. Hallett, M.: Overview of human tremor physiology. Mov. Disord. **13**(S3), 43–48 (1998). https://doi.org/10.1002/mds.870131308
15. McDonald, A.D., Ferris, T.K., Wiener, T.A.: Classification of driver distraction: a comprehensive analysis of feature generation, machine learning, and input measures. Hum. Factors J. Hum. Factors Ergon. Soc. 001872081985645 (2019). https://doi.org/10.1177/0018720819856454

16. Fernández, A., Usamentiaga, R., Carús, J.L., Casado, R.: Driver distraction using visual-based sensors and algorithms. Sensors **16**(11), 1805 (2016). https://doi.org/10.3390/s16111805
17. Wierwille, W.: Visual and manual demands of in-car controls and displays. In: Peacock, B., Karwowski, W. (eds.) Automotive Ergonomics, pp. 99–320. Taylor & Francis, Washington (1993)
18. Sodhi, M., Reimer, B., Llamazares, I.: Glance analysis of driver eye movements to evaluate distraction. Behav. Res. Methods Instrum. Comput. **34**(4), 529–538 (2002). https://doi.org/10.3758/bf03195482
19. Kircher, K., Ahlström, C.: Issues related to the driver distraction detection algorithm AttenD, p. 15
20. Engström, J., Johansson, E., Östlund, J.: Effects of visual and cognitive load in real and simulated motorway driving. Transp. Res. Part F Traffic Psychol. Behav. **8**(2), 97–120 (2005). https://doi.org/10.1016/j.trf.2005.04.012
21. Kountouriotis, G., Spyridakos, P., Carsten, O., Merat, N.: Identifying cognitive distraction using steering wheel reversal rates. Accid. Anal. Prev. **96**, 39–45 (2016). https://doi.org/10.1016/j.aap.2016.07.032
22. Yekhshatyan, L., Lee, J.D.: Changes in the correlation between eye and steering movements indicate driver distraction. IEEE Trans. Intell. Transp. Syst. **14**(1), 136–145 (2013). https://doi.org/10.1109/tits.2012.2208223
23. T. 22 ISO: Road vehicles - Ergonomic aspects of transport information and control systems - Calibration tasks for methods which assess driver demand due to the use of in-vehicle systems ISO/TC 22 N 2898 (2012)
24. Macdonald, W.A., Hoffmann, E.R.: Review of relationships between steering wheel reversal rate and driving task demand. Hum. Factors **22**(6), 733–739 (1980)
25. Cousineau, D.: Confidence intervals in within-subject designs: a simpler solution to Loftus and Masson's method. Tutor. Quant. Methods Psychol. **1**(1), 42–45 (2005). https://doi.org/10.20982/tqmp.01.1.p042
26. Morey, R.D.: Confidence intervals from normalized data: a correction to cousineau. Tutor. Quant. Methods Psychol. **4**(2), 61–64 (2008). https://doi.org/10.20982/tqmp.04.2.p061

Development of a Driving Model That Understands Other Drivers' Characteristics

Shota Matsubayashi[1]([⊠])[iD], Hitoshi Terai[2][iD], and Kazuhisa Miwa[1]

[1] Nagoya University, Nagoya, Aichi 4648601, Japan
shota.matsubayashi@nagoya-u.jp
[2] Kindai University, Iizuka, Fukuoka 8208555, Japan

Abstract. In this study, a driving model that observes others' behaviors and that makes its decisions based on the estimated characteristics of the other driver in the situation where two lanes merge on a highway was developed. Additionally, drivers' probabilities of decisions in relation to their primary characteristics was interpreted. We presumed various drivers have different characteristics such as aggression and caution that affect their making decisions. We simulated the merging behaviors of two drivers in a merging lane and in a main lane after the driver in the merging lane had estimated the characteristics of the driver in the main lane as a typical case. The results of the estimation-success case revealed that two drivers changed lanes immediately after selecting and canceling their decisions several times. However, the results of the estimation-failure case revealed that if a standoff between two drivers occurred, it would take longer to change lanes than in the estimation-success case. Furthermore, the lateral swaying of the cars was worse in the estimation-failure case than in the estimation-success case because the two drivers allocated much cognitive resources and time to monitor of the other car. The importance of understanding others and building a model that understands others in traffic is discussed.

Keywords: Driving behaviors · ACT-R · Understanding others

1 Introduction

1.1 Understanding Others in Traffic

It is imperative to understand others in traffic so as to ensure safety as well as flow of the traffic. When in traffic, drivers tend to behave in accordance with their own individual characteristics and make decisions by observing others' behaviors. For example, in an intersections, when drivers observe pedestrians' behaviors, they may assume that they will cross the street at the intersection. Consequently, the drivers will decide their own behaviors such as vehicle routing and speed in

Supported by JSPS KAKENHI Grant Number 1918H05320.

H. Krömker (Ed.): HCII 2020, LNCS 12213, pp. 29–39, 2020.
https://doi.org/10.1007/978-3-030-50537-0_3

accordance with this estimated probability. Driving behaviors depend strongly on the characteristics of the drivers [10]. Therefore, if drivers make incorrect estimations, discrepancies between them are likely to occur, which may lead to serious accidents. Therefore, it is imperative to clarify the cognitive mechanism underlying how drivers understand others and make their own decisions so as to ameliorate traffic.

Clarifying this process of understanding and building a model of driving may also contribute to the development of advanced driving assistance systems. If driving assistance systems comprehend drivers' processes of understanding others, they will be able to provide appropriate assistance that is suitable for every driver.

Previous studies have revealed the importance associated with different assistance that advanced driving assistance systems provide to drivers in relation to their characteristics. Various empirical studies on the relationship between driving assistance systems and human drivers have revealed that the information provided by the system and its behaviors influence drivers' subjective evaluation of the system including trustworthiness and acceptance [2,9]. Additionally, the usability evaluation items—effectiveness and understandability—are related to changes in driving behaviors such as speed [4,5]. Therefore, such assistance has to be designed by taking drivers' cognitive aspects such as trustworthiness, usability evaluation, and behavioral aspects including changes in speed into consideration. Accordingly, it is imperative for assistance systems to comprehend drivers' underlying cognitive mechanisms and mental state. If the system comprehends drivers' mental states, it can predict how the drivers understand others and what they are likely to do next. Providing proactive assistance to drivers based on these predictions may contribute to the flow of traffic and safety of all.

It is imperative to represent the mechanism of understanding others as a descriptive model so as to incorporate it into an assistance system. Cognitive architecture such as the Adaptive Control of Thought–Rational (ACT-R) [1] is beneficial in describing the cognitive mechanisms of individuals because they are designed by tanking cognitive characteristics of human into account. This architecture can implement various cognitive tasks and their behavioral data can be measured. Furthermore, data about mental states, which are difficult to obtain in psychological experiments such as a preliminary state for making decisions, are also available. Therefore, the internal process with cognitive architecture can be verified in detail.

1.2   Objective

In this paper, a driving model that observes others' behaviors and that makes decisions based on the estimated characteristics of the other drivers in the merging situation (Fig. 1) was developed by employing ACT-R [1]. Understanding others in such situation is imperative to ensure safety in traffic. When two drivers in two adjacent lanes observe the other's behaviors, they have to decide who should go ahead and who should follow behind. Subsequently, one driver will try to change lanes by observing the other car. In the real situations, drivers often

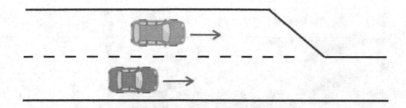

Fig. 1. Merging situation in this study

have to change lanes by the end of a section of the road. Therefore, drivers often try to understand the other driver within a short span of time.

We also interpreted the probabilities of decisions regarding merging as drivers' primary characteristics. For example, when two cars are driving adjacent to one another, some drivers may be aggressive in their attempts to either speed up or slow down and subsequently, may change lanes immediately. However, other drivers may be cautious and thus, unlikely to make such decisions. We presumed that various drivers have different characteristics and accordingly, we simulated their driving behaviors. In particular, we verified merging behaviors after driver estimated one other's characteristics correctly or incorrectly as a typical case.

2 Simulation Method

2.1 Merging Behaviors

We built an environment where a two-lane road merges into a one-lane road in Unity (Fig. 2). While the lane on the drivers' left was the merging lane, the lane on their right was the main lane. We simulated the merging behaviors until the car in the merging lane finished changing lanes (Fig. 1).

At the outset, two cars were driving adjacent to one another. The drivers attempted to make LEAD or FOLLOW decisions according to the merging flows (Details in Sect. 2.2). The driver who selected a LEAD decision sped up his/her car so as to get ahead of the other car. Contrastingly, the driver who selected a FOLLOW decision slowed down his/her car in order to follow the other car. If there was a lot of space or a great difference in speed between the two cars, the driver in the merging lane tried to change lanes so as drive in the main lane. Therefore, if the two drivers did not make the same LEAD/FOLLOW decision, they would be able to change lanes quickly and easily.

2.2 Model Structure

Our merging model was developed by employing ACT-R, which is a leading cognitive architecture [1]. ACT-R has various modules including visual and memory modules. Furthermore, ACT-R can implement various tasks with production rules in an if-then rule format. The design of each module is based on individuals'

Fig. 2. Overview of the running situations in this study

cognitive characteristics. For example, ACT-R cannot see two objects simultaneously and it takes a certain amount of time to move its eyes toward an object and read the information. These constraints enabled us to simulate the internal cognitive processes of individuals. Recently, driving behaviors have been simulated by employing ACT-R in many studies; these include in-car interface use that influences driving performance and lateral deviations [6].

We extended the running model developed by previous studies [7,8] to a merging model. Specifically, we added the production rules of monitoring other's behaviors and changing lanes to the existing production rules in relation to driving straight.

In Fig. 3, the merging flow of the driver in the merging lane is depicted. At first, the driver in the merging lane watches the car in the main lane and acquires relative distance (hereinafter RD) between the two cars (Read-car_1st). Subsequently, the driver monitors the other driver again so as to gain relative velocity (hereinafter RV) between the two cars (Read-car_2nd). The driver makes a LEAD/FOLLOW decision by referring to the values of RD and RV. However, the driver's selection of LEAD/FOLLOW depends on the circumstances. The driver may have to clear the values of RD and RV and monitor the other car again. If drivers select a LEAD decision, they will speed up the car and slow it down if they select a FOLLOW decision. Thereafter, the driver updates the value of RD and RV (Read-car_3rd) and tries to make a lane-changing decision by referring to the updated values. If drivers decided to change lanes, they will turn the steering wheel and move into the main lane. However, if driver perceive that changing lanes is impossible or do not take action after making

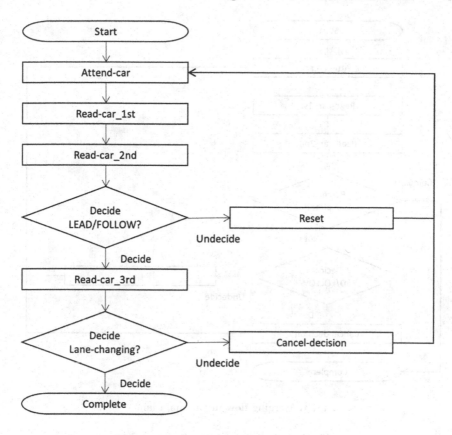

Fig. 3. Merging flow in the merging lane

a LEAD/FOLLOW decision, they will cancel the decisions, resume the speed
they were traveling at, and start monitoring the other driver from the beginning
(Cancel-decision).

The core aspects in this flow are the LEAD/FOLLOW decisions where the
drivers select one of the decisions and the change-lane decision where they decide
whether to change lanes or to cancel their LEAD/FOLLOW decision. These
decisions are defined by the logistic curve surface functions, which refer to the
RD and RV between the two cars (Details in Sect. 2.3). This means that drivers
are likely to make LEAD or FOLLOW decisions and change lanes when the RD
and RV is greater than zero and unlikely to select the LEAD/FOLLOW decision
and likely to cancel their decision when the values are closer to zero, in which
case the cars will be driving next to one another.

The merging flow in the main lane was the same as the one in the merging
lane except for the following two aspects (Fig. 4). First, the production rules of
the lane-changing decision were removed because the drivers in the main lane do
not change lanes. Second, the production rules, which enable drivers to perceive

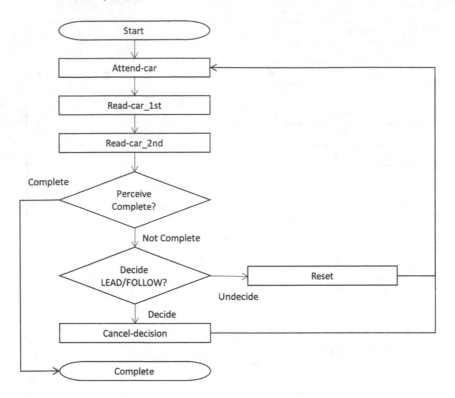

Fig. 4. Merging flow in the main lane

whether the car in the merging lane has crossed the center line and to finish their own monitoring of the other drivers, were added (Perceive Complete).

It is noteworthy that the production rules of merging constructed in this study are independent of those of driving straight [7,8]. However, both production rules require the use of the visual module in ACT-R. Therefore, conflict between the merging and driving processes may occur and interrupt each other. This leads to a delay in changing lanes or swaying while driving.

2.3　Representation of Individual Characteristics

Individual driving characteristics in merging behaviors are represented by manipulating the three constant terms of logistic curve surfaces (Fig. 5) in the LEAD/FOLLOW decision and lane-changing decision (Eq. 1).

$$p = \frac{1}{1 + exp(-b_0 + b_1 \times RD + b_2 \times RV)} \tag{1}$$

In Eq. 1, b_0 determines the intercept of the curve surface and b_1 and b_2 determine the sensitivities to the RD and RV between the two cars, respectively. The higher b_0 becomes, the more aggressively drivers try to make LEAD/FOLLOW

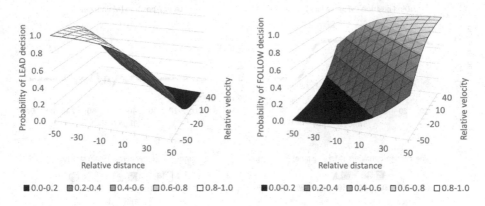

Fig. 5. Overview of the logistic curve surfaces of the LEAD/FOLLOW decision and lane-changing decision

decisions and change lanes. This means that they have aggressive characteristics and want to make decisions themselves. Meanwhile, the lower b_0 becomes, the more unlikely drivers will make any decisions and change lanes. This means that they are characterized by caution and allow the other to make the decisions. In this study, b_1 and b_2 representing the sensitivities to the RD and RV are fixed.

In cases when an aggressive driver and a cautious driver for LEAD decisions are matched, the aggressive driver is likely to select the LEAD decision and the cautious driver the FOLLOW decision. Consequently, the RD and RV between the two cars are likely to become greater and less time will be needed to change lanes. However, if both drivers are cautious, they are likely to select the same FOLLOW decision and the two cars will continue to drive next to each other. Therefore, it will take longer to change lanes.

3   Simulation Results

In this study, we simulated 90-s merging behaviors after the driver in the merging lane estimated the characteristics of the driver in the main lane. Two kinds of typical cases were simulated. First, two drivers attempted to finish changing lanes after the driver in the merging lane correctly estimated the other driver as aggressive for LEAD decisions and become more cautious. The parameter of b_0 regarding LEAD decisions of the driver in the merging lane was set to $- 2$ and that of the driver in the main lane was set to $+2$. Second, two drivers attempted to finish changing lanes after the driver in the merging lane incorrectly estimated the driver in the main lane to be cautious for LEAD decisions; in other words, both drivers became cautious. The parameters of b_0 regarding LEAD decisions of both drivers were set to $- 2$. The parameters of b_0 for FOLLOW decisions were set to zero and all other parameters were set to default.

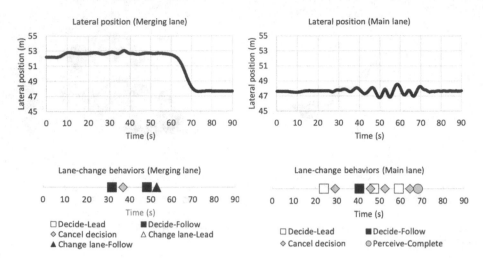

Fig. 6. Results of the estimation-success case

3.1 Estimation-Success Case

In Fig. 6, the results of the estimation-success case are depicted where two drivers attempted to change lanes after the driver in the merging lane had estimated the aggressive characteristics of the driver in the main lane successfully. The upper figures represent the lateral positions of each car and the lower figures represent the time points when the major production rules of merging behaviors were implemented. The figure on the left represent the driver in the merging lane and those on the right represent the driver in the main lane. In relation to the car in the merging lane, the result of the lateral position revealed that the driver started to change lanes after 60 s and finished changing lanes 10 s later. However, the results of the production rules revealed that the driver selected a FOLLOW decision within 30 s, but canceled five seconds thereafter. Subsequently, the driver selected a FOLLOW decision again within 50 s and decided to change lanes a few seconds later.

Although the car in the main lane drove straight throughout, it swayed laterally from 40 to 70 s. The results of the production rules revealed that the driver were engaging in merging behaviors during this period. Moreover, the driver selected both LEAD and FOLLOW decisions and canceled them repeatedly. The swaying stopped after 70 s and the subsequent driving was smooth because the driver in the main lane perceived that the car in the merging lane had finished changing lanes and accordingly, aborted his/her merging behaviors, which enabled him/her to allocate more cognitive resources and time to drive straight.

In essence, in the estimation-success case, two drivers selected LEAD/FOLLOW decisions and canceled them several times in the beginning. Subsequently, the driver in the merging lane selected a FOLLOW decision while the driver in the main lane selected a LEAD decision. They eventually finished changing lanes smoothly.

3.2   Estimation-Failure Case

In Fig. 7, the results of the estimation-failure case where two drivers attempted to change lanes after the driver in the merging lane incorrectly estimated the cautious characteristics of the driver in the main lane are presented. Both drivers were cautious. The results of the lateral positions revealed that the two drivers had not finished changing lanes within 90 s. According to the production rules, the driver in the merging lane decided to start changing lanes after 75 s, but did not complete it. As time passed, the lateral swaying of the car in the merging lane got significantly worse because the driver selected four FOLLOW decisions and canceled it three times before starting to change lanes. Similarly, the driver in the main lane allocated his/her cognitive resources in five decisions and cancelations while driving straight, which resulted in the lateral swaying after 30 s.

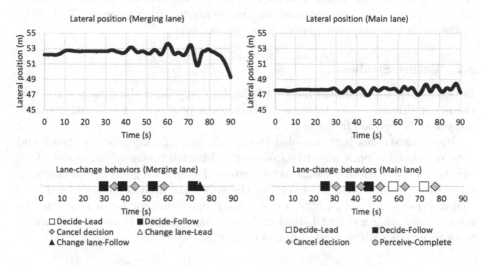

Fig. 7. The simulation results of the estimation-failure trial.

In essence, in the estimation-failure case, there was a standoff between the two drivers and it took a longer time to finish changing lanes than in the estimation-success case because they selected the same FOLLOW decisions. Furthermore, the lateral swaying of the cars was worse than in the estimation-success case because the two drivers allocated more cognitive resources in relation to merging behaviors.

4   Discussions and Conclusions

4.1   Effects of Understanding Other's Characteristics

We developed a driver model that monitors others' driving behaviors and makes its own decisions based on estimated characteristics in situations where lane merge. Additionally, we showed that the success or failure of the estimation influences driving behaviors significantly.

In the estimation-success case, that is, when one driver is aggressive and the other cautious for LEAD decisions, the aggressive driver is likely to select a LEAD decision and the cautious driver a FOLLOW decision. When such decisions are made, the RD and the RV are not near zero and subsequently, the two cars change lanes easily and quickly.

However, in the estimation-failure case, that is, when both drivers are cautious for LEAD decisions, they are both likely to select the same FOLLOW decisions. The RD and the RV between the two cars are almost zero and the two cars drive next to one another because both drivers slow down their car in relation to their FOLLOW decisions. In such circumstances, the driver in the merging lane judges that he/she cannot change lanes safely and may monitor the other car several times. Lateral swaying occurs because the driver cannot allocate enough cognitive resources to operate his/her steering wheel. The driver in the main lane may experience similar phenomena.

4.2   Future Studies

In this study, we simulated merging behaviors only after the drivers estimated the other drivers' characteristics. However, the processes of estimating others are yet to be verified in detail.

Previous studies have revealed the complexity of understating others and have suggested various ways to implement estimation processing in models. One way to estimate another's state is to assume that other's goals are the same as one's own [3]. Although we employed the way that drivers changes their own characteristics in favor of other drivers, they may compromise or conform. We intend simulating merging behaviors by employing multiple models in various ways to enhance an understanding of others.

4.3   Conclusion

We developed a driving model that observes others' behaviors and that makes its decisions based on the estimated characteristics of the other driver in situations where lanes merge. Furthermore, we simulated the merging behaviors after the driver estimated the other's characteristics correctly or incorrectly as a typical case. The results revealed that the success or failure of the estimation influences driving behaviors significantly.

Driving requires drivers to execute many complex tasks simultaneously and sometimes these tasks have strict temporal limitations. Understanding others is a part of these tasks, but it is imperative to ensure the safety of all as well as the flow of the traffic. Clarifying the cognitive process of understanding and building a model of driving will contribute to the social development. In the future, it is hoped that advanced driving assistance systems will provide adequate assistance that are suitable for each driver by employing our descriptive model of understanding others.

References

1. Anderson, J.R.: How Can the Human Mind Occur in the Physical Universe?. Oxford University Press, New York (2007)
2. Beggiato, M., Krems, J.F.: The evolution of mental model, trust and acceptance of adaptive cruise control in relation to initial information. Transp. Res. Part F: Traffic Psychol. Behav. **18**, 47–57 (2013). https://doi.org/10.1016/j.trf.2012.12.006
3. Laird, J.E.: It knows what you're going to do: adding anticipation to a Quakebot. In: Proceedings of the Fifth International Conference on Autonomous Agents, AGENTS 2001, pp. 385–392 (2001). https://doi.org/10.1145/375735.376343
4. Matsubayashi, S., et al.: Empirical investigation of changes of driving behavior and usability evaluation using an advanced driving assistance system. In: The Thirteenth International Conference on Autonomic and Autonomous Systems (ICAS 2017) (c), pp. 36–39 (2017)
5. Matsubayashi, S., et al.: Cognitive and behavioral effects on driving by information presentation and behavioral intervention in advanced driving assistance system. Cogn. Stud. **25**(3), 324–337 (2018). https://doi.org/10.11225/jcss.25.324
6. Salvucci, D.D.: Predicting the effects of in-car interface use on driver performance: an integrated model approach. Int. J. Hum. Comput. Stud. **55**(1), 85–107 (2001). https://doi.org/10.1006/ijhc.2001.0472
7. Salvucci, D.D.: Modeling driver behavior in a cognitive architecture. Hum. Factors J. Hum. Factors Ergon. Soc. **48**(2), 362–380 (2006). https://doi.org/10.1518/001872006777724417
8. Salvucci, D.D., Gray, R.: A two-point visual control model of steering. Perception **33**(10), 1233–1248 (2004). https://doi.org/10.1068/p5343
9. Verberne, F.M., Ham, J., Midden, C.J.: Trust in smart systems: sharing driving goals and giving information to increase trustworthiness and acceptability of smart systems in cars. Hum. Factors **54**(5), 799–810 (2012). https://doi.org/10.1177/0018720812443825
10. Vichitvanichphong, S., Talaei-Khoei, A., Kerr, D., Ghapanchi, A.H.: What does happen to our driving when we get older? Transp. Rev. **35**(1), 56–81 (2015). https://doi.org/10.1080/01441647.2014.997819

Voice User-Interface (VUI) in Automobiles: Exploring Design Opportunities for Using VUI Through the Observational Study

Fangang Meng[1], Peiyao Cheng[1(⊠)], and Yiran Wang[2]

[1] Design Department, Faculty of Social Science, Harbin Institute of Technology, Shenzhen, People's Republic of China
chengpeiyao@hit.edu.cn
[2] Internet of Vehicles (IoV), Intelligent Driving Group (IDG), Baidu.com Times Technology Co., Ltd., Beijing, People's Republic of China
wangyiran01@baidu.com

Abstract. The integration of voice interaction in automobiles has become a trend for current automobile development. This study aims to explore the design opportunities of using voice interaction in automobiles. To achieve this goal, we conducted an observational study to observe different behavior during driving. Based on video and voice recording collected through observation, we categorized the different behaviors and different sounds during the driving process. Next, by analyzing the relationships between the sound and the drivers' behavior, we discussed the design opportunities for involving VUI.

Keywords: Voice interaction · VUI · Driving · Observation

1 Introduction

Voice interaction is becoming a hotly discussed issue nowadays. As a novel way of human-computer interaction, it becomes increasingly common and integrated into many electronic devices, such as mobile phones, tablets, consumer electronic products, and cars. In Chinese markets, Roewe and NIO spent extensive efforts on designing a vehicle-mounted interactive system. These voice interaction systems provide the basic supports for driving, such as opening the sunroof through voice, using voice to control phone calls, and using voice to listen to music. In academia, extant research has been conducted with a focus on investigating the characteristics of VUI. For example, voice interaction may disclose personal information, therefore, researchers conducted a study on the security of personal information in voice interaction in public space (Easwara Moorthy and Vu 2015), and researchers also expand the application scope of voice interaction to intelligent classification system, provides a good example for processing and categorizing voice messages (Ji and Rau 2019).

Thus far, involving VUI in automobiles is still in the early stage. Many problems and opportunities remain unknown. Specifically, although there are several successful examples of involving VUI in automobiles, we still lack a comprehensive overview of the possible opportunities as well as the influences of these opportunities on the driving experience.

© Springer Nature Switzerland AG 2020
H. Krömker (Ed.): HCII 2020, LNCS 12213, pp. 40–50, 2020.
https://doi.org/10.1007/978-3-030-50537-0_4

Therefore, this paper aims to explore the opportunities in the onboard interaction in the automobile through observing drivers' driving process. Specifically, we observed five drivers' driving activities and the sounds related to these activities. Next, we categorized the driving activities and analyzed driving activities through the consumer journey map. We also discussed the possibilities of involving VUI based on these analyses.

2 Literature Review

Currently, researches on VUI focus on the relationship between the user and the VUI system. This often involves the user's psychological characteristics and the internal structure of the voice interaction system. According to an integrated model, the attitudes of VUI users are affected by the proposed driving factors, including trust, perceived risks, perceived enjoyment, and mobile self-efficacy, significantly (Nguyen et al. 2019). Another study shows three finds on how does user resistance influences user experience in the context of the voice user interface of in-vehicle infotainment (IVI) system (Kim and Lee 2016).

Many studies have focused on voice human-computer interaction interface of smart products, smart homes, and cars. The design of a smart home voice interface for the elderly should give them more ability to control the whole system to support daily living based on voice command (Portet et al. 2013). For the sake of driving safety, researches record the activities of driving and reveal the details when drivers using phones (Esbjörnsson et al. 2007). Another research shows voice interaction during driving plays an essential role in the effectiveness and safety in the car (Peissner and Doebler 2011). Some researches concentrate on the details of driving control. By investigating the effects when switching between touch and speech input on task efficiency and driver distraction in a dual-task setup, it shows that the sequential combination of adequate modalities for subtasks reduced the duration of the entire interaction in the car (Özcan and van Egmond 2012).

Current research on automobile interaction focuses on specific aspects such as driving safety and switch control. It can effectively tackle on the particular problems related with VUI. However, to integrate VUI in automobiles, we lack a comprehensive understanding of different opportunities. This paper focuses on the whole driving process. The results provide avenues for integrating VUI in automobiles.

3 Methods

3.1 Participants

We collected 5 drivers to participate in this study. These participants are ordinary drivers whose driving experience is mainly in urban areas. They differed in ages and driving experience. Their driving experience ranged from 1 to 20 years. The detailed information of drivers, cars and driving routes in the five observation is shown in Table 1:

Table 1. Table captions should be placed above the tables.

Participant	Information
Participant 1	33-year-old male, 8 years of driving experience, 3 months of vehicle age. Car model: Roewe RX5
Participant 2	43-year-old male, 20 years of driving experience, 4 years of vehicle age. Car model: Mercedes C200L
Participant 3	33-year-old female, 5 years of driving experience, 5 years of vehicle age. Car model: Hyundai ix35
Participant 4	31-year old female, 1 year driving experience, 1 year of vehicle age. Car model: Jeep Compass
Participant 5	63-year-old male, 15 years of driving experience, 1 year of vehicle age. Car model: Roewe RX5

3.2 Procedure

During the observation, each driver's driving process was recorded. The observer sat in the back seat, using video-camera to film driving activities and using the recorder to record all the voice during the driving process. The camera's shooting ranged from the steering wheel to the main visual interactive panel on the right side of the driver. In the process of filming, when the driver's operation went beyond the screen, the observer adjusted the shooting range to track the driver's activities. In addition, the recorder was placed in the vehicle's compartment near the driver to record the sound of the whole process. Moreover, in the observation process, the observer attempted to keep silent to avoid direct communication with the driver. In this way, the driver can keep his or her normal driving habits. Each observation ranged from half an hour to an hour.

After the observation of the driving process, a short interview was conducted with each driver to obtain their personal information, including age, driving experience, vehicle model, and driving time. They were also asked about the sounds they noticed while driving and their views on onboard voice interaction.

3.3 Data Processing

After the observation and acquisition of video and audio data, each driving process was played back and analyzed one by one. While analyzing data, we firstly coded drivers' activities according to the timeline. Along with the activities, the corresponding sound was also coded. For example, while intending to turn right, the driver turned on flashing light with the sound 'click-click' playing. Moreover, other sources of sound were also considered, such as voice navigation.

After initial coding, we intended to look for patterns and themes. This coding process was conducted observation by observation and resulted in a number of codes related to drivers' activities. We followed the prior research to categorize these activities. Finally, this process resulted in activities with three different levels.

4   Results

After initial coding, different driving activities emerged. We followed Bubb's (2011) framework to analyze the driving activities, which resulted in an overview of different activities categorized in three levels (see Table 2).

Table 2. The three levels of driving activities.

Level	Activity
The primary driving task (Keep the car on the course)	Take the initiative to check the map to confirm navigation Passive voice listening to confirm navigation (once a minute on average)
The secondary driving task (Shows driving intention)	Use the horn Open the light Make a turn
The tertiary tasks (Make the driver comfortable)	Adjust the visor The air-conditioning Open the skylight Adjust the radio

In addition to the three-level activities, five themes resulted from our analysis: showing driving intention, improving driving comfort, navigation through smart-phones, and encountering traffic jams.

4.1   Showing Driving Intention

In general, there are two main ways for drivers to show their intentions to other vehicles and pedestrians: horns and lights. The lamplight includes turn signal, flood-light, double flashing light. The most frequently used light is the turn signal. According to the driving regulations, many operations in the driving process need to turn on the turn signal in advance to indicate their intentions, such as turning the corner, changing lanes, overtaking, turning around, entering and exiting the roundabout, stopping and starting.

Using the horn is another way for drivers to show their intention, as shown in Fig. 1. During the observation, the driver used the horn when the car in front slows down in the middle of the road. In this process, the vehicle in front slows down for some reason, fails to notice and blocks the vehicle in the rear, so the driver of the vehicle is observed to sound the horn for a warning, indicating that the horn is a very direct and efficient way of vehicle-to-vehicle interaction.

Fig. 1. Horning (by P1).

4.2 Improving Driving Comfort

We found that drivers tend to do some activities to improve their comfort while driving, include adjusting the air environment (air conditioning, windows, sunroof), radio, and sun visor.

Adjust the Air Conditioner. During the observation, we found that adjusting the car's air environment needs a series of operations. Before turning on air conditioning, drivers need to make sure that the window and the skylight have been closed. After open air conditioning, the driver needs to adjust the wind patterns, the temperature, wind location, blowing direction. These actions need to conduct several times. Next, the driver will also put hands on the diffuser to feel the wind, as shown in Fig. 2. This action was used as feedback to adjust again to achieve the most comfortable state.

During the adjustment process, the driver can receive the feedback sound of the equipment. When the window is closed, there will be the sound of the window rising and locking. When the air conditioner is turned on, there will be the sound of the air conditioner running.

Fig. 2. Adjust air conditioner blowing direction (by P1, P4).

Open the Visor. When the sunlight affected the driver's eyesight, the driver opened the visor and adjusted it to the appropriate position in order to eliminate the influence of sunlight (see Fig. 3). This process also requires adjustments several times. While adjusting the visor, the driver needs to free a hand, which could increase the risks of driving. During this process, no audio information was involved.

Fig. 3. Open the visor (by P3)

Switch Radio. The drivers can listen to the radio while driving. The drivers need to turn on the radio and switch the radio channel. As shown in Fig. 4, for the same driver, the shortcut key is used when switching radio channels, and the knob on the interactive interface is used to adjust the volume. Therefore, variables need to be further controlled to determine whether the current driving state or the difference between volume and channel switching affects the use of different interactive keys. While switching the radio, no audio feedback was provided. The interaction mainly based on physical interaction.

Fig. 4. Left-channel switching; Right-volume adjustment (by P4)

4.3 Navigation Through Smartphone

Participants tend to use voice navigation when the driving route is unfamiliar.

After getting on the car and before departure, drivers make the navigation route, place the smartphone on the smartphone bracket around the steering wheel, and start the navigation system and departure. Among the five participants, three of them did not have in-car voice navigation. Thus, they used the voice navigation system installed in smartphones.

Because of using smartphones to navigate, participants need to fix the smartphone prior to driving. As shown in Fig. 5 below, drivers tend to place the smartphone closer to their eyesight and the steering wheel.

While using a navigation system, drivers not only need to hear the voice instructions but also watch the visual interface to confirm the route. Thus, the position to view the visual screen is crucial. Also due to this, drivers would not prefer to using the onboard navigation system. They felt the position of the screen was not optimal, which was not convenient enough during driving.

Fig. 5. The use of voice navigation on mobile phones (by P1, P4)

Among the five participants, two participants had in-car voice navigation, but they still used smartphone voice navigation. They seemed to get used to using the navigation system installed in smartphones. Participants used navigation mainly for the following three purposes:

Route Planning. Participants tend to use smartphone navigation when they were unsure about the route to a destination. While using the navigation system, participants need to enter the destination information before driving. The navigation system can plan and inform the driver of the route information in advance. For example, drivers can be informed about the congested road in front and replan the route.

Safety Instruction. The voice navigation also provides instruction for safety. The voice reminder of voice navigation provides information to the driver according to the road conditions, and the corresponding reminder information is provided based on the geographical location of the car and the intersection, school, sidewalk, illegal photo point, etc. The frequency of such information is about once a minute on average.

Route/Visual Confirmation. Moreover, the navigation system on the smartphone can also provide visual confirmation. As is shown in the Fig. 6b, in the case of a road that is difficult to recognize, for example, there are two similar turning directions at the intersection, the smartphone voice navigation will use the arrow to display the specific choice direction for visual assistance. In this way, the drivers can get the route information clearly.

Fig. 6. Check the intersection map (by P2)

4.4 Driving in a Traffic Jam

During the observation, we found that drivers showed slight anxiety when they encountered traffic jams. While waiting for red lights, they usually checked their phones repeatedly. They checked navigation routes, and they also checked other applications that not related to driving, such as social media Apps like WeChat. Participants repeatedly checked their smartphones for the purpose of releasing their anxiety rather than getting useful information (Fig. 7).

Fig. 7. Checking your phone while stuck in traffic (by P2) (Color figure online)

5   Exploring Design Opportunities for Integrating VUI in Automobile Through Journey Map

In order to explore the opportunities for integrating VUI, we created a typical journey map based on the main findings from the observation study. Along with this journey map, the main activities during driving were created. The drivers' emotions were also analyzed (Fig. 8).

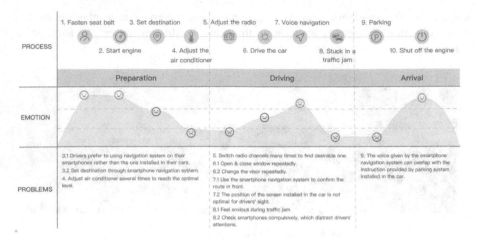

Fig. 8. Journey map

Based on the journey map above as well as the findings from the observation study, several design opportunities of involving VUI can be recognized.

Before driving, in the preparation stage, drivers often make some changes to the inner environment. While making these changes, drivers tend to adjust it several times to reach the optimal level. In this process, drivers tend to use physical buttons and knobs. For example, drivers use the buttons to adjust the air conditioner. Such physical interaction can provide immediate feedback to drivers, which helps drivers take further actions, such as improve the air volume or change the airflow direction. Through physical interactions, this process is fluent and efficient. Therefore, involving VUI seems not to be an optimal choice. VUI cannot respond as quickly as physical interaction. Moreover, through physical interaction, drivers can make subtle and accurate changes, which are difficult to achieve for VUI.

When encountering traffic jams, we found that drivers tend to suffer from anxiety. They try to deal with the anxiety through checking their smartphone compulsively. These can be risky for driving. Specifically, while drivers' views focused on their smartphone screen, their attentions on road-situation can be largely eliminated. Thus, this can be an opportunity to involve VUI. Particularly, involving VUI can be a way to release drivers' negative emotions triggered by traffic jams. For example, VUI system can play some jokes or music that can clam down drivers.

Currently, drivers have the habit of using smartphones to navigation although cars can equip with navigation systems. When we asked drivers why they choose their smartphone to navigate rather than the automobile equipped one, they tend to already get used to their smartphone. They also mentioned that the screen equipped in the car was not arranged in the optimal position for viewing. They have more flexibility while using their own smartphone. However, there can be risks for heavily relying on the navigation system in the smartphone. For example, during parking, the instruction provided by the navigation system in the smartphone can be overlapped with the one installed in the car. Thus, while involving VUI in the car, we should control other sources of voice instruction to prevent possible conflicts.

6 Discussion

6.1 Summary of Findings

This study explores the design opportunities for involving VUI in automobiles. Specifically, we conducted an observational study. Through observing 5 drivers, we concluded the different driving activities: 1) show driving intention, 2) improve driving comfort, 3) navigation on the way, 4) encountering traffic jams. Next, we created a journey map based on these findings, analyzed drivers' emotions, and explored the relevant design opportunities for involving VUI.

Generally speaking, VUI is a novel and powerful way to integrate into automobiles. But it can only create a desirable driving experience while the characteristics of VUI match with the goals of driving activities. For example, for showing driving intention, drivers use horn and light to achieve. These activities have been thoroughly rooted in driving habits and driving regulations. Moreover, drivers currently can press the horn and turn on the light through physical interactions effectively and efficiently. Thus, there is little space for involving VUI in for showing driving intention. Similarly, for improving driving comfort, drivers use buttons or rotary knob to adjust certain settings, such as air conditioner, visor, radio, etc. While changing these settings to reach the optimal driving environment, driver often need several rounds of adjustments. They need to firstly change the settings, felt the changes, and reset the settings until satisfied. In this process, current physical interaction seems to be superior to VUI because physical interaction can provide subtle and accurate adjustments and provide immediate feedback.

However, VUI has potentials for navigation. We found that drivers are getting used to voice navigation. But they used the navigation system in smartphones more often than the ones installed in the automobiles.

We also found that during traffic jams, drivers suffer from some negative emotions. At present, drivers attempted to deal with these negative emotions through checking smartphones. But this carries the risks for driving. Thus, involving VUI can be a promising way to help drivers deal with negative emotions in the traffic jam. When the car sensed the traffic jam, the VUI system can be activated to play some jokes or music for drivers. In this way, drivers' negative emotions are released while their sights are not occupied.

6.2 Limitations and Future Research

There are several opportunities to strengthen this research. First, in this research, we only collected five drivers to participate in the observation study. Due to the small sample, there could be activities that we did not capture. Thus, future research could continue this study by collecting more participants. The results might provide further validity to this research. Second, this research uses the qualitative research methods to explore design opportunities. Although the identified opportunities are promising, they still need further validation. How to realize these opportunities still requires further investigation. Third, in this research, we only considered the situation with one driver. But in the real situation, there can be passengers on the car and they may also be users of VUI. Thus, future research could address this.

Acknowledgements. This work was partly supported by China National Social Science Fund [grant number 19CYY052].

References

Easwara Moorthy, A., Vu, K.-P.L.: Privacy concerns for use of voice activated personal assistant in the public space. Int. J. Hum. Comput. Interact. **31**(4), 307–335 (2015). https://doi.org/10.1080/10447318.2014.986642

Ji, X., Rau, P.-L.P.: Development and application of a classification system for voice intelligent agents. Int. J. Hum. Comput. Interact. **35**(9), 787–795 (2019). https://doi.org/10.1080/10447318.2018.1496969

Esbjörnsson, A.M., Juhlin, O., Weilenmann, A.: Drivers using mobile phones in traffic: an ethnographic study of interactional adaptation. Int. J. Hum. Comput. Interact. **22**(1–2), 37–58 (2007). https://doi.org/10.1080/10447310709336954

Bubb, H.: Traffic safety through driver assistance and intelligence. Int. J. Comput. Intell. Syst. **4**(3), 287–296 (2011)

Nguyen, Q.N., Ta, A., Prybutok, V.: An integrated model of voice-user interface continuance intention: the gender effect. Int. J. Hum. Comput. Interact. **35**(15), 1362–1377 (2019). https://doi.org/10.1080/10447318.2018.1525023

Kim, D., Lee, H.: Effects of user experience on user resistance to change to the voice user interface of an in-vehicle infotainment system: implications for platform and standards competition. Int. J. Inf. Manag. **36**(4), 653–667 (2016). ISSN 0268-4012

Portet, F., Vacher, M., Golanski, C.: Design and evaluation of a smart home voice interface for the elderly: acceptability and objection aspects. Pers. Ubiquit. Comput. **17**, 127–144 (2013). https://doi.org/10.1007/s00779-011-0470-5

Özcan, E., van Egmond, R.: Basic semantics of product sounds. Int. J. Des. **6**(2), 41–54 (2012)

Measuring Driver Distraction
with the Box Task – A Summary
of Two Experimental Studies

Tina Morgenstern[1]([✉]), Daniel Trommler[1], Yannick Forster[2],
Frederik Naujoks[2], Sebastian Hergeth[2], Josef F. Krems[1],
and Andreas Keinath[2]

[1] Chemnitz University of Technology, Chemnitz, Germany
tina.morgenstern@psychologie.tu-chemnitz.de
[2] BMW Group, Munich, Germany

Abstract. The evaluation of the distraction potential of secondary task activities while driving has traditionally been focused on visual-manual tasks. In previous years, different test protocols have been developed and standardized to evaluate the distraction effects of in-vehicle information systems while driving. However, the assessment of cognitive distraction has not received much attention in this context. In the present paper, a new method, that combines a two-dimensional tracking task (the so called 'Box Task') with the Detection Response Task, is proposed. Thus, visual-manual as well as cognitive distraction effects can be assessed. Two evaluation studies are summarized that confirm the ability of this new evaluation method to distinguish between different types and levels of distraction.

Keywords: Box Task · Lane Change Task · Evaluation methods · In-vehicle information systems

1 Introduction

The use of advanced in-vehicle information systems (IVIS) and smartphones while driving has increased in recent years. Research indicates that using IVIS while driving can distract drivers from their primary driving task of guiding the vehicle safely and reacting to road hazards in a timely manner [1]. The use of nomadic electronic devices (e.g., smartphones or tablets) creates an even greater risk as they provide additional opportunities for drivers to perform complex everyday tasks, such as checking e-mails, browsing or watching videos. Complex visual-manual interactions while driving increase the risk of safety-critical events substantially, mainly because they require long off-road glances [2].

To relieve the user, most automobile manufacturers offer voice-based user interfaces (UIs) in their vehicles. In recent years, these voice-based UIs cover more and more in-vehicle tasks, making visual-manual interaction virtually obsolete. However, voice-based interaction can cause cognitive load [3, 4], which in turn can increase visual distraction due to impaired gaze behavior [5]. The goal of manufacturers, especially interface designers, is, therefore, to guarantee that the use of IVIS does not,

H. Krömker (Ed.): HCII 2020, LNCS 12213, pp. 51–60, 2020.
https://doi.org/10.1007/978-3-030-50537-0_5

or only little, interfere with the primary driving task. To assess this, easy-to-use methods for measuring distraction caused by IVIS are necessary.

Currently, several standardized test methods are available to assess the distraction potential of IVIS, such as the lane Change Task (LCT) [6, 7], the Alliance of Automobile Manufacturers (AAM) distraction protocol [8] or the NHTSA distraction protocol [9]. However, driver distraction is a multi-dimensional construct, and there are indications that these test protocols cannot distinguish between *different* dimensions of driver distraction [10]. A relatively new method that is theoretically able to measure both visual-manual as well as cognitive distraction is the so-called "Box Task" (BT) combined with the Detection Response Task (DRT) [11]. Visual-manual load induced by a secondary task is measured via a two-dimensional tracking task, while cognitive load is assessed simultaneously by a haptic DRT [12]. This approach is based on the Dimensional Model of Driver Demand, which consists of the two components – "physical demand" and "cognitive demand" [13].

This paper presents a summary of two experimental studies that were used to evaluate the BT+DRT. The validation studies consisted of

(1) a comparison of the BT+DRT with established test protocols using a range of artificial and naturalistic secondary tasks and
(2) an in-depth study investigating the BT+DRT's capability to distinguish between different dimensions of driver distraction.

2 Experiment 1

In a first experiment, the potential of the BT+DRT to assess secondary task demand while driving was supposed to be examined by comparing this method with two already established test protocols. Specifically, the LCT and a driving simulation task (implemented according to the NHTSA test protocol [9]) were chosen as comparison methods. Moreover, different artificial as well as naturalistic secondary tasks were used to evaluate the BT+DRT in comparison to the LCT and the driving simulation task.

2.1 Method

Participants and Experimental Design. The sample of the first experiment consisted of 29 participants (12 female, 17 male) with a mean age of 32 years ($SD = 7.29$). All participants held a driving license. During the experiment, the participants had to engage in secondary tasks while performing the BT+DRT, the LCT and a driving simulation task. A 3×7 within-subject design with two independent variables was used (factor 1: "test method", factor 2: "secondary task"). The factor "secondary task" consisted of seven different secondary task conditions (including a baseline without secondary task engagement).

Material. Within the BT, a two-dimensional tracking task was simulated. Participants were instructed to hold a blue box within two yellow boundaries (see Fig. 1). The blue box was changing its lateral position and size continuously within a sinusoidal pattern.

Participants had to react through using a steering wheel (to adjust the lateral position of the box; representing lane maintenance) and a foot pedal (to adjust the size of the box, representing headway). For the presentation of the BT, we used a 23″ flat screen monitor. Box size and box position could be controlled using a Logitech MOMO force-feedback steering wheel with foot pedal. In parallel to the BT, participants had to react to haptic stimuli that were presented on their right shoulders every three to five seconds by pressing a button on the steering wheel. The implementation of the DRT followed the ISO norm [12].

Within the LCT, participants were asked to drive at a constant speed of 60 km/h on a three-lane road. Drivers were instructed to change lanes according to signs appearing at the roadside. The LCT was implemented according to the ISO standard [6]. For the presentation of the LCT, we used the same hardware as for the BT.

Within the driving simulation task, participants had to drive with a speed of approximately 80 km/h while maintaining the lane and keeping a constant distance to a lead vehicle. The driving simulation task was implemented according to the NHTSA test protocol with benchmark (DS-BM) in a dynamic car following scenario (DFD-BM) [9]. The driving simulator was a static simulator that covered a 135-degree field of view using three projections of 220 cm × 200 cm each.

Fig. 1. Example screen of the Box Task: The two yellow boxes represent the guide boxes (i.e. inner and outer boundaries) (Color figure online)

Six secondary tasks that differed in respect to their resource demands and their type (i.e., artificial or naturalistic) were employed:

- Counting Task (artificial, cognitive): Participants had to count backwards in steps of 7 starting from 581, 583 or 585 [14].
- Rapid Serial Visual Presentation (RSVP; artificial, visual-manual): Participants had to react to numbers that randomly appeared between serially presented letters by touching a tablet device [15].
- Surrogate Reference Task (SuRT; artificial, visual-manual): Participants had to search for a target circle (8 mm in diameter) in a collection of distractors (6 mm in diameter) that are presented on a tablet device [16].
- Radio Tuning (naturalistic, visual-manual): Participants had to adjust radio frequencies on a smartphone. The radio tuning task was implemented with the AAM radio tuning app of the Technical University of Munich [17].

- Destination Entry Task (naturalistic, visual-manual): Participants had to enter an address consisting of city, street and street number in different input fields on a tablet device. The destination entry task was implemented according to the specifications of the NHTSA [9].
- Texting Task (naturalistic, visual-manual-cognitive): Participants had to enter missing words of well-known proverbs on a smartphone. The texting task was implemented according to Ranney et al. [18].

Procedure. The experiment consisted of three test blocks (BT+DRT, LCT and driving simulation task). At first, participants were introduced to the respective test method (i.e., BT+DRT, LCT or driving simulation task). After a familiarization phase, a baseline run was conducted (i.e., the test method was performed without secondary task engagement). Before each dual task trial, participants received a detailed instruction to the respective secondary task and had the opportunity to practice the task. Test methods and secondary tasks were (partially) balanced.

Dependent Measures. Table 1 provides an overview of the performance measures that were used for the analyses.

Table 1. Performance measures of the test methods.

Test method	Performance measure	Description
BT	Variability of box size (SDLatP)	Standard deviation from the ideal box size
	Variability of box position (SDLongP)	Standard deviation from the ideal box position
DRT	Hit rate	The rate for correct hits (i.e., participants reactions to the stimulus within 2.5 s)
	Mean reaction time	The time participants needed to react to the stimulus
LCT	Mean path deviation (MDEV)	Mean path deviation from the ideal lane change path (using the adaptive model; i.e., the reference path was participants' baseline run)
Driving simulation task	Number of lane exceedances	The number of instances the vehicle crossed the lane boundaries
	Standard deviation of lane position (SDLP)	The variability of the deviation from the lane center
	Mean distance headway	The mean distance of the vehicle to the lead vehicle

2.2 Results

Significant differences across secondary tasks regarding the variability of box position and box size were found. Participants showed the highest variability of box position and box size during the texting and destination entry condition. Regarding DRT performance measures, hit rate differed significantly across the secondary tasks. Participants showed the lowest hit rate during the counting condition. In addition, there were significant differences in mean reaction time across the secondary tasks. Here, the lowest mean reaction time was found during the texting condition.

The MDEV values of the LCT also differed significantly across the secondary tasks. On average, participants showed the highest MDEV values during the destination entry condition.

The mean number of lane exceedances in the driving simulation task differed only slightly across the secondary task conditions. The lowest numbers of lane exceedances were found during the counting condition, the highest during the texting condition. The SDLP did not differ significantly between the secondary task conditions. Regarding the mean distance headway, a significant difference across secondary tasks was found. The greatest mean distance headway occurred in the texting condition, the lowest in the counting condition.

In Fig. 2, the performance measures of the test methods across the secondary task conditions are depicted. To make the performance measures across the different test methods more comparable, z-standardized mean values were calculated.

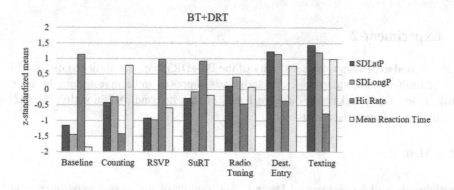

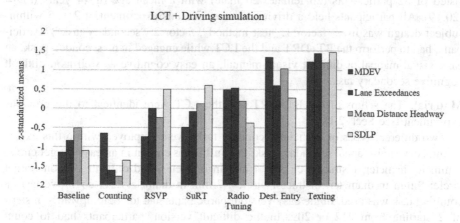

Fig. 2. Z-standardized performance measures across the test methods and secondary tasks in Experiment 1.

2.3   Discussion

The results of the first experiment indicate that the BT+DRT is an effective method to assess secondary task demand. The performance measures of the BT were particularly sensitive to visual-manual secondary task demand. The results were comparable to those of the LCT. In addition, the results regarding the DRT performance measures showed that the DRT is able to cover cognitive distraction effects. Participants showed the lowest hit rate during the counting task and the highest mean reaction time during the texting task. Both tasks are cognitively distractive. Surprisingly, the performance measures of the driving simulation task, which was implemented according to the NHTSA guidelines [9], were less sensitive to different kinds of distraction. Only for the mean distance headway, a significant difference across secondary task conditions could be found. The commonly used SDLP as well as the mean number of lane exceedances could not accurately differentiate between the secondary tasks. Moreover, participants showed the best driving performance during the counting condition (and not during the baseline condition), indicating that drivers might have enough spare capacities to engage in secondary tasks while driving in a simple car following scenario.

3   Experiment 2

In the second experiment, the sensitivity of the BT+DRT (i.e., the ability to distinguish between different secondary task demands) was supposed to be investigated by using artificial secondary tasks with different levels of task demand. Additionally, the LCT was used as a comparison method.

3.1   Method

Participants and Experimental Design. The sample of the second experiment consisted of 52 participants (26 female, 26 male) with a mean age of 44 years ($SD = $ 20.19). All participants held a driving license. For this experiment, a 2×5 within-subject design was used (factor 1: "test method", factor 2: "secondary task"). Participants had to perform the BT+DRT and the LCT while engaged in no secondary task, an easy visual-manual, a difficult visual-manual, an easy cognitive as well as a difficult cognitive secondary task.

Material. The setups of the BT+DRT and the LCT were identical to the previous experiment (see Sect. 2.1).

Two different types of artificial secondary tasks were employed within this experiment. As visual-manual secondary task, the SuRT was chosen in an easy (target circle: 8 mm in diameter, distractor circles: 4 mm in diameter) and difficult version (target circle: 8 mm in diameter, distractor circles: 7 mm in diameter). As cognitive task, a counting task was used. In the easy version, participants had to count upwards in steps of 2 starting from 212 or 209. In the difficult version, participants had to count backwards in steps of 7 starting from 585 or 582 [14].

Procedure. In Experiment 2, we followed the same procedure as in Experiment 1. There were two test blocks (BT+DRT and LCT). Participants received an instruction to the primary driving task (i.e., BT+DRT or LCT). After a practice trial, participants were asked to perform the primary driving task without secondary task engagement. Before each dual task trial, the respective secondary task was explained. Test methods and secondary tasks were balanced.

Dependent Measures. The dependent measures regarding BT, DRT and LCT were identical to Experiment 1 (see Sect. 2.1).

3.2 Results

We found significant differences across the secondary tasks for the variability of box position and box size. Participants showed the highest variability during the difficult SuRT condition for both box size and box position. When considering the DRT performance measures, significant differences across secondary tasks were found for hit rate and mean reaction time. Participants' hit rate was the lowest and mean reaction time the highest in the difficult counting condition.

The MDEV in the LCT also differed significantly across secondary task conditions. Participants showed the highest MDEV in the difficult SuRT condition.

In Fig. 3, the z-standardized mean values across the test methods and secondary task conditions are presented.

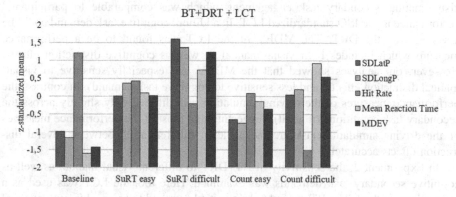

Fig. 3. Z-standardized performance measures across the test methods and secondary tasks in Experiment 2.

3.3 Discussion

The results of the second experiment indicate that the BT+DRT is able to distinguish between visual-manual as well as cognitive secondary tasks of different task difficulty. Whereas the BT is especially sensitive to visual-manual secondary task demand, the DRT is able to cover cognitive distraction effects. The results of the LCT were comparable to the results of the BT, indicating that the LCT is especially sensitive to visual-manual secondary task demand. Moreover, the results showed that the difference in participants' BT and LCT performance between the easy and difficult SuRT condition was significant. This indicates that the performance measures of the BT and the LCT are sensitive to different visual-manual task demand. However, we could also find significant differences between the easy and difficult condition of the counting task for both BT and LCT. Hence, BT and LCT are also sensitive to cognitive distraction effects (albeit to a lesser extent). However, the overall BT and LCT performance during the cognitive secondary tasks was higher than during the visual-manual secondary tasks. In contrast, during the DRT, participants' hit rate was higher and mean reaction time lower when engaged in the visual-manual compared to the cognitive secondary tasks. In addition, the performance measures of the DRT were able to distinguish between the easy and difficult counting task significantly. However, there were also significant differences between the easy and difficult SuRT conditions.

4 Conclusions

The aim of the present paper was to summarize the results of two experimental studies on the evaluation of the Box Task method. In the first study, the BT+DRT was compared with established test protocols (i.e., LCT and NHTSA test protocol), whereas in the second study the sensitivity of the BT+DRT was examined by focusing on secondary tasks with different levels of task difficulty.

The results of Experiment 1 indicate that the BT+DRT can assess several levels of secondary task demand. In particular, participants' BT performance decreases during visual-manual secondary task engagement, which was comparable to participants' performance in the ISO standardized LCT. In addition, cognitive task demand could be assessed with the DRT. The MDEV of the LCT was found to be a performance measure which includes both visual-manual as well as cognitive distraction effects. However, our analyses showed that the MDEV was especially sensitive to visual-manual distraction effects (and less sensitive to cognitive task demand). In contrast, the performance measures of the driving simulation task differed only slightly across the secondary task conditions (if at all), suggesting that most of the performance measures of the driving simulation task are incapable of differentiating between several distraction effects accurately.

In Experiment 2, the sensitivity of BT+DRT to different visual-manual as well as cognitive secondary task demands was examined. Here too, the LCT was used as a comparison method. The BT proved to be a suitable method to assess different levels of visual-manual secondary task demand. Due to the integration of the DRT, cognitive distraction effects could also be assessed. The results regarding the LCT suggest that this test method can be used to investigate visual-manual task demand. However, the LCT is less able to cover cognitive distraction effects. This is in line with the finding that cognitive distraction has only little impact on lane keeping performance [3].

In sum, the two experimental studies have shown that the BT+DRT seems to be a cost-effective and easy-to-implement method to assess in-vehicle system demand. Compared to previous methods, the BT+DRT has the advantage that visual-manual as well as cognitive distraction effects can be assessed separately.

Acknowledgment. This research was funded by the BMW Group. Statements in this paper reflect the authors' views and do not necessarily reflect those of the funding body.

References

1. Tijerina, L., Parmer, E., Goodman, M.J.: Driver workload assessment of route guidance system destination entry while driving: a test track study. In: Proceedings of the 5th ITS World Congress, Seoul, pp. 1–8 (1998)
2. Simmons, S.M., Hicks, A., Caird, J.: Safety critical event risk associated with cell phone tasks as measured in naturalistic driving studies: a systematic review and meta-analysis. Accid. Anal. Prev. **87**, 161–169 (2016). https://doi.org/10.1016/j.aap.2015.11.015
3. Engström, J., Markkula, G., Victor, T., Merat, N.: Effects of cognitive load on driving performance: the cognitive control hypothesis. Hum. Fact. **59**(5), 734–764 (2017). https://doi.org/10.1177/0018720817690639
4. Strayer, D.L., Cooper, J.M., Turrill, J., Coleman, J., Medeiros-Ward, N., Biondi, F.: Measuring Cognitive Distraction in the Automobile. AAA Foundation for Traffic Safety, Washington, D.C. (2013)
5. Trbovich, P., Harbluk, J.L.: Cell phone communication and driver visual behavior: the impact of cognitive distraction. In: Proceedings of CHI 2003 Extended Abstracts on Human Factors in Computing Systems, Ft. Lauderdale, pp. 728–729 (2003)
6. ISO 26022: Road vehicles - Ergonomic aspects of transport information and control systems - Simulated lane change test to assess in-vehicle secondary task demand (2010)
7. Mattes, S.: The lane-change-task as a tool for driver distraction evaluation. In: Strasser, H., et al. (eds.) Quality of Work and Products in Enterprises of the Future, pp. 57–60. Ergonomia, Stuttgart (2003)
8. Alliance of Automobile Manufacturers Driverfocus Telematics Working Group: Statement of Principles, Criteria and Verification Procedures on Driver Interactions with Advanced In-Vehicle Information and Communication Systems (2006). https://autoalliance.org/wp-content/uploads/2018/08/Alliance-DF-T-Guidelines-Inc-2006-Updates.pdf
9. National Highway Traffic Safety Administration: Visual-Manual NHTSA Driver Distraction Guidelines for In-Vehicle Electronic Devices. Report No. 37, vol. 77. National Highway Traffic Safety Administration, Washington D.C. (2012)
10. Engström, J., Markkula, G.: Effects of visual and cognitive distraction on lane change test performance. In: Proceedings of the 4th International Driving Symposium on Human Factors in Driver Assessment, Training and Vehicle Design, Washington D.C. (2007)
11. Hsieh, L., Seaman, S.: Evaluation of the two-dimensional secondary task demand assessment method. Unpublished Report, Department of Communication Sciences and Disorders, Wayne State University (n.d.)
12. ISO 17488: Road vehicles—Transport information and control systems—Detection-response task (DRT) for assessing attentional effects of cognitive load in driving (2016)
13. Young, R., Seaman, S., Hsieh, L.: The dimensional model of driver demand: visual-manual tasks. SAE Int. **4**(1), 33–71 (2016). https://doi.org/10.4271/2016-01-1423
14. Petzoldt, T., Krems, J.F.: How does a lower predictability of lane changes affect performance in the Lane change task? App. Ergon. **45**(4), 1218–1224 (2014). https://doi.org/10.1016/j.apergo.2014.02.013
15. Broadbent, D.E., Broadbent, M.H.P.: From detection to identification: response to multiple targets in rapid serial visual presentation. Percept. Psychophys. **42**(2), 105–113 (1987). https://doi.org/10.3758/bf03210498

16. Mattes, S., Hallén, A.: Surrogate distraction measurement techniques: the lane change test. In: Regan, M., Lee, J., Young, K. (eds.) Driver Distraction 2006, pp. 107–122. CRC Press (2009)
17. Krause, M., Angerer, C., Bengler, K.: Evaluation of a radio tuning task on Android while driving. Procedia Manuf. **3**, 2642–2649 (2015). https://doi.org/10.1016/j.promfg.2015.07.334
18. Ranney, T.A., Baldwin, G.H.S., Parmer, E., Martin, J., Mazzae, E.N.: Distraction effects of manual number and text entry while driving. Report No. DOT HS 811 510. National Highway Traffic Safety Administration, Washington D.C. (2011)

I Care Who and Where You Are – Influence of Type, Position and Quantity of Oncoming Vehicles on Perceived Safety During Automated Driving on Rural Roads

Patrick Rossner[✉] and Angelika C. Bullinger

Chair for Ergonomics and Innovation, Chemnitz University of Technology,
Chemnitz, Germany
patrick.rossner@mb.tu-chemnitz.de

Abstract. There is not yet sufficient knowledge on how people want to be driven in a highly automated vehicle. Currently, trajectory behaviour as one part of the driving style is mostly implemented as a lane-centric position of the vehicle in the lane, but drivers show quite different preferences, especially with oncoming traffic. A driving simulator study was conducted to investigate seemingly natural reactive driving trajectories on rural roads in an oncoming traffic scenario to better understand people's preferences regarding driving styles. 30 subjects experienced a static and a reactive (based on manual driving) trajectory behaviour on the most common lane widths in Germany: 2.75 m and 3.00 m. There were twelve oncoming traffic scenarios with vehicle variations in type (trucks or cars), quantity (one or two in a row) and position (with or without lateral offset to the road centre) in balanced order. Results show that reactive trajectory behaviour and wider lane widths lead to significantly higher perceived safety. We also identified quantity, type and position of oncoming vehicles as factors that influence perceived safety during automated driving. Trucks and vehicles with lateral offset to the road centre lead to significantly lower perceived safety. We recommend an adaptive driving trajectory, which modifies trajectory behaviour on different lane widths and adjusts its behaviour on type and position of oncoming vehicles. The results of the study help to design an accepted, preferred and trustfully trajectory behaviour for highly automated vehicles.

Keywords: Automated driving · Trajectory behaviour · Perceived safety · Rural roads

1 State of Literature and Knowledge

Sensory and algorithmic developments enable an increasing implementation of automation in the automotive sector. Ergonomic studies on highly automated driving constitute essential aspects for a later acceptance and use of highly automated vehicles [1, 2]. In addition to studies on driving task transfer or out-of-the-loop issues, there is not yet sufficient knowledge on how people want to be driven in a highly automated vehicle [3–5]. First insights show that preferences regarding the perception and rating

© Springer Nature Switzerland AG 2020
H. Krömker (Ed.): HCII 2020, LNCS 12213, pp. 61–71, 2020.
https://doi.org/10.1007/978-3-030-50537-0_6

of driving styles are widely spread. Many subjects prefer their own or a very similar driving style and reject other driving styles that include e.g. very high acceleration and deceleration rates or small longitudinal and lateral distances to other road users [6, 7]. Studies show that swift, anticipatory, safe and seemingly natural driving styles are prioritized [8, 9]. In existing literature, trajectory behaviour as one part of the driving style is mostly implemented as a lane-centric position of the vehicle in the lane. From a technical point of view this is a justifiable and logical conclusion, but drivers show quite different preferences, especially in curves and in case of oncoming traffic [10, 11]. In manual driving, subjects cut left and right curves and react on oncoming traffic by moving to the right edge of the lane. When meeting heavy traffic, subjects' reactions are even greater [12–14]. The implementation of this behaviour into an automated driving style includes high potential to improve the driving experience in an automated car. Previous studies [15, 16] showed tendencies to higher perceived safety, significantly higher driving comfort and driving joy as well as preferences for a seemingly natural reactive trajectory behaviour based on manual driving. The paper shows further inferential statistical analyses to identify essential characteristics for trajectory design and modification and gives an outlook on further planned studies.

2 Method and Variables

The aim of the study was to investigate seemingly natural reactive driving trajectories on rural roads in an oncoming traffic scenario to better understand people's preferences regarding driving styles. A fixed-based driving simulator (Fig. 1) with an adjustable automated driving function was used to conduct a within-subject design experiment. 30 subjects experienced a static and a reactive trajectory behaviour on the most common lane widths in Germany: 2.75 m and 3.00 m. This resulted in four experimental conditions that were presented in randomized order to minimize potential systematic biases. All subjects were at least 25 years old and had a minimum driving experience of 2.000 km last year and 10.000 km over the last five years (see Table 1 for details). The static trajectory behaviour kept the car in the centre of the lane throughout the whole experiment whereas the reactive trajectory behaviour moved to the right edge of the lane when meeting oncoming traffic. There were twelve oncoming traffic scenarios that varied in type (trucks and cars), quantity (one or two in a row) and position (cars in the middle of the oncoming lane and cars with lateral offset to the road centre) in balanced order – see Fig. 2. The participants were required to observe the driving as a passenger of an automated car.

Fig. 1. Driving simulator with instructor centre (left) and an exemplary subject (right)

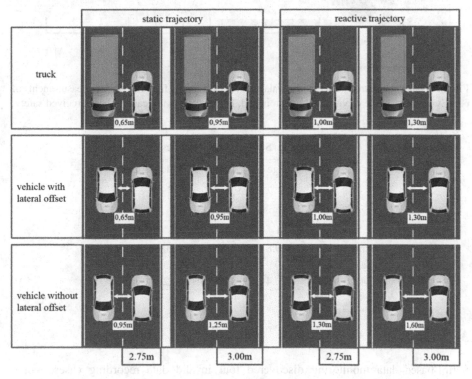

Fig. 2. Variations of oncoming traffic, resultant lateral distances to the ego-vehicle on two different lane widths (2.75 m, 3.00 m) in two different trajectory behaviour models (static, reactive)

During the drive subjects' main feedback tool was an online handset control to measure perceived safety as shown in Fig. 3. This tool provides information about the occurrence of safety concerns in each location of the track and could be recorded in sync with video, eye-tracking, physiological or driving data [9]. After each experimental condition subjects filled in questionnaires regarding acceptance [17], trust in automation [18] and subjectively experienced driving performance [19] and ere interviewed at the end of the study. The questionnaire also required single item ratings regarding perceived safety, driving comfort, driving joy and driving style on a 11-point Likert scale with values from 0 (very low) to 10 (very high).

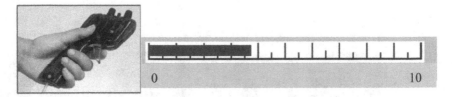

Fig. 3. Handset control (left) and visual feedback (right) for the online-measurement of perceived safety while driving highly automated. Higher values indicate higher perceived safety.

Table 1. Subjects characteristics

	Number	Age		Driver's license holding [years]		Mileage last five year [km]	
		M	SD	M	SD	M	SD
Female	12	29.8	7.9	10.6	4.2	40,083	32,745
Male	18	30.9	6.8	11.9	6.1	68,333	43,661
Total	30	30.4	7.1	11.3	5.3	54,208	41,501

3 Results

Script-based data monitoring discovered four invalid data recording cases, which needed to be excluded for further analysis. In a first step, handset control data was reversed and cumulated to analyse peaks of low perceived safety – hereinafter referred to as perceived safety concerns – during each drive depending on number, type and position of oncoming traffic (see Fig. 4). Based on this overview, perceived safety concerns were further analysed performing four-factor ANOVAs with repeated measurements (see Fig. 5, Table 2 and Table 3).

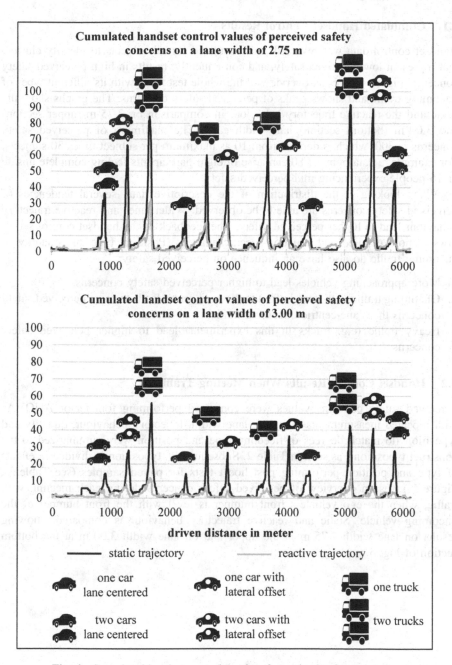

Fig. 4. Cumulated handset control results of perceived safety concerns

3.1   Cumulated Handset Control Results

Handset control data was reversed and cumulated for all subjects to identify clusters that represent low perceived safety, and consequently, results in high perceived safety concerns. Figure 4 gives an overview of the whole test route with its different types of oncoming traffic and shows peaks of perceived safety concerns. The graphs show the static and the reactive trajectory behaviour in comparison on 2.75 m (upper section) and 3.00 m (bottom section) lane width each. The maximum of perceived safety concerns is 300, which is derived from 10 as maximum per subject times 30 subjects. For example, a data point of 80 can arise of eight participants feeling complete unsafe or 16 people experiencing mid perceived safety.

When looking at the distribution of the descriptive data, several tendencies of perceived safety concerns are able to be observed. Wider lanes and reactive trajectory behaviour lead to higher perceived safety. The feedback of the handset control set allows a more detailed and situation-specific analysis. Position, type and quantity of oncoming traffic do also have an influence on perceived safety:

1. More approaching vehicles lead to higher perceived safety concerns.
2. Oncoming traffic with lateral offset to the road centre leads to more perceived safety concerns than lane-centric oncoming traffic.
3. Heavy traffic (e.g. trucks in this experiment) lead to higher perceived safety concerns.

3.2   Handset Control Results When Meeting Traffic

Perceived safety concerns values were compared performing four-factor ANOVAs with repeated measurements including lane width, trajectory behaviour, quantity and typosition (to match degrees of freedom, type and position were summarized to the construct typosition), as seen in Table 2. Subsequently, typosition is divided in effects of type and position performing post hoc t tests for paired samples (see Table 3). Figure 5 gives an overview of perceived safety concerns during each meeting with traffic, when the ego vehicle's front bumper is level with the front bumper of the oncoming vehicle. Static and reactive trajectory behaviour is compared, showing results on lane width 2.75 m in the upper and on lane width 3.00 m at the bottom section of Fig. 5.

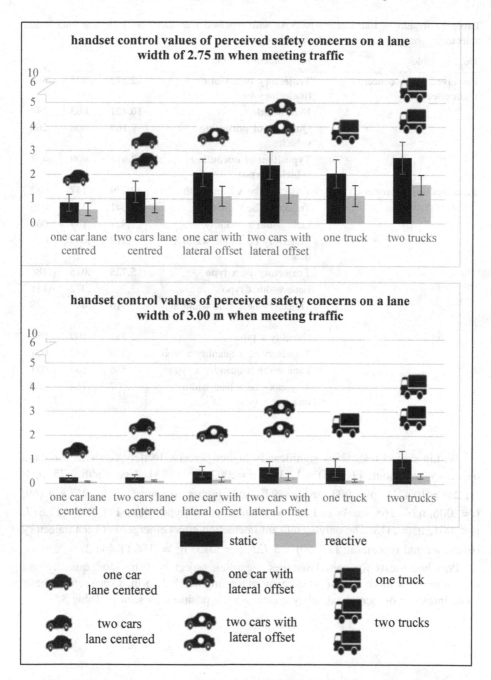

Fig. 5. Handset control results of perceived safety concerns when meeting traffic

Table 2. Results of four-factor ANOVAs with repeated measurements including lane width, trajectory behaviour, quantity and typosition

Dep. variables	Independent variables	F	p	η_p^2
Perceived safety concerns main effects	**Trajectory behaviour (trajectory be)**	**12.372**	**.002**	**.331**
	Lane width	**10.421**	**.003**	**.294**
	Quantity of oncoming vehicles	**9.169**	**.006**	**.268**
	Typosition of oncoming vehicles (typo)	**17.617**	**.000**	**.413**
Perceived safety concerns interaction effects	Trajectory be x lane width	2.239	.147	.082
	Trajectory be x quantity	1.041	.317	.040
	Lane width x quantity	1.827	.189	.068
	Trajectory be x lane width x quantity	.212	.649	.008
	Trajectory be x typo	**5.725**	**.013**	**.186**
	Lane width x typo	3.262	.072	.115
	Trajectory be x lane width x typo	.716	.439	.028
	Quantity x typo	2.385	.109	.087
	Trajectory be x quantity x typo	.035	.952	.001
	Lane width x quantity x typo	.536	.564	.021
	Trajectory be x lane width x quantity x typo	.207	.754	.008

Within-subject tests show significantly higher perceived safety concerns for static trajectory behaviour, $F(1, 25) = 12.372$, $p = .002$, $\eta_p^2 = .331$, lane width 2.75 m, $F(1, 25) = 10.421$, $p = .003$, $\eta_p^2 = .294$, and higher quantity, $F(1, 25) = 9.169$, $p = .006$, $\eta_p^2 = .268$, trucks and cars with lateral offset (typosition), $F(1, 25) = 17.617$, $p < .001$, $\eta_p^2 = .413$. The only significant interaction effect emerges between trajectory behaviour and typosition, $F(2, 50) = 5.725$, $p = .013$, $\eta_p^2 = .186$ (Table 2).

Post hoc t-tests for paired samples regarding trajectory behaviour, quantity and typosition show more significant differences for lane width 2.75 m. Type seems to have more influence on perceived safety concerns than position, as seen in Table 3.

Table 3. Results of post hoc t-tests for paired samples

	t	*p*	*t*	*p*	*t*	*p*	*t*	*p*	*t*	*p*	*t*	*p*
	one car lane centred		two cars lane centred		one car with lateral offset		two cars with lateral offset		one truck		two trucks	
Trajectory behaviour: static – reactive												
2.75 m	1.45	.161	**2.75**	**.011**	**2.15**	**.042**	**3.44**	**.002**	**2.51**	**.019**	**2.12**	**.044**
3.00 m	**2.07**	**.049**	2.37	.026	1.45	.161	1.71	.099	1.33	.196	2.37	.026
Lane width: 2.75 m – 3.00 m												
static	1.80	.085	**2.39**	**.025**	**2.87**	**.008**	**3.30**	**.003**	**2.26**	**.032**	**2.67**	**.013**
reactive	1.86	.075	**2.24**	**.034**	**2.25**	**.033**	**2.51**	**.019**	**2.29**	**.031**	**3.10**	**.005**

Quantity: one – two

	2.75 m			3.00 m		
	car lane centred	one car with lateral offset	truck	car lane centred	one car with lateral offset	truck
static	**-2.62** **.015**	-1.03 .314	**-2.45** **.021**	-.10 .919	-.70 .489	-1.37 .184
reactive	**-2.10** **.046**	-.32 .752	-1.52 .142	-1.14 .267	-.70 .489	-1.98 .058

	t	*p*	*t*	*p*	*t*	*p*	*t*	*p*
Typosition – part type: car with lateral offset – truck								
	2.75 m				3.00 m			
	one		two		one		two	
static	**-2.95**	**.007**	**-2.85**	**.009**	-1.23	.230	**-2.66**	**.013**
reactive	**-2.41**	**.023**	**-2.74**	**.011**	-1.33	.194	-1.90	.060
Typosition – part position: car lane centred – car with lateral offset								
	2.75 m				3.00 m			
	one		two		one		two	
static	**-2.76**	**.011**	**-3.96**	**.001**	-1.64	.113	**-2.14**	**.043**
reactive	**-3.02**	**.006**	-1.77	.089	-1.313	.201	-1.24	.255

4 Conclusion and Outlook

The aim of the study was to investigate seemingly natural reactive driving trajectories on rural roads in an oncoming traffic scenario to better understand people's preferences regarding driving styles. The use of manual drivers' trajectories as basis for implementing highly automated driving trajectories showed high potential to increase perceived safety [10, 11, 15, 16]. Data revealed significantly higher perceived safety concerns for the static trajectory behaviour and the 2.75 m lane width. We also identified quantity, type and position of oncoming vehicles as factors that influence perceived safety during automated driving. We isolated the effect of vehicle type by comparing trucks and cars with identical lateral distances to the ego vehicle. Additionally, we analysed the effect of vehicle position by comparing cars with different lateral differences to the ego vehicle. Both evaluations led to significant differences for the described cases. Trucks lead to significantly higher perceived safety concerns than cars. Cars with lateral offset to the road centre also lead to significantly higher perceived safety concerns than lane centred cars. Based on the results so far, it is concluded that factors which influence perceived safety in manual driving [11–14] are also factors influencing perceived safety during highly automated driving. As drivers cannot react to oncoming traffic by shifting to the right edge of the lane, the automated vehicle has to do so to increase perceived safety and driving comfort of the passenger. A possible conclusion is an adaptive driving trajectory, which modifies trajectory behaviour on different lane widths and adjusts its behaviour on type and position of oncoming vehicles. Therefore, it seems most relevant to investigate manual trajectory behaviour in more detail to implement better reactive trajectories that include less negative side effects and lead to a better driving experience. It is important to note that a positive driving experience has the potential to improve the acceptance of highly automated vehicles [5, 9] and therefore has both ergonomic and economic benefits.

Acknowledgements. This research was partially supported by the German Federal Ministry of Education and Research (research project: KomfoPilot, funding code: 16SV7690K). The sponsor had no role in the study design, the collection, analysis and interpretation of data, the writing of the report, or the submission of the paper for publication. We are very grateful to KonstantinFelbel Marty Friedrich and Maximilian Hentschel for their assistance with data collection and analysis.

References

1. Banks, V.A., Stanton, N.A.: Keep the driver in control: automating automobiles of the future. Appl. Ergon. **53**, 389–395 (2015)
2. Elbanhawi, M., Simic, M., Jazar, R.: In the passenger seat: investigating ride comfort measures in autonomous cars. IEEE Intell. Transp. Syst. Mag. **7**(3), 4–17 (2015). https://doi.org/10.1109/mits.2015.2405571
3. Gasser, T.M.: Herausforderung automatischen Fahrens und Forschungsschwerpunkte, vol. 6. Tagung Fahrerassistenz, München (2013)
4. Mayr, J., Bengler, K.: Literaturanalyse und methodenauswahl zur gestaltung von systemen zum hochautomatisierten fahren. FAT-Schriftenreihe **276**, 1–57 (2015)

5. Siebert, F.W., Oehl, M., Höger, R., Pfister, H.-R.: Discomfort in automated driving – the disco-scale. In: Stephanidis, C. (ed.) HCI 2013. CCIS, vol. 374, pp. 337–341. Springer, Heidelberg (2013). https://doi.org/10.1007/978-3-642-39476-8_69

6. Festner, M., Baumann, H., Schramm, D.: Der Einfluss fahrfremder Tätigkeiten und Manöverlängsdynamik auf die Komfort- und Sicherheitswahrnehmung beim hochautomatisierten Fahren. 32nd VDI/VW- Gemeinschaftstagung Fahrerassistenz und automatisiertes Fahren, Wolfsburg (2016)

7. Griesche, S., Nicolay, E., Assmann, D., Dotzauer, M., Käthner, D.: Should my car drive as I do? What kind of driving style do drivers prefer for the design of automated driving functions? In: Proceedings of Contribution to 17th Braunschweiger Symposium Automatisierungssysteme, Assistenzsysteme und eingebettete Systeme für Transportmittel (AAET), Its Automotive Nord E.V., pp. 185–204 (2016). ISBN 978-3-937655-37-6

8. Bellem, H., Schönenberg, T., Krems, J.F., Schrauf, M.: Objective metrics of comfort: Developing a driving style for highly automated vehicles. Transp. Res. Part F: Traffic Psychol. Behav. **41**, 45–54 (2016)

9. Hartwich, F., Beggiato., M., Dettmann., A., Krems., J.F.: Drive me comfortable: Customized automated driving styles for younger and older drivers. 8. VDI-Tagung ,Der Fahrer im 21. Jahrhundert (2015)

10. Bellem, H., Klüver, M., Schrauf, M., Schöner, H.-P., Hecht, H., Krems, J.F.: Can we study autonomous driving comfort in moving-base driving simulators? Validation study. Hum. Fact. **59**(3), 442–456 (2017). https://doi.org/10.1177/0018720816682647

11. Lex, C., et al.: Objektive erfassung und subjektive bewertung menschlicher trajektoriewahl in einer naturalistic driving study. VDI-Ber Nr. **2311**, 177–192 (2017)

12. Dijksterhuis, C., Stuiver, A., Mulder, B., Brookhuis, K.A., de Waard, D.: An adaptive driver support system: user experiences and driving performance in a simulator. Hum. Fact. **54**(5), 772–785 (2012). https://doi.org/10.1177/0018720811430502

13. Mecheri, S., Rosey, F., Lobjois, R.: The effects of lane width, shoulder width, and road cross-sectional reallocation on drivers' behavioral adaptations. Accid. Anal. Prev. **104**, 65–73 (2017). https://doi.org/10.1016/j.aap.2017.04.019

14. Schlag, B., Voigt, J.: Auswirkungen von Querschnittsgestaltung und laengsgerichtet Markierungen auf das Fahrverhalten auf Landstrassen. Berichte der Bundesanstalt fuer Strassenwesen. Unterreihe Verkehrstechnik, no. 249 (2015)

15. Roßner, P. Bullinger, A.C.: Drive me naturally: design and evaluation of trajectories for highly automated driving manoeuvres on rural roads. In: Technology for an Ageing Society, Postersession Human Factors and Ergonomics Society Europe Chapter 2018 Annual Conference, Berlin (2018)

16. Rossner, P., Bullinger, A.C.: Do you shift or not? Influence of trajectory behaviour on perceived safety during automated driving on rural roads. In: Krömker, H. (ed.) HCII 2019. LNCS, vol. 11596, pp. 245–254. Springer, Cham (2019). https://doi.org/10.1007/978-3-030-22666-4_18

17. Van der Laan, J.D., Heino, A., De Waard, D.: A simple procedure for the assessment of acceptance of advanced transport telematic. Transp. Res. Emerg. Technol. **5**(1), 1–10 (1997)

18. Jian, J.Y., Bisantz, A.M., Drury, C.G.: Foundations for an empirically determined scale of trust in automated systems. Int. J. Cogn. Ergon. **4**(1), 53–71 (2000)

19. Voß, G., Schwalm, M.: Bedeutung kompensativer fahrerstrategien im kontext automatisierter fahrfunktionen. Berichte der Bundesanstalt für Straßenwesen, Fahrzeugtechnik Heft F 118 (2017). ISBN 978-3-95606-327-5

Evaluation of Driver Drowsiness While Using Automated Driving Systems on Driving Simulator, Test Course and Public Roads

Toshihisa Sato[✉] ⓘ, Yuji Takeda ⓘ, Motoyuki Akamatsu,
and Satoshi Kitazaki ⓘ

Human-Centered Mobility Research Center,
National Institute of Advanced Industrial Science and Technology (AIST),
Tsukuba, Japan
toshihisa-sato@aist.go.jp

Abstract. This paper describes an investigation of evaluation indices for assessing driver conditions when using an automated driving system. We focused on a driver drowsiness in the automated mode. A driving simulator experiment was conducted to identify evaluation indices which were sensitive to the subjective evaluation of the driver's drowsiness. The following indices were calculated based on the driver's eye movement data recorded for 60 s before the RtI (Request to Intervene): number of blinks, duration of blinking, PERCLOS (Percent of Eyelid Closure), pupil diameter, number of saccade, amplitude of saccade, and velocity of saccade. We also measured the driver's driving performance after a transition from the automated driving to the manual driving mode. The results of the driving simulator experiment suggested that PERCLOS was sensitive to the subjective assessment of the reduction of the driver's alert level. And this index was highly related to the time to initiate driver's steering operation after the RtI presentation. We have developed a prototype of the driver monitoring system that detects drivers' eyelid movements. The findings obtained from a test course experiment and a public road experiment indicated the effectivity of the driver monitoring system for evaluating quantitatively the driver's drowsiness in the automated driving condition. The results of the public road experiment imply that the duration of blinking as well as PERCLOS might be necessary to estimate the delay of the steering response time after the transition to manual driving.

Keywords: Driver monitoring · Drowsiness · Automated driving systems

1 Introduction

When drivers use automated driving systems (especially, the level 3 in the SAE definition [1]), they might be engaged in non-driving related activities, such as reading texts, matching a movie, gaming, and listening to the music [2]. However, "sleeping" will not be allowed, because the drivers should be in the loop (physical control of the vehicle and monitoring the driving situation) [3] when the operational design domain of the automated driving systems exceeds its functional limitation. Monitoring of a

© Springer Nature Switzerland AG 2020
H. Krömker (Ed.): HCII 2020, LNCS 12213, pp. 72–85, 2020.
https://doi.org/10.1007/978-3-030-50537-0_7

driver's drowsiness and predicting of the influences of the lower arousal level on a transition from the automated to manual driving are one of the fundamental functions to develop driver monitoring systems installed in future automated driving systems. This paper describes an investigation of evaluation indices for assessing the driver drowsiness both in virtual and real driving environments. We measured driver conditions while using an automated driving system and driver behaviors after the transition to manual driving, and then analyzed correlations between the driver states and the transition behaviors.

Several indices have been proposed to assess the driver's arousal level in the general experimental conditions as well as in the automotive human factor domains [4]. In summary, the following changes of the evaluation indices, focusing on eye-related metrics, were suggested when a human felt sleepy:

- Number of blinks increases or decreases [5, 6].
- Blinking duration prolongs and blinking duration ratio increases [7, 8].
- Blinking speed decreases [9].
- PERCLOS (Percent of Eyelid Closure) increases [10].
- Pupil diameter decreases [11].
- Pupil low frequency fluctuation occurs [11].
- Velocity and amplitude of saccadic eye movements decrease [12].
- Slow eye movement occurs [13].

The driver's blinking frequency increased and the driver's blinking duration ratio increased when he/she uses an adaptive cruise control system and a lane-keeping system simultaneously in comparison with a manual driving condition [8, 14]. PERCLOS also increased in the highly automated driving condition than that in manual driving condition, which were suggested based on a driving simulator experiment [15]. When the authors compared the eye-related measurement indices between in a driving seat and in a passenger seat (assuming a human condition under an automated driving mode), the blinking duration was longer in the passenger seat than that in the driving seat [16].

Previous studies were conducted either in a simulator or in a real road condition, and little has investigated the evaluation metrics for assessing a driver drowsiness through virtual and real road traffic environments. In this study, we investigated indices of the driver drowsiness assessment that can be adapted to non-wearable detection under real road traffic environments. Non-wearable detection technique is an important to contribute to practical implementation of the driver monitoring systems.

First, we conducted a driving simulator experiment to select the metrics for assessing the driver's alert level within a variety of eye-related measurement indices. In the driving simulator experiment, we used wearable eye-mark device to detect the driver's eye and eyelid movements. A prototype of a driver monitoring system that can use non-wearable device (camera) has been developed based on the findings obtained from the simulator experiment. Second, a test course experiment was conducted to confirm a possibility of the driver state assessment through the driver monitoring system in a real sunlight environment. Finally, we carried out a field operational test to confirm whether the evaluation indices obtained from the driving simulator and test course experiments were valid for assessing a driver drowsiness for longer travel duration in real road traffic environments. We have also investigated an influence of the driver lower alertness on a transition behavior after a driver received information on RtI (Request to Intervene).

2 Driving Simulator Experiment

2.1 Methods

Driving simulator experiments were performed to evaluate the driver's state, in terms of recognition capabilities, while he/she is using an automated driving system. Scenario assumed a situation where the vehicle was running on a four-lane highway (Fig. 1).

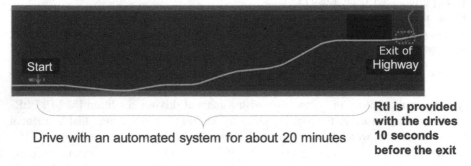

Fig. 1. Driving course in driving simulator experiment. Operational design domain of automated systems is "on highway", and transition from automated to manual driving occurs before the exit of the highway.

It was running alone on the left lane in the automated mode. After driving about 20 min, the vehicle approached a junction (a simulated highway exit), where the participant was instructed to suspend automatically the automated mode (RtI, visual icon and verbal message, occurred) and switch to manual driving before changing lanes to enter the junction. The RtI was presented 10 s before reaching the highway exit. This scenario simulates the situation where a switch to manual driving is required, while running in level 3 automatic driving, to exit from a highway. Assuming the possibility of gradual change in the driver's arousal level before reaching the junction, the arousal level at the time of RtI presentation was evaluated from the facial expression and the subjective evaluation using KSS (Karolinska Sleepiness Scale) [17].

The participant was instructed to exit from the highway when he/she comes to an intersection junction. The following instruction was given to the participant: there is no need to monitor the automated systems and the surrounding driving conditions, and the automated driving system is not capable of handling procedures of exit from the highway. Because the automated drive mode had been turned off before the intersection came into view (the mode change took place before the "Exit Sign" was given), the participant was instructed to drive manually by himself to exit from the highway after he/she received the RtI. The contents of RtI, including presentation timing, interface design, and automatic shutdown of the automated systems just after the RtI were also instructed to the participants before the experimental trial.

The driving simulator consisted a real vehicle cabin, a 6 degrees-of-freedom electric motion system, and a 300-degree field of view screen. 44 drivers (26 females and 18 males, average age: 39.1 years old from 21 to 73 years old) participated in the driving simulator experiment.

Eye movements and Eyelid movements were measured, and we calculated the following indices: number of saccade, amplitude of saccade, velocity of saccade, pupil diameter, number of blinks, blinking duration, PERCLOS (Percent of Eyelid Closure). Eyelid motions were recorded using a wearable eye camera (eye-mark recorder EMR-9, nac Image Technology Inc.) to determine the percentage length of time of eyelid closure in one minute, where eye closure was defined as the state pupil diameter becomes unmeasurable. Eye movement data recorded by the same eye-mark recorder in two minutes before RtI were used to calculate the indices for saccadic eye movements.

2.2   Results

Driver States Indices in Automated Driving. We divided the participants into two groups based on the results of the KSS: low arousal group (29 participants, average KSS: 7.93, SD: 0.92) and high arousal group (15 participants, average KSS: 4.47, SD: 1.81). Figure 2 presents the comparison results of the evaluation indices for assessing the participants' drowsiness before the RtI. A significant difference was found in PERCLOS. PERCLOS was higher in the low arousal group than that in the high arousal group. The difference between the two groups in the pupil diameter suggests significant tendency. The pupil diameter decreased when the participants felt subjectively a drowsiness.

Relationship Between Driver States Before RtI and Driving Performance After RtI. Figure 3 presents the results of a relationship between PERCLOS and the steering operation response time. PERCLOS was calculated before RtI and the steering response was the driving performance index after RtI. The time to onset steering operation was significantly correlated with PERCLOS (correlation coefficient: 0.33, $p < 0.05$). The steering response time was later when the participant's arousal level was lower quantitatively. The pupil diameter was also significantly correlated with the time to initiate the steering operation (correlation coefficient: -0.40, $p < 0.01$).

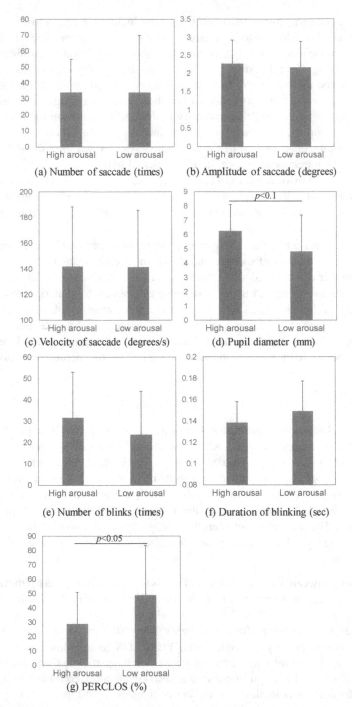

Fig. 2. Results of evaluation indices for assessing driver's drowsiness while using the automated driving system. Each graph shows average and standard deviation (error bar) of the assessment metrics.

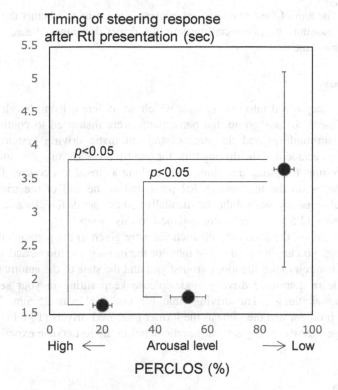

Fig. 3. Results of a relationship between PERCLOS before RtI and time to initiate steering operation after RtI. The data were divided into three categories based on PERCLOS (from 0 to 30%, from 30 to 65%, and higher than 65%). The error bar presents standard errors of PERCLOS and the steering response time respectively.

The results obtained from the driving simulator experiment indicate that PERCLOS and pupil diameter were sensitive to the driver's drowsiness, and these two indices were significantly correlated with the driving performance after the transition from the automated to manual driving. Pupil diameter is influenced by the brightness in surrounding environments. In a driving simulator, the brightness does not change, and thus the pupil diameter could be applied to the driver drowsiness assessment. The brightness is dramatically changed under real road traffic environments due to a time period of a day, weather conditions, shadows by several kinds of structures.

3   Test Course Experiment

A prototype of the driver monitoring systems has been developed to detect the driver states in the automated driving mode. The prototype consists of a camera that detects the position and direction of driver's face, driver's eye movements, and driver's eyelid movements. The aim of the test course experiment is to estimate whether the prototype of the driver monitoring system can measure the driver drowsiness based on the driver's eyelid movements.

3.1 Methods

Drivers were categorized into two groups, which are different from the viewpoint of driving conditions. In one group, the participants were instructed to continue monitoring your surroundings and the state of the automatic driving system to guard yourself against accidents. The driving time for one trial was 15 min. An automatically driven vehicle runs following a leading vehicle along a closed test course. There were no other vehicles on the test course. RtI presented at the end of the trial, and the participant subsequently steered the car manually. 20 drivers (9 females and 11 males, average age was 42.5 years old) were assigned to this group.

In another group, the following instructions were given to the participants: "While you are driving, you are not required to monitor the driving situations, and you do not need to pay attention to the situations around you and the state of the automatic driving system. While in automatic driving mode, please keep sitting in your seat quietly without doing anything." The driving time for one trial was 25 min. The other experimental protocol was the same as the former group. 31 drivers (15 females and 16 males, average age was 47.7 years old) participated in the test course experiment with this scenario.

3.2 Results

Figure 4 presents the results of the driver state metrics measured via the driver monitoring system in the test course experiment. Road environments of a test course have little variety, and driving on a test course could be boring. The instruction of "without the need of monitoring" and the longer driving time could lead to the decrease of the driver's arousal level while using an automated driving system. The number of blinks, the blinking duration, and PERCLOS suggest significant differences in the comparison between the monitoring and no monitoring conditions.

PERCLOS increased in the no monitoring group, suggesting a similar tendency to the driving simulator experiment. The number of blinks was lower and the blinking duration was higher in the no monitoring group. Although there were not statistically significant differences, the changes of the two indices when the driver might feel a drowsiness were the same as the driving simulator experiment.

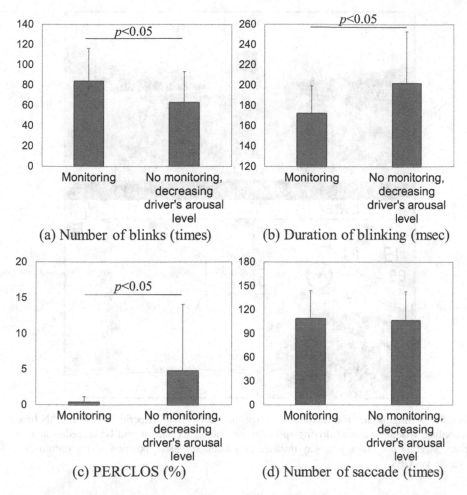

Fig. 4. Results of evaluation indices for assessing driver's drowsiness in the test course experiment. (a), (b), (c) were calculated from the eyelid movement data for 1 min per one participant, and (d) was from the eye movement data for 2 min per one participant. Each graph shows average and standard deviation (error bar) of the assessment metrics.

4   Public Road Experiment

4.1   Methods

Instrumented vehicles for the field operational test of the driving monitoring systems were a Tesla Model S and a Benz E-class both equipped with Level 2 automated driving systems. 42 drivers (18 females and 24 males, average age: 33.5 years old, average driving experience: 14.5 years, non-participation in the simulator and test course experiments) drove each of the instrumented vehicles. The driving route was on Tomei and Shin-Tomei expressways in Japan. Time length of driving was about 3.5 h per one participant.

80   T. Sato et al.

Fig. 5. Driving scene and recording image in public road experiment. The CAN-based recording system measured driving speed, accelerations, accelerator and brake pedals applications, steering operation, headway distance to a leading vehicle, position of the instrumented vehicle via GPS.

An on-site questionnaire was carried out in 15 min time interval by an operator who rode in the instrumented vehicle: the operator asked the participants about their arousal level using the KSS. The driver states were detected by the driver monitoring system, and the vehicle states were measured via a CAN-based recording system during the measurement trial.

Pseudo-RtI (visual icon and auditory message) presented 10 times per one driving route. The automated driving mode terminated after the driver received the pseudo-RtI, and he/she drove manually for about 5 min. Then the driver turned on the automated driving and continued the automatic driving mode. We recorded the driver's hand movements installed in the instrumented vehicle (see Fig. 5) and calculated a time for griping the steering wheel again after he/she received the pseudo-RtI (the driver grasped the lower part of the steering wheel while the automated system was active).

4.2 Results

Figure 6 and Fig. 7 present the results of the relations between KSS and PERCLOS and between KSS and duration of blinking in the public road experiment, respectively.

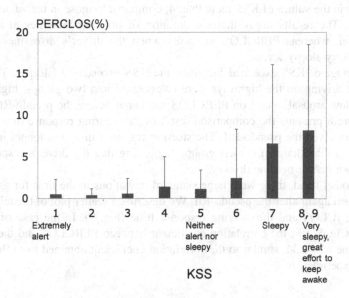

Fig. 6. Results of PERCLOS based on KSS in the public road experiment. Average and standard deviation (error bar) of PERCLOS were calculated per each KSS score, using all of the data from the on-site questionnaires of 42 participants.

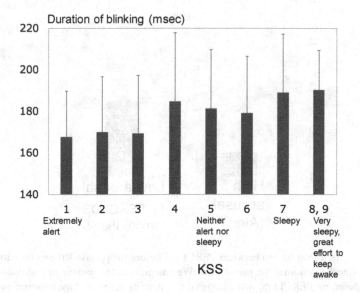

Fig. 7. Results of duration of blinking based on KSS in the public road experiment. Average and standard deviation (error bar) of the duration of blinking were calculated per each KSS score, using all of the data from the on-site questionnaires of 42 participants.

PERCLOS increased as the KSS became higher. Especially, the averages of KSS were more than 5% when the values of KSS exceeded 7 (sleepy). The duration of blinking increased when the participants subjectively felt sleepy, indicating a similar tendency to the relation between the KSS and PERCLOS. The durations of blinking were higher in the values of KSS more than 4, compared to those in the values of KSS less than 3. The results imply that the duration of blinking increases at an earlier drowsy level, whereas PERCLOS increases when the driver's drowsiness reaches sleepy and very sleepy levels.

The average of KSS exceeded 5% when the KSS exceeded 7 (sleepy). The driver states while driving on the highways were categorized into two groups: high arousal (alert) and low arousal, based on PERCLOS measured before the pseudo-RtI presentation. Figure 8 presents the comparison result of the steering response time after the participant received the pseudo-RtI. The steering response time was longer in the low arousal group than that in the alert group, suggesting that the driver's response was delayed when he/she became drowsy.

On the other hand, there were large standard deviations in the time for griping the steering wheel again after the pseudo-RtI. We describe a scatter plot of the relationship between PERCLOS and the steering response time (Fig. 9). In the case of the data above PERCLOS 5%, the correlation coefficient between PERCLOS and the steering response time was 0.34, similar to the correlation coefficient obtained from the driving simulator experiment.

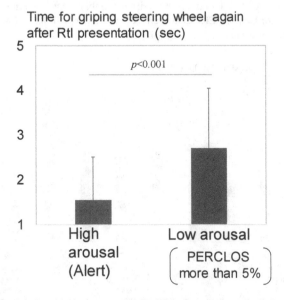

Fig. 8. Results of a relationship between PERCLOS before the pseudo-RtI and time for griping the steering wheel again after the pseudo-RtI. We categorized the participants' drowsiness into two groups based on PERCLOS, and compared the steering response time between in the high arousal (alert) group and in the low arousal group.

When we focused on the steering response time, which were measured in PER-CLOS below 5%, we found out large variation of the response time from about 0 to 6 s. Then, we divided the response time data into two groups, based on the duration of blinking in which the threshold value was 180 ms from Fig. 7. The major part of longer steering response time more than 3 s was found in longer duration of blinking (more than 180 ms). We analyzed the images of the driver's face recorded in the longer steering response. The video images implied that the participants were fighting off drowsiness.

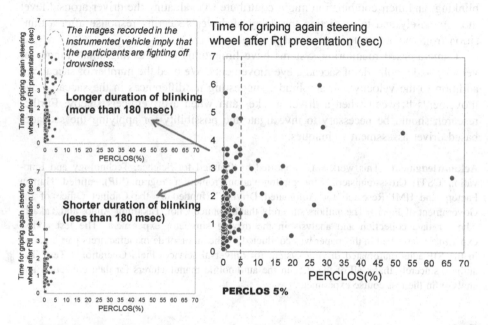

Fig. 9. Scatter plot of the relationship between PERCLOS before the pseudo-RtI and steering response time after the pseudo-RtI. Large variation was found in the steering response time below PERCLOS 5%. The duration of blinking might contribute to estimating longer steering response time in the data below PERCLOS 5%.

5   Discussion

The results obtained from the driving simulator, test course, and public road experiments indicate that PERCLOS is highly related to driver drowsiness in the automated driving systems, and the camera-based driver monitoring system could detect such eyelid movements in real road traffic environments. PERCLOS is well used to estimate a human awake level and to predict consequential behavioral delays or errors in several kinds of experimental settings [18, 19]. Our findings suggest that PERCLOS can be applied to evaluating the driver's drowsiness while an automated driving system is active and to predicting the delay of the driver's response after a transition from the automated to manual driving.

One limitation of measuring PERCLOS concerns a depth of the estimated driver's drowsiness level. PERCLOS could not detect a light sleeping condition due to the necessity for detecting fully eye closures. The results of our public road experiment implied such limitation: Longer time to respond to RtI presentation was found in the lower PERCLOS region (when PERCLOS was less than 5%). The duration of blinking could be applied to an estimation of the lower drowsiness level: The duration of blinking increased at lower KSS values compared to PERCLOS, and the longer steering response time could be categorized based on the average duration of blinking corresponding to the KSS value of 4. Our findings imply that PERCLOS, duration of blinking, and their combination might contribute to evaluating the driver arousal level more precisely and to predicting the delay of driver's steering response after a transition from the automated to manual driving.

Camera-based monitoring systems have limitations for detecting small ranges of velocity and amplitude of saccadic eye movements. We used the number of saccade in addition to the velocity and amplitude, suggesting no differences in the saccadic eye movements between when a driver awakes and when a driver feels sleepy. Further research should be necessary to investigate the possibility for applying the saccade-based driver assessment techniques.

Acknowledgment. This work was supported by Council for Science, Technology and Innovation (CSTI), Cross-ministerial Strategic Innovation Promotion Program (SIP), entitled "Human Factors and HMI Research for Automated Driving" (funded by the Cabinet Office of the Government of Japan). The authors sincerely thank Damee Choi, Takafumi Ando, and Takashi Abe for data collection and analysis in the driving simulator experiment. The test course experiment described in this paper was conducted by the automobile manufacturers participating in the SIP-adus (automated driving system for universal service) Field Operational Test. The authors sincerely thank all of the staff in the automobile manufacturers for data collection and analysis in the test course experiment.

References

1. SAE International: Taxonomy and definitions for terms related to driving automation systems for on-road motor vehicles J3016_201806 (2018)
2. Naujoks, F., Befelein, D., Wiedemann, K., Neukum, A.: A review of non-driving-related tasks used in studies on automated driving. In: Stanton, N. (ed.) AHFE 2017. AISC, vol. 597, pp. 525–537. Springer, Cham (2018). https://doi.org/10.1007/978-3-319-60441-1_52
3. Merat, N., et al.: The "Out-of-the-Loop" concept in automated driving: proposed definition, measures and implications. Cogn. Technol. Work **21**, 87–98 (2019). https://doi.org/10.1007/s10111-018-0525-8
4. Kaida, K., Akerstedt, T., Kecklund, G., Nilsson, J.P., Axelsson, J.: Use of subjective and physiological indicators of sleepiness to predict performance during a vigilance task. Ind. Health **45**, 520–526 (2007)
5. Crevits, L., Simons, B., Wildenbeest, J.: Effect of sleep deprivation on saccades and eyelid blinking. Eur. Neurol. **50**(3), 176–180 (2003)
6. Atienza, M., Cantero, J.L., Stickgold, R., Hobson, J.A.: Eyelid movements measured by Nightcap predict slow eye movements during quiet wakefulness in humans. J. Sleep Res. **13**(1), 25–29 (2004)

7. Tucker, A.J., Johns, M.W.: The duration of eyelid movements during blinks: changes with drowsiness. Sleep **28**, A122 (2005)

8. Jamson, A.H., Merat, N., Carsten, O.M.J., Lai, F.C.H.: Behavioural changes in drivers experiencing highly-automated vehicle control in varying traffic conditions. Transp. Res. Part C Emerg. Technol. **30**, 116–125 (2013)

9. Johns, M.W., Tucker, A.J.: The amplitude-velocity ratios of eyelid movements during blinks: changes with drowsiness. Sleep **28**, A122 (2005)

10. Wierwille, W.W., Ellsworth, L.A., Wreggit, S.S., Fairbanks, R.J., Kim, C.L.: Research on vehicle-based driver status/performance monitoring: development, validation, and refinement of algorithms for detection of driver drowsiness. National Highway Traffic Safety Administration Final Paper, DOT HS 808 247 (1994)

11. Wilhelm, B., Wilhelm, H., Ludtke, H., Streicher, P., Adler, M.: Pupillographic assessment of sleepiness in sleep-deprived healthy subjects. Sleep **21**(3), 258–265 (1998)

12. Di Stasi, L.L., Catena, A., Canas, J.J., Macknik, S.L., Martinez-Conde, S.: Saccadic velocity as an arousal index in naturalistic tasks. Neurosci. Biobehav. Rev. **37**(5), 968–975 (2013)

13. Shin, D., Sakai, H., Uchiyama, Y.: Slow eye movement detection can prevent sleep-related accidents effectively in a simulated driving task. J. Sleep Res. **20**(3), 416–424 (2011)

14. Cha, D.: Driver workload comparisons among road section of automated highway systems. SAE Technical Paper 2003-01-0119 (2003)

15. Dinges, D.F., Mallis, M.M., Maislin, G., Powell, J.W.: Evaluation of techniques for ocular measurement as an index of fatigue and as the basis for alertness management. National Highway Traffic Safety Administration Final Paper, DOT HS 808 762 (1998)

16. Takeda, Y., Sato, T., Kimura, K., Komine, H., Akamatsu, M., Sato, J.: Electrophysiological evaluation of attention in drivers and passengers: toward an understanding of drivers' attentional state in autonomous vehicles. Transp. Res. Part F Traffic Psychol. Behav. **42**(1), 140–150 (2016)

17. Akerstedt, T., Gillberg, M.: Subjective and objective sleepiness in the active individual. Int. J. Neurosci. **52**, 29–37 (1990)

18. Chua, E.C., et al.: Heart rate variability can be used to estimate sleepiness-related decrements in psychomotor vigilance during total sleep deprivation. Sleep **35**(3), 325–334 (2012)

19. Abe, T., et al.: Detecting deteriorated vigilance using percentage of eyelid closure time during behavioral maintenance of wakefulness tests. Int. J. Psychophysiol. **82**(3), 269–274 (2011)

Conflict Situations and Driving Behavior in Road Traffic – An Analysis Using Eyetracking and Stress Measurement on Car Drivers

Swenja Sawilla[✉], Christine Keller, and Thomas Schlegel

Institute of Ubiquitous Mobility Systems, Karlsruhe University of Applied Sciences, Moltkestr. 30, 76131 Karlsruhe, Germany
iums@hs-karlsruhe.de
http://iums.eu

Abstract. Car drivers constantly have to assess and evaluate conflict situations in traffic. Current technology used in autonomous vehicles to detect dangerous traffic situations is not as safe as decisions made by human drivers. We analyzed the behavior of drivers in conflict situations in a real-life study. We used head mounted eyetracking glasses and a pulse bracelet to collect data on driver's behavior. We used eyetracking and video data to analyze conflict situations and driver's gaze in this situations. The collected pulse data was used to determine driving stress. As a result of our work, we derived a classification method to analyze conflict situations in road traffic. We also developed a model of influencing factors of driving behavior in those conflict situations. Furthermore, our real-life study gives an overview of the potential and limits of the use of eyetracking in real traffic situations.

Keywords: Eyetracking · Driving behavior · Real-life study · Driver stress

1 Introduction

The perception of surrounding traffic is crucial for car drivers. Drivers rely on their senses to control their vehicle in a goal-oriented and reliable manner [1]. Compared with other sensations, the visual perception is of particular importance for car traffic. About 90% of driving information is detected via the eyes [2]. In real life traffic, it is important for the driver to perceive crucial information at the right time, in order to react correctly. Many road users interact in road traffic with each other. Every day there is a multitude of different conflict situations. The aim of our research is to get an overview of these conflicts and humans reaction. To receive more information about eye movements, we realized an eyetracking study on the road. In order to gain more conclusions about the eye- and driving behavior, we looked more closely on the situation when participants turned their car. We also investigated whether conflicts causes stress for the driver.

H. Krömker (Ed.): HCII 2020, LNCS 12213, pp. 86–103, 2020.
https://doi.org/10.1007/978-3-030-50537-0_8

1.1 Driving Stress

Everyone knows stress and experiences it in various forms while driving a car. In the following, we discuss stress factors and look how they relate to driving behavior.

Evers (2011) said that stress is caused by a stimulus that can be either physical or psychological in nature. A person's response to such a stimulus can trigger intense stress or emotional tension. Factors such as overload or time pressure also influence stress [3]. Kaluza (1991) stated that especially driving beginners experience stress. For example in the driving test, they have unpredictable situations, which they cannot influence or control themselves. These demanding situations can trigger stress reactions. Particularly when they are unsure, whether the skills they have learned are sufficient to successfully cope with the challenges [4]. According to Matthews (2002), Fig. 1 shows the triggers of stress while driving [5].

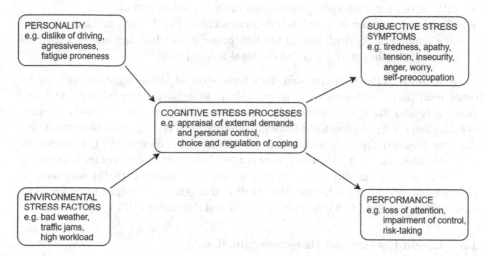

Fig. 1. An outline transactional framework for driver stress [5].

Environmental stress factors like visibility or poor roads, can affect our actions and personal abilities. The personality factor describes the characteristics of a person, such as an aversion to driving. Another example is that a driver, who is prone to frustration, finds a red light signal annoying. In contrast, another driver finds this to be insignificant. These stress factors act in certain situations and influence the cognitive stress process. The cognitive stress process describes the driver's assessment of stressful situations and their ability to cope with them in different ways. For some drivers, the thought of an extreme situation is sufficient to trigger stress and panic. These stress thoughts are controlled and perceived depending on the person. The cognitive stress process includes two results. On the one hand, the subjective stress, such as anxiety, anger or fatigue. On the other hand, the drivers action, such as changes in speed or loss of attention [5].

1.2   Responsible Driving and Self-assessment of Drivers

The VINCI Autoroutes Foundation for responsible driving has launched a European barometer survey on responsible driving. Its aim is to change driving behavior in order to improve road safety. The study surveyed people (N = 11,038) from eleven different countries. Most of the people admit that they violate basic traffic rules [6].

The following shows a list of dangerous and uncivil driving behavior on European roads.

- 89% confirm that they exceed the speed limit. In Germany and Sweden this is even confirmed by 93%.
- 63% admit that they do not keep sufficient distance from the vehicle in front.
- 56% forget to flash when overtaking another road user or turning.
- 55% forget to drive more slowly when construction workers are on the road.
- 21% say they drive without wearing a seat belt.
- 11% drive on the emergency lane when there is a lot of traffic.
- 4% drive even when they feel the effects of alcohol. For 7%, this had a direct effect on a conflict or accident. One in ten Europeans admits to being behind the wheel knowing that they have exceeded the legal alcohol limit.

In addition, 81% of drivers said, they were worried about aggressive driving by other road users. Although most people admit to behave "dangerously", they are positive behind the wheel. Almost all respondents (97%) use the adjectives attentive (74%), calm (57%) and courteous (26%) to describe their driving style. In contrast, few describe their driving style as stressed (11%), never aggressive (3%), irresponsible (1%) and dangerous (1%). However, they are less indulgent when it comes to assessing the behavior of other drivers. At least one negative adjective (83%) was used to describe another drivers behavior. The result is that others are judged to be irresponsible (47%), stressed (34%), aggressive (31%) and dangerous (30%) [6].

1.3   Known Conflicts and Dangerous Situations

In order to find out about conflict and dangerous situations, it is helpful to have a closer look at the development of accidents. A study on accidents on German roads by the German Federal Statistical Office shows that accidents cannot always be traced back to a single cause. In 2017, there were 302.656 accidents with personal injury in Germany with an average of 1.4 causes. Most accidents are due to human error. In 2017, 88% of all recorded accidents were caused by misconduct. About 8% were general causes, such as road conditions or weather conditions. Technical deficiencies corresponded to about 1% [7].

According to the Federal Statistical Office (2017) 16% of accidents occur when turning off, turning around and reversing, as well as when starting up and retracting. 15% of the accidents took place when the right of way or the priority of other road users was disregarded. Other personal injuries were caused if the distance to another vehicle was not sufficiently maintained. In addition, 12% of accidents are caused by non-adapted speed. Incorrect use of the road, such as not keeping to the lane, also caused 12% of the accidents. Wrong behavior towards pedestrians, influence of alcohol and overtaking of another road user correspond to 4% of accidents. The worst consequences of accidents are caused by inappropriate speed. 1.077 people lost their lives in speeding accidents and 60.079 people were injured. About one in three people that died in accidents involving cars, died in accidents caused by non-adapted speed. On the other hand, about 6% were injured when turning or turning around [7].

1.4 Driver Distraction

Several studies concluded that drivers spend up to 50% of their driving time on secondary tasks, that are not relevant to the driving task [1]. On average, these are every six minutes [8]. Objects, persons or events outside the vehicle often distract drivers. It is important that secondary tasks of low complexity, such as talking to a passenger or operating the volume control for music, do not increase the risk of an accident. Visual distraction, on the other hand, such as advertising on the roadside, leads to a greater distraction of the actual driving task. The more eye-catching the advertisement, the higher the probability of attracting attention [9].

According to the Chair of Engineering and Traffic Psychology at the Technical University in Braunschweig, three forms of distraction are distinguished. Each form of distraction has its characteristic features. The first form is the visual distraction. The driver's gaze is not directed towards the road for a certain time. Causes can be conscious actions or events that tempt the driver to turn his gaze there. Another form is the mental distraction. In this case, the driver's gaze is directed to the street. A dangerous situation is seen, but is not recognized by digressive thoughts or, for example, a telephone call. In motor distraction, both the gaze and concentration are directed towards the road. It is not possible for the driver to act appropriately on what is happening on the road. An example for motor distraction is that the driver is drinking or eating while driving. In general, several forms of distraction occur simultaneously. A suitable example is searching for an object in the glove compartment. The driver is visually, mentally and motorically distracted [10].

2   User Study in Real Traffic

Perception processes of sensory input are important for human information processing. Through their sensory organs, humans can interpret the prevailing environmental conditions and adapt their actions accordingly. Vehicle drivers are dependent on their senses in order to steer their vehicle in a target-oriented and reliable manner [1]. Compared to other sensations, visual perception is of particular importance for road traffic. We performed an eyetracking study to gather more information about driving behavior and eye movements of drivers in real traffic.

2.1   Participants

Participants (N = 10) with valid German driver licenses participated in our study. The participants drove their own car and were free to choose when they wanted to drive. Driving their own car reduced possible discomfort. This way, we could rule out any distractions caused by an unfamiliar car. Among all ten participants, two persons were female (participant 1 and 2). The participants were on average 23 years old, eight of them were students, and two working. They had an average driving experience of 5,8 years. Regarding to the driving behavior, we asked the participants about driving practice and traffic tickets. Eight participants reported that they had a speeding ticket before and three participants had received a ticket for illegal parking. Most of the participants stated that they did not have to travel long distances by car during the week. In average they drive 129,17 km per week. According to Kontogiannis (2006), past accidents have a direct impact on stress and driving behavior [11]. For this reason, we asked the participants about their experienced accidents. These we divided into accidents as a driver and those, experienced as a passenger. Six of ten participants had an accident as a driver. One participant was involved in an accident as a passenger. It is remarkable that four accidents were parking accidents. As can also been seen in Sect. 1.3, in Germany, most misconduct occurs during parking [7].

2.2   Study Structure

In road traffic, unpredictable conflicts occur every day. Drivers often have to make decisions. Most information from the surrounding traffic is perceived through vision. Therefore, we used eyetracking glasses and cameras and evaluated occurring conflicts (Fig. 2).

Fig. 2. A participant during our study.

We used the head mounted Tobii Glasses 2.0, that were calibrated with a Lenovo ThinkPad (T460s) and the Tobii Pro Glasses Controller program for analysis (version 1.83.11324-RC1). During our study, we measured the participant's pulse to find out how and if it changes. Therefore, we used a Fitbit Charge 3, as pulse bracelet. Additionally, we equipped the participant's cars with two cameras (Sony 4K Splashproof

Exmor R). One of the cameras recorded the front view and another one focused at the participant. Our study leader, who drove as a passenger, noted conspicuous situations in a notebook.

We gave all participants a destination in advance and told them to reach this destination independently. In general, we observed the participant's driving behavior. We were specifically interested in how the participant behaved towards other road users and in situations and whether patterns are recognizable. Unforeseeable conflicts that can generate stress were also relevant. In the city, we had the focus on changing interaction with other road users and turning processes. During the study, we gave the participants two tasks. One was parking in a parking space. We chose different parking spaces, depending on the route taken by the participants and did not give any instructions whether the participants should park backwards, sideways or forwards. We let the participants choose the way they felt was appropriate for the situation. Our second task for the participant was turning. We allowed the participants to use aids, such as navigation systems. The participants were told to behave as usual. We allowed the participants to listen to music, sing, comments about others, as they would normally do while driving. The study leader documented conspicuous situations and provided an overall impression of the participant's behavior.

3 Results

In order to be able to compare the behavior of the participants in the conflicts, we developed a categorization.

Table 1. Scheme for the analysis of conflict situations while driving a car.

Used Assistance		Pulse			Gaze behavior		Driving behavior		Driving style	
	Meaning	Ranking	(bpm)		Meaning		Meaning		Meaning	
I	Navigation device (fixed)		60-70	1	Observe	11	Flashing	22	Calm	
II	Navigation device (holder attached)		71-80	2	Search	12	Light braking	23	Impatient	
III	Navigation via Smartphone		81-90	3	Gaze not on the street	13	Strongly braking	24	Nervous	
IV	Map (online)		91-100	4	Side mirror	14	Evasion	25	Attentive	
V	Automatic		101-110	5	Rear mirror	15	Evasive onto the oncoming lane	26	Aggressive	
VI	Parking assistance		<130	6	Signpost noticed	16	Accelereate	27	Dangerous	
VII	Lane assistance			7	Not perceived	17	Drive up	28	Stressed	
VIII	Distance sensor			8	Navigation checked	18	Stopping	29	Courteous	
				9	Speed controlled	19	Commenting			
				10	Shoulder look	20	Annoyed			
						21	Demand			

Table 1 defines classifications for the analysis of the participant's conflicts. Number I to VIII define assistance systems used during the study. Number II refers to a navigation system attached to the windscreen or by an external holder. In addition, we distinguished between navigation devices and smartphones. The participants placed the smartphones on their laps or in a rack under the radio. As a result, the eye movements towards the smartphone were different. We encoded the pulse measurements with a color ranking from grey to dark red. The highest measured pulse of a participant determined the limit. Women over 20 years have an average resting pulse of 74 bpm and men a resting pulse of 71 bpm [12]. Since the difference between the genders is very small and only two women took part in the study, the resting pulse of an adult was taken as a reference point. It is between 60–80 bpm [13]. We illustrated the resting pulse in grey tones. Grouping the pulse data in ten steps is sufficient to obtain meaningful findings. Table 1 also lists our categories of gaze behavior, driving behavior and driving style. In the first category, we list the gaze behavior, which was determined by evaluating the eyetracking data. The difference between observing and searching is that the participants fixed a certain object while observing. On the other hand, the gaze behavior was rather erratic while searching. This was especially the case when searching for a parking or turning possibility. Number 3 stands for situations in which the view was directed to instruments in the car instead of the road. Numbers 4 to 6 describe if the objects side mirrors, rear-view mirrors and signposts were perceived. Number 7 indicates whether the participant did not perceive an important object in the traffic environment. The speed was recorded by the eyetracking glasses through the look of the participants at the speedometer display. Number 10 describes a shoulder look where the head was turned and the view was not focused on the front traffic environment. We noted the driving behavior by analyzing the video and sound recordings. Some classifications were recorded by observations of our study leader. The notes of our study leader helped to describe the driving behavior and the reaction in conflict situations. We identified flashing either by the flashing symbol on the speedometer or by the tone of the flashing (number 11). The difference between the categories light and strongly braking is that the participant reduced the speed in a short time during heavy braking (numbers 12 and 13). Evasion means that the participants evaded to the edge on their roadway. In contrast, when switching to the opposite lane, the center lane was crossed (number 14, 15). The driving characteristic "accelerate" was noted as soon as the participant increased their speed in a short time. This was recognizable by the conspicuous engine noise (see number 16). The participant's comments and the impression of the passenger (number 17) lead to notes about the impact on another vehicle. Number 18 describes when the participant stopped their car due to a particular situation. We noted normal comments (number 19) and annoyed comments (number 20) that were made by participants and distinguished normal from annoyed comments by the sound of their voice and the reactions of the participant. For example, an orange traffic light was commented by the participant (categorized as number 19) and complaints were made about a car driving too slowly (categorized as number 20). This was different for each participant and situation and could be registered by the assessment of the study leader in the passenger's seat. The category driving style describes the driver's skill in the conflict situations. For this classification, we used the same adjectives that were used in the questions towards participants in the European barometer [6].

3.1 Conflict Situations in the City

This section deals with the study results. The legends to the tables can be found in the classification discussed in the previous section (Table 2).

Table 2. Information of the participants (scheme: see Table 1).

ID	Local knowledge	Used assistance	Weather				Road		Time (total)	Pulse (Ø)
			Sun	Rain	Cloudy	Snow	Wet	Dry	mm:ss	bpm
1	Local	IV	Yes				Yes		19:10	77,44
2	Local	II, VI		Yes	Yes		Yes		15:59	86,22
3	Local	III, V	Yes				Yes		16:41	88,01
4	Local	I, V, VI		Yes	Yes		Yes		17:56	69,19
5	Known	III	Yes			Yes		Yes	19:40	80,56
6	Unknown	III					Yes		21:16	84,90
7	Unknown	III		Yes	Yes		Yes		20:25	77,05
8	Local	IV					Yes		20:32	73,16
9	Non-local	III	Yes				Yes		31:43	101,31
10	Non-local	II		Yes	Yes		Yes		25:12	103,05

This assessment was noted by the study leader during the study, based on comments and impressions in certain situations, as well as by the information from our survey the participants did before the study.

The participants 1 to 4 and 8, are familiar with driving in our study city. Participants 1 and 8 used Google Maps on their smartphones to find out where the target is. All other participants used a device with a navigation aid. Parking aids were used by participant 2 and 4. The weather conditions were mostly rainy and cloudy due to the winter period. One day there was snowfall. For almost all participants, the road was wet. On average, the participants spent about 21 min in city traffic. Six participants were above the resting pulse. Especially the participants 9 and 10 had a conspicuously permanently increased pulse. Since the study was conducted in real traffic, conflicts occurred individually and were noted by the study leader and analyzed in the video data.

Table 3 shows conflict situations, which the participant caused. For participants 2, 5 and 8, conflicts occurred when turning left and not keeping enough distance. As can be seen in 1.4, this is where most of the misconduct occurs. The participants were impatient and aggressive in these situations. Participant 7 did not notice a green traffic light and pedestrians waiting at the crossing. In both cases, their eyes were not directed at the decisive components. In particular, stopping at a green traffic light was dangerous and the participant was alerted by another road user honking his horn. Participant 3 experienced a severe braking during the threading of a narrowing of the roadway because another road user was braking. The participant had observed this attentively and reacted foresightedly. Due to a narrowing of the roadway, a backlog formed at an intersection. Participant 8 blocked the roadway. He reacted attentively and behaved calmly. In general, the pulse of the participants was mostly in the resting pulse or slightly increased.

Table 3. Conflict situations caused by the driver (scheme: see Table 1).

ID	Turn left		Overlooked		Distance	Road constriction	
	Lane blocked	Drive up another vehicle	Green light	Pedestrians (zebra crossing)	Drive up	Threading	Backwater junction
2		1, 11, 12, 17, 23, 26					
3						1, 13, 19, 25, 29	
5	1, 11, 12, 14, 16, 18, 19, 26, 27						
7			3, 7, 12, 18, 27	7, 22			
8					1, 16, 17, 20, 23, 26		1, 12, 18, 22, 25

ID	Parking		Overlooked		Navigation		
	Forward	Backward	Proceed	Wrong turn	Route unclear	Forgot statement	Wrong turning lane
1	2, 4, 11, 12, 19, 21, 24		7, 19				
2	1, 2, 27						
3	1, 2, 22, 27						
4	1, 2, 22						
5	1, 2, 5, 18, 19, 23, 28			1, 2, 3, 8, 12, 19, 24, 28			
6	2, 12, 18, 19, 24			1, 4, 8, 11, 19, 22	2, 3, 8, 12, 19, 21, 24, 28		1, 2, 8, 11, 12, 14, 19, 24, 28
7	1, 2, 4, 11, 18, 19, 23, 25				1, 2, 3, 6, 8, 14, 19, 24, 28		
8		2, 4, 10, 23, 26	1, 2, 19, 22				
9		2, 4, 10, 22		1, 8, 19, 23	1, 2, 3, 8, 19, 21, 24, 28		
10	2, 11, 22					1, 4, 11, 19, 28	

Table 3 also shows the parking task. When performing the task, eight participants chose to park forwards. It was noticeable that two participants had an elevated pulse. Participant 1, 6, 8, 9 and 10 focused on the search for a parking space and did not pay attention to their traffic environment. The participants behaved differently during parking. Participant 1 and 6 seemed insecure and nervous. The participants 3, 4, 9 and 10 appeared calm. It is striking that, apart from participant 1, 7 and 10, parking was not indicated by a flashing light. This could be an indication of parking misconduct (see Sect. 1.3). It also confirms the result of the VINCI survey (see Sect. 1.2). A frequent conflict was that the participants got lost or had unclear instruction of the navigation aid. At least one conflict occurred among the participants which were locals in the city, but unfamiliar with the area they were driving in. All participants commented on the situation or the instruction of the navigation aid. The pulse of the participants was above the resting pulse during the conflicts with the navigation device and during the wrong turn. When the route was unclear, the eye was not on the road for a few moments, but on the navigation device.

Table 4 shows the conflict situations caused by other road users. The behavior and reaction of the participants was evaluated. Truck and car drivers who did not keep their lane when turning two lanes, caused the most conflict situations in the city. In one case, a lorry restricted vision when turning left at an intersection. Participant 1 behaved calmly and attentively observed the traffic environment.

In addition, the vehicles blocked the road by stopping or unloading and the participants 8 and 10 had to move to the opposite lane. They did not flash during the oncoming lane switch, as was the case with some left turns.

Table 4. Conflict situations caused by other road users – part I (scheme: see Table 1).

ID	LKW			PKW		
	Track is not kept	Visual restriction	Road Blocked	Track is not kept	Road blocked	Parking
1	1, 12, 14, 19, 24	1, 11, 12, 18, 19, 22, 25, 29	1, 4, 5, 11, 15, 22, 25	1, 11, 19, 25		
3	1, 13, 19, 25, 26			1, 12, 19, 23, 26		
4				1, 11, 12, 19, 22, 25		
5					1, 12, 18, 19, 22, 25	
8			1, 11, 12, 14, 18, 23		1, 15, 22	1, 2, 14, 26
10					1,2,15,16,27,28	

Table 5 shows conflict situations caused by pedestrians and cyclist. A frequent conflict situation was when pedestrians were on the street. In one case, pedestrians walked across the street despite the red pedestrian lights. These pedestrians were observed and participant 7 behaved calmly and attentively. Participant 5, on the other hand, did not see the pedestrians on the street. In the case of participant 9, cyclists met in a one-way street. The participant was in a narrow traffic area, where he could hardly evade. He seemed attentive, but at the same time stressed. He kept enough distance and stopped to let the cyclist and his dog pass. Participant 9 was the only one who stated in the survey that he is usually stressed while driving.

Table 5. Conflict situations caused by other road users – part II (scheme: see Table 1).

ID	Pedestrians		Cyclist		
	On the street	Walk over the street by red light	Opposite lane	With dog on a leach	Covered by parking cars
1	1, 14, 22, 25				
5	7, 27				
6	1, 14, 22, 25				
7		1, 12, 18, 19, 22, 25			
8	1, 12, 22, 25				
9			1, 14, 25, 28	1, 2, 12, 18, 22, 25	1, 18, 19, 24, 28, 29
10	1, 14, 22				

3.2 Findings on Driving Behavior

The weather conditions during the study led to visibility restrictions. If the sun shone while driving, it blinded the participants and they put the panel down. This restricted the field of vision. Participant 2 experienced fog on the country road, which also restricted visibility. She commented on this and continued attentively at a slower pace. Some of the participants had restricted visibility due to rain and the windscreen wiper was activated. In addition, the fogged windows of participant 7 led to a reduced visibility. Due to new snow, participant 6 lost control of his car for a short moment when changing lanes. We therefore could observe repeatedly that weather conditions influence driving behavior.

While driving, participant 4 drank his coffee and made a phone call via his car's hands-free system. This means the driver was engaged in secondary activities while driving (see Sect. 1.4). His gaze was not on the road for short moments while drinking coffee and when operating the hands-free car kit. To what extent the participant was distracted from his driving task as a result cannot be determined from the gaze data alone. His pulse was not elevated and he seemed relaxed.

All participants listened to music during the study, either via radio or via smartphone. It was noticeable that some of the participants sang along familiar traffic environment and seemed relaxed. In many situations where glances over the shoulder would have been necessary, the participants did omit them. We got the impression that the participant flashed only when they thought it was necessary. If they had to concentrate, for example when parking, the traffic was observed and they stopped talking.

To sum up all the conflicts, it is clear that the participants observe other road users and comment on their behavior. Some of the participants were annoyed about the behavior of others. When turning, parking, turning in side streets or at roundabouts, however, flashing was often forgotten. When using the navigation device, the participants usually listened to the instructions. Only when the instructions about the route were unclear did the participants look at the navigation device. The participants usually commented the navigation instructions. Especially when they got lost, some participants reacted irritably and stressed. It was noticeable that especially the non-local participants were lost or drove into the wrong lane. Of the five participants, who were familiar with the city, two did not use a navigation device. The others were using it, but were driving "their" route and followed not the navigation information. A shoulder glance was only made in individual cases like parking. A further result is that weather conditions can restrict visibility when driving. We also confirmed that drivers are distracted from driving by secondary activities. We determined that the driver's eyes were not on the road for a few moments when engaging in secondary activities.

3.3 Analysis of Turning a Car with Eyetracking

One task during the study was to turn the car around. According to Federal Statistical Office (2017) turning is one of the accidents with the highest misconduct by a driver [7]. We used the eyetracking data, to analyze the turning process. Our aim was to find out, how misconduct occurs in this situation.

Paragraph 9 of the Road Traffic Regulations (StVO) contains the guidelines for turning off, turning around and reversing. There is an increased duty of care if you want to turn around by car. Turning may be carried out, if the danger to traffic can be excluded. It is necessary to check the traffic before turning the vehicle. Turning should not be done in places where there is no clarity [14]. Driving schools give advice on how to turn the car safely. First, the driver must check the traffic from both directions. In particular, rearward traffic should be monitored. Parking vehicles and the curb should be observed. During the procedure, one should behave calmly and always pay attention to suddenly appearing road users. In order to turn properly, traffic signs should be taken into account [15]. During turning, anything that is not forbidden is permitted. The difficulty is that there are many possibilities and the driver has to make a driving decision [16].

We visualized the recorded eyetracking data with the program Tobii Pro Lab (version:1.102.16417) as heat maps to show the general process of turning. The heat maps illustrate the eye movements of the participants in the form of dots and highlight particularly long looked at Areas of Interests in color.

We used the description of the driving school and the StVO to divide the turning into three steps. First, the driver searches a suitable place for turning and during this time, the traffic environment will be observed (step 1). The driver makes a decision and turns (step 2). Afterwards, the driver places in road traffic again (step 3).

Area of Interests (AOI) allow numerical and statistical analysis based on objects of particular interest to the stimulus [17].

Figure 3a shows, which AOIs we chose to analyze the eyetracking data for the turning task. We kept the design abstract to standardize the turning steps. It shows a view of the car from above. The green rectangles represent the traffic environment and the side space. The yellow rectangles represent instruments of the car. We illustrated the view from the shoulder with blue rectangle and the road users with orange rectangle. We also divided the AOIs into right (R), left (L), front (F) and back (B).

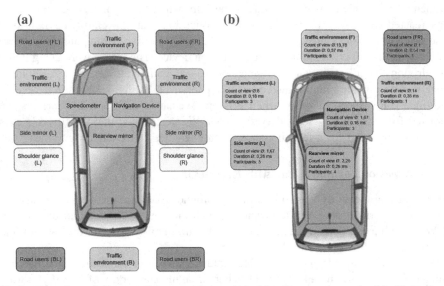

Fig. 3. Overview of all areas of interest (a), step 1: looking for a turning point (b). (Color figure online)

During the first step, all participants with an average of 13 views observed the traffic environment (Fig. 3b). Rear traffic was observed by four participants through the rearview mirror. Three participants looked into the left side mirror and to the traffic environment on the left. On average, the participants looked with 1,67 views in the side mirrors. Compared to the left traffic environment, the side mirror was looked at longer (0,35 ms). One participant had a road user in the right front field of vision, who was observed with one view. Three participants looked at the navigation device with a duration of 0,16 ms.

Figure 4a illustrates the second and third step of the turning. In the second step, all participants looked to the left traffic environment as well as the front traffic environment during the turning. The duration of the views was between 0,30–0,32 ms. Compared to the first step, fewer views were used for the traffic environment in front. One participant looked into the rearview mirror with a duration of 0,08 ms. Five participants used the left side mirror with an average of 1,6 views. The participants did not use the right mirror, during turning.

In the third step, six participants did a right shoulder look when they got themselves into road traffic back (Fig. 4b). On average, at least four views with a duration of 0,19 ms were made. Almost all participants looked back into the left side room and the traffic environment in front. Compared to the other two steps, the participants observed the left field of vision more frequently.

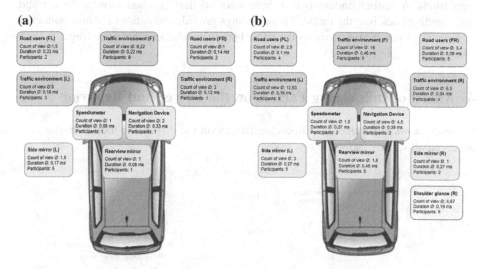

Fig. 4. Step 2: turning (a), Step 3: get back in lane with traffic (b).

If one compares these three steps, the focus of the participants is on the front field of vision. In general, rearward traffic is only partially observed by individual participants. This can be seen in the first and second step, as our participants rarely looked into the rearview mirror. The focus in the first step is on finding a suitable place to turn. The rear traffic is therefore hardly noticed. In addition, when turning in the second step, no participants made a shoulder look. In comparison, the left traffic environment is observed more often than the right. This could be related to how the drivers had to get into the traffic again and how much traffic was on the roads. Some participants looked at the navigation device while turning, especially when the participant was lost. In none of the steps were road users perceived from behind or a left shoulder glance. Driving schools particularly point to observing rear traffic (see Sect. 3.3). Too little observation of the rear traffic field can lead to mistakes and conflicts with other road users.

We also had a closer look on the participant's decision, where they wanted to turn and when they decided to flash. We evaluated this through the video and the sound recordings. It is interesting that many participants used the confluence area, for example a parking lot, for turning. Participant 6 and 8 both turned in a confluence area of a one-way street. Driving on one-way streets in the forbidden opposite direction is gross negligence against § 49 Para. 3 No. 4 StVO and contrary to regulations [18]. Both participants were aware of it through the seen one-way street sign and commented on the turning. Most of the participants indicated their intention by flashing shortly before turning the car, in the second step. Two participants reported this towards the end of the first step, when a spot became visible early on. When they reassigned to road traffic, three participants reported this by flashing. Four out of ten participants did not use the flashing light when turning.

The analysis of the turning shows that the rear traffic environment in particular was hardly observed. In addition, the driver turned to places that were in violation of the regulations. A further finding is that there were no flashes when turning the car and when getting back into the traffic. These findings provide indications for misconduct in turning accidents. It also confirms the VINCI study's statement that 56% forget to flash when turning (see Sect. 1.2).

4   Model of Influencing Factors on the Driver and Behavior

Figure 5 shows our model of influencing factors on the driver and their behavior. We observed all shown factors in our study.

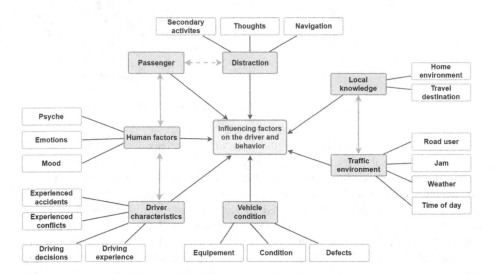

Fig. 5. Influencing factors on the driver and behavior.

Traffic environment can change any time and can hardly be influenced by the driver, but must be taken into account while driving. Examples are weather conditions, other road users, congestion and road conditions. Relevant for the driving behavior is whether the driver is familiar with those factors. Usage of a navigation device can lead to a distraction from the street, for example. Distraction is also caused by secondary activities such as making phone calls while driving or by searching for music. The interaction between passengers is included as influencing factor. On one hand, passengers can distract the driver (represented by dashed line), but they can also provide helpful information. The mood, psyche, and emotions of a driver are individual factors, but affect driving behavior (see Sect. 1.4). This is likely related to driving character-istics such as driving experience and decision-making. For example, previously experienced accidents can trigger stress (see Sect. 1.1). Important influence factors on driving behavior are car characteristics. Each car offers varying equipment for driver's assistance, such as navigation systems and parking assistants. The condition and the functionality of all instruments of the car has influenced driving behavior in our study.

5 Limitations and Conclusion

Some participants did not have a holder for their smartphone and placed it on their laps or on a shelf under the radio. Gaze data on this section are missing, because the participants looked under the eyetracking glasses. It is therefore advantageous for eyetracking evaluations, if the navigation device is installed or attached to an external holder on the windscreen. During pulse tracking, data may be lost if the wristband slips. Therefore, we checked whether the pulse measured before starting the study. It should also be noted that all functions that cause the wristband to blink or vibrate should be, and were, switched off.

Due to bad weather conditions and the danger of storms, one participant cancelled the appointment. Another participant had to be towed away due to a technical defect in the car shortly before the start of the study. It is advisable to include alternative dates in the planning.

As we conducted the study in a real traffic environment, none of the evaluated situations is the same. In an attempt to estimate the driver's different visual tasks, a calculation by Fastenmeier (1995) resulted in 26,460 fundamentally different traffic situations [19]. Therefore, it would be advisable to increase the number of participants, in order to gather more data on similar situations, or to use simulations, if specific situations should be examined. On the basis of the designed tables in Sect. 3.1, an impression was given of the conflicts that had occurred between the participants in traffic, as well as the driving and viewing behavior of the participants. We analyzed conflict situations with the help of the allocation tables and provide indications of important characteristics for recognizing conflicts. Many conflicts occurred in urban traffic, as opposed to rural traffic. This can be explained by the fact that in urban spaces many road users interact with each other in a confined space. What is striking, is the behavior of non-local participants who used a navigation device to reach their desti-nation. Unclear instructions from the navigation device were not understood and looked up on the device. This often led to misinterpretations, which led to a deviation of the route. Further conflicts were caused by the weather and influenced the driving behavior. The sun, for example, restricted the participants view and snow caused the vehicle to behave uncontrollably. In addition, side activities such as drinking coffee averted the view from the road. As Pilgerstorfer and Boets (2015) already surveyed in a study, drinking distractions occur while driving [20].

The evaluation of the pulse data shows that it is not sufficient to obtain a mean-ingful stress level when driving and in future work, additional stress measurements should be taken. However, it can be seen whether the participants pulse was above the resting pulse. It was noticeable that in situations with the navigation device, the pulse was increased and the participants appeared insecure and stressed.

Our study gave an initial insight into the driving behavior and the potential to evaluate it, using modern technologies such as eyetracking glasses and a pulse wristband. The result of the work provides an insight into driving behavior and gives hints on how to identify conflicts. In addition, the work gives an impression of the stress level of participants The turning task was examined more closely with regard to gaze behavior and shows misbehavior on the part of the participants. Drivers are able to react to unpredictable conflicts situations created by other road user. The behavior of other road users is observed, evaluated and assessed. It also becomes clear that there are many components that need to be considered in traffic, since drivers make individual decisions and react differently to different stimuli.

References

1. Kettwich, C.: Ablenkung im Straßenverkehr und deren Einfluss auf das Fahrverhalten, Karlsruher Institut für Technologie: Spektrum der Lichttechnik (2014)
2. Godfrey, N.: DriverMetrics. https://www.drivermetrics.com/blog/the-anatomy-of-driver-behaviour-infographic/. Accessed 9 Feb 2019
3. Evers, C.: Auswirkungen von Belastungen und Stress auf das Verkehrsverhalten von Lkw-Fahrern. VKU Verkehrsunfall und Fahrzeugtechnik, vol. 49, no. 1 (2011)
4. Kaluza, G.: Gelassen und sicher im Stress. Springer, Heidelberg (1991). https://doi.org/10.1007/978-3-662-45807-5
5. Matthews, G.: Towards a transactional ergonomics for driver stress and fatigue. Theor. Issues Ergon. Sci. 3(2), 195–211 (2002)
6. Giobbe, E.: Ferron und Ludovica, Connected Objects, Rudeness and Drowsiness: Scientists Explain Our Driving Behaviour. VINCI Autoroutes, Rueil Malmaison Cedex (2018)
7. Bundesamt, S.: Unfallentwicklung auf deutschen Straßen, Wiesbaden (2018)
8. McEvoy, S., Stevenson, S., Woodward, M.: The impact of driver distraction on road safety: results from a representative survey in two Australian states. Inj. Prev. 12(4), 242–247 (2006)
9. Klauer, S., Dingus, T.A., Neale, V.L., Sudweeks, J.D., Ramsey, D.J.: The impact of driver inattention on nearcrash/crash risk: an analysis using the 100-car naturalistic driving study data. Technischer Bericht, National Highway Traffic Safety Administration (2006)
10. Polizei Rheinland Pfalz: Ablenkung im Straßenverkehr
11. Kontogiannis, T.: Patterns of driver stress and coping strategies in a Greek sample and their relationship to aberrant behaviors and traffic accidents. Accid. Anal. Prev. 38(5), 913–924 (2006)
12. U.S. Department of Health and Human Service: Resting pulse rate reference data for children, adolescents, and adults. National Health Statistics Reports number 41, 1999–2008 United States (2011)
13. American Heart Association: National center. https://www.heart.org/en/health-topics/high-blood-pressure/the-facts-about-high-blood-pressure/all-about-heart-rate-pulse. Accessed 12 Feb 2019
14. Bußgeldkataolg, Wenden mit dem Auto: Neuer Bußgeldkatalog (2019). https://www.bussgeldkatalog.de/wenden/. Accessed 17 Jan 2019

15. Umkehren – Grundfahraufgaben. http://www.grundfahraufgaben.com/umkehren.html. Accessed 17 Jan 2019
16. Bußgeldkatalog: Wenden Sie das Auto bitte - Wie geht das eigentlich? (2019). https://www.bussgeldkatalog.org/wenden-auto/. Accessed 17 Jan 2018
17. Kunze, U.: Evaluation einer Focus+Context Visualisierungstechnik mit Eyetracking. Institut für Visualisierung und Interaktive Systeme, Stuttgart (2018)
18. Fachanwaltskanzlei Verkehrsrecht Hamburg, Ratgeber: Verkehrsregeln. http://fachanwaltskanzlei-verkehrsrecht-hamburg.de/start/bussgeld/pkw-kraftraeder/strassenverkehrsregeln/. Accessed 13 Feb 2019
19. Fastenmeier, W.: Die Verkehrssituation als Analyseeinheit im Verkehrssystem. TÜV Rheinland, Bonn (1995)
20. Pilgerstofer, M., Boets, S.: Wie wirkt sich Ablenkung auf das Fahrverhalten aus?. Kuratorium für Verkehrssicherheit, Wien (2015)

Decision-Making in Interactions Between Two Vehicles at a Highway Junction

Asaya Shimojo[1](✉), Yuki Ninomiya[1], Shota Matsubayashi[1],
Kazuhisa Miwa[1], Hitoshi Terai[2], Hiroyuki Okuda[3],
and Tatsuya Suzuki[3]

[1] Graduate School of Informatics, Nagoya University, Nagoya, Japan
{shimojo,ninomiya}@cog.human.nagoya-u.ac.jp,
shota.matsubayashi@nagoya-u.jp,
miwa@is.nagoya-u.ac.jp
[2] Faculty of Humanity-Oriented Science and Engineering, Kindai University,
Fukuoka, Japan
teraihitoshi@gmail.com
[3] Graduate School of Engineering, Nagoya University, Nagoya, Japan
{h_okuda,t_suzuki}@nuem.nagoya-u.ac.jp

Abstract. Driving often involves situations where interaction between drivers is required, for instance in situations where two lanes merge. In previous studies, models have been proposed wherein the environmental relationship, such as relative distance between two cars determines the drivers' driving behavior (Hiramatsu, Jang, Naemoto, Ito, Yamazaki, and Sunda, 2017). However, according to Simulation Theory in the Theory of Mind, the driver may think "If I were you, I would drive in this way," and so their behavior is determined by referring to this simulation. In this study, this hypothesis was examined in the merging scenario. Results show that a driver would drive according to traffic norms or the driving tendency (e.g., the degree of acceleration/deceleration, or average speed). This suggests that the Simulation Theory proposed in the Theory of Mind is unlikely to be adopted in decision-making process in the merging scenario.

Keywords: Decision-making · Theory of Mind · Lane-changing

1 Introduction

There are many situations wherein the activity of driving requires drivers to interact with each other via a car system. This activity includes merging and overtaking on a highway, or giving way at an intersection or a narrow road. It is possible to perform the interactions even when drivers cannot see each other. To ensure a smooth flow of traffic, drivers determine their own driving behaviors using external information, such as the environmental relationship between themselves, other cars and traffic norms, which are systematically or empirically accumulated. For instance, the environmental relationship could include such things as relative distance and relative speed between two cars, and the traffic norms that include right of way and traffic signs.

© Springer Nature Switzerland AG 2020
H. Krömker (Ed.): HCII 2020, LNCS 12213, pp. 104–113, 2020.
https://doi.org/10.1007/978-3-030-50537-0_9

Previous studies on the interactional behavior between drivers have mainly examined the principle that a driver makes a decision based on the kind of external information they receive. For instance, Hiramatsu, Jang, Naemoto, Ito, Yamazaki, & Sunda (2017) predicted driving behaviors based on the relative distance between a car and the preceding car when following a car in front [1]. They concluded that people's driving behaviors could be predicted by referring to the relative distance between the two cars when two cars are driving in convoy.

Additionally, it has been shown that people's driving behaviors characteristics can be estimated by the relative speed and distance to the car in front, and the model could predict what drivers would do next based on this estimation [2].

However, some studies argue that consideration of other cars affects the decision of the driving behaviors in interactive scenarios between several drivers, such as following a car in front or merging [3]. Yoshikawa & Takagi (2008) examined what factors affect driving behaviors by analyzing what participants said when driving with a talk aloud protocol. Consequently, the utterances about the behaviors of other cars and the decision-making based on the behaviors accounted for about 40% of the total utterances. For instance, it included a report that "I decided to decelerate in consideration of the oncoming car because the road is narrow." Thus, it is suggested that the driving behaviors is determined based on not only objectively external information, but also an estimation of the driving tendency and intentions of the other driver involved.

In psychology, humans can estimate mental states of another people that cannot be directly observed from the outside (hereinafter, "the Model of Others"; e.g., purpose, intention, and belief) to determine their own behaviors. Such mental functions are called Theory of Mind [4]. Simulation Theory has been proposed as one of the mechanisms for estimating the Model of Others [5]. According to Simulation Theory, humans simulate what they would do if they were in another person's situation and convert their own state of mind to that of the other person's state of mind. Then, they would estimate the Model of Others by assuming that the other's mental state is the same as their own and make decisions based on considering this.

In this study, a merging scenario on a highway was used as an experimental situation. Specifically, in order to achieve merging while considering the other car in the adjacent lane, participants decide to accelerate so as to come out ahead of the other car (hereinafter, "Lead decision"), or decelerate to join the lane behind the other car (hereinafter, "Follow decision"). In such a situation, participants need to estimate the Model of the Others and consider the state of the other driver and make a decision based on their simulation. They might think "If I were the driver of the other car, would I lead or follow in this situation?" and make a Lead/Follow decision based on their estimated answer. If decisions are made in this way, the Lead/Follow decision in the merging lane is determined according to their assumption of Lead/Follow decision in the main lane, and vice versa. For instance, a driver who pulled out in front of the other car in the main lane when merging would follow the other car when merging from the merging lane.

If the driver decides their own behaviors based on the simulation that "If I were a driver of the other car, would I lead in front of or follow behind the car for merging in this situation?," their decisions in the merging lane are dependent on the assumed behaviors in the main lane.

2 Method

In this experiment, we used a highway merging junction as an experimental task and examined how drivers make lane-change decisions when moving from the merging lane to the main lane. This experiment was performed with the approval of the research ethics committee at the Institute of Innovation for Future Society (MIRAI) of Nagoya University (approval number: 2019–17).

2.1 Participants

Twenty-four participants (13 women and 11 men, mean age 43.92, range 22–60, $SD = 11.40$) were paid $50 for the 90-min experiment.

2.2 Task

Participants drove in a merging scenario on the highway that consisted of two lanes; a main lane and a merging lane (see Fig. 1). On the highway, two cars were adjacent to one another: a car driven by participants (hereinafter, "self car"), and a car driven in the adjacent lane to the self car (hereinafter, "other car"). Until the car reached 360 m from the starting point, a wall separated the lanes so that participants could not see the other car in the adjacent lane. During this period, participants were required to keep driving at 80 km/h.

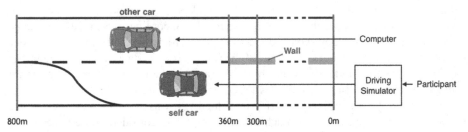

(a) Merging Lane condition

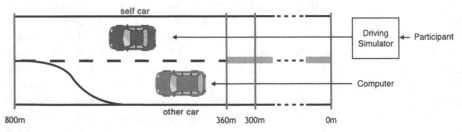

(b) Main Lane condition

Fig. 1. The task situation. Participants drive a car on a road consisting of two lanes, the main lane and the merging lane. On the road, there are two cars: a car driven by participants; and an autonomous car driven on the adjacent lane to the road the self car is being driven on.

Participants drove in the following two conditions; they drove the self car in the merging lane and merged the self car into the main lane (hereinafter, "merging lane condition"), and drove the self car in the main lane and let the other car merge from the merging lane into main lane (hereinafter, "main lane condition"). At that time, participants made Lead/Follow decisions.

This experiment was performed using a driving simulator (see Fig. 2). As shown in Fig. 2-a, the road created by Unity (ver.2019.2.17) was projected on five screens (three in front and one on each side). Participants sat in the driver's seat set in front of the screen and drove in the same way as driving an actual car (see Fig. 2-b).

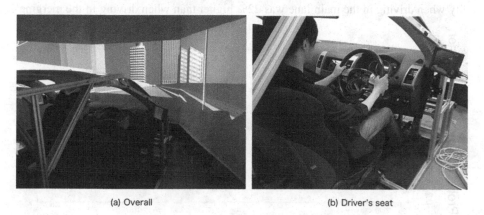

(a) Overall   (b) Driver's seat

Fig. 2. The experimental environment using a driving simulator.

2.3 Procedure

Participants drove under the following two conditions: 5 times in the practice trial, and 25 times in the actual trial. In the actual trial, when participants completed a lane-change, whether the self car lead in front of or follow behind the other car was recorded as a data point. The order of the trials was counterbalanced among participants.

- The merging lane condition:

Participants drove a car in the merging lane and then merged into the main lane. Participants then decided whether to lead in front of (Lead) or follow behind the other car (Follow decision). The other car was set to run autonomously at a constant speed.

- The main lane condition:

Participants drove a car in the main lane and decided whether to lead in front of (Lead) or follow behind the other car coming from the merging lane (Follow decision). The other car ran at a constant speed and merged smoothly into the main lane only when there was a certain amount of distance between the two cars.

Five levels of relative distances were set between the self car and the other car. Specifically, when the self car reached 300 m (shortly before the 360 m mark where other car could be observed), the other car was located +10 m, +5 m, 0 m, −5 m, or −10 m from the 300 m mark. Participants performed five trials under each relative distance for a total of 25 trials. The order of the 25 trials was randomized.

3   Results

3.1   Overall Result

To understand the overall tendency of how decisions are made in the main lane and the merging lane, we conducted logistic regression analysis to predict lead probability at the relative distance (−10 m, −5 m, 0 m, +5 m, +10 m). Figure 2 shows the results of this analysis.

The lead probability at the intercept of the model (in short, the lead probability when the relative distance is 0) was calculated. The results show that the lead probability when driving in the main lane was 22% higher than when driving in the merging

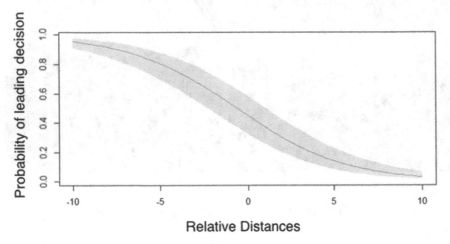

(a) In the merging lane condition

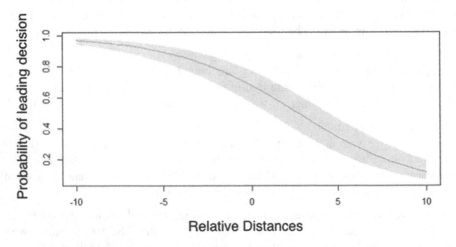

(b) In the main lane condition

Fig. 3. Logistic regression models to predict lead probability with relative distance in each lane.

lane (the probability of a leading decision in the main lane condition = .68, and the probability of a leading decision in the merging lane condition = .46) (Fig. 3).

3.2   Result of Individual Behaviors

The following analysis was conducted to verify whether participants determined their own driving behaviors according to the Model of Others based on Simulation Theory. We used logistic regression analysis to calculate the probability that the self car would lead in front of the other car when participants drove in each lane based on the relative distance (−10 m, −5 m, 0 m, +5 m, +10 m). We defined the intercept of the regression equation when the relative distance was 0 as the "probability of making a leading decision." Figure 4 shows the relationship of the probability of making a leading decision.

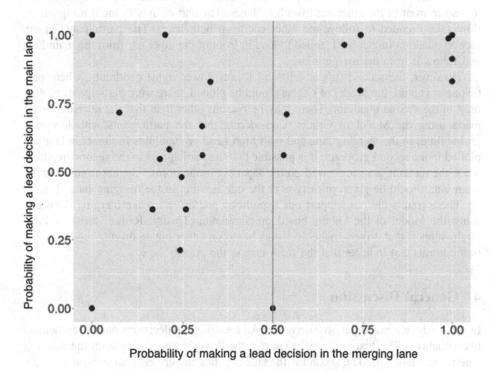

Probability of making a lead decision in the merging lane

Fig. 4. Correlations for the probability of making a leading decision between the main lane and merging lane conditions. The vertical axis indicates the probability that the self car would lead in front of the other car in the main lane condition. The horizontal axis indicates the probability that the self car would lead in front of the other car in the merging lane condition.

If participants estimated the Model of the Others through use of Simulation Theory and decided their own actions based on the estimated model, the following result is predicted: when participants drive in the merging lane, the Model of the Others is estimated based on their main lane driving behaviors. In this case, if the participant tends to follow the other car in the main lane, they presume that the other car in the main lane will also follow the other car. Therefore, the participant would pull out in front of the other car. The participants who displayed this behavior are plotted in the fourth quadrant (lower right) in Fig. 4. Conversely, the participants who tended to pull out in front of the other car in the main lane would tend to follow the other car in the merging lane; such participants are plotted in the second quadrant (upper left) in Fig. 4. In short, if the participants make merging decisions using the Model of the Others based on Simulation Theory, their decisions should be plotted in the second and fourth quadrants.

Consequently, ten participants (45% of the total) were plotted in the first quadrant (upper right), eight (30%) were plotted in the second quadrant (upper left), and six (25%) were plotted in the third quadrant (lower left). No participants were plotted in the fourth quadrant (lower right). The participants in the first quadrant (upper right) tended to lead in front of the other car from both lanes. The participants in the third quadrant (lower left) tended to follow the other car from both lanes. The participants in the second quadrant (upper left) tended to lead in front of the other car from the main lane and follow it from the merging lane.

However, the second (upper left) and fourth (lower right) quadrants, where participants who use the Model of Others would be plotted, there were participants plotted only in the second quadrant. There may be reasons other than that the decision being made using the Model of Others. Considering that the participants with low lead probabilities in the merging lane and with high Lead probabilities in the main lane are plotted in the second quadrant, it is possible that the participants in the second quadrant made the merging decision based on the Japanese traffic norm, "the car running in the main lane should be given priority over the one running in the merging lane" [6].

These results did not support our hypothesis that drivers make merging decisions using the Model of the Others based on Simulation Theory. Rather, these findings might support that drivers make decisions based on either innate driving tendencies or traffic norms that indicate that the main lane is the priority lane.

4 General Discussion

In this study, we examined how drivers make Lead/Follow decisions on the highway in two situations. The first is the situation that the driver's car merges from the merging lane to the main lane. The second is the situation that the driver's car runs in the main lane, and the other car merges from the merging lane to the main lane. As a result, it was shown that the driver would drive according to their own driving tendency or traffic norms. This suggests that Simulation Theory proposed in the Theory of Mind is unlikely to be used in decision-making while merging from one lane to another.

It is assumed that participants thought that other drivers do not always make the same inference as they might make. Simulation Theory assumes that the self and the other make the same inference to estimate the Model of Others [7]. Therefore, if a driver deems the other driver not to be identical as him- or herself, the driver would be unlikely to estimate the mental status of the other driver based on simulation. This leads to the following question; do drivers make merging decisions mechanically on their own driving habits or traffic norms?

Regarding the estimation of the Model of Others, apart from Simulation Theory, "Theory Theory" has been proposed [8]. According to Theory Theory, humans estimate the Model of Others by interpreting behaviors a person took in a situation according to the individual's knowledge, such as stereotypes. Humans can estimate the Model of Others of an opponent based on the impression formed by the opponent's behaviors even in a scenario where it is their first interaction with the opponent [9, 10].

It is also possible that the drivers form and use impressions toward the driver of the other car during driving. Hosokawa, Shino, Kamata, Kanamori, Fuwamoto, & Umemura (2008) propose a mathematical model that estimates the driving tendency of the driver based on driving behaviors, such as the degree of deceleration and the timing of indicating his or her intention using blinkers [11]. For instance, this model estimates that drivers with frequent deceleration and early use of their blinkers tend to drive cautiously. In this way, the impression of the other driver is formed by collating the driver's attribution of the other car with prior knowledge (e.g., stereotype, patterns of behaviors). The driver's attribution is determined by his or her sex, social status, driving experience or driving tendency (e.g., the degree of acceleration/deceleration, average speed, and timing of the use of their blinkers). For instance, considering the drivers' social status of the other car, a driver might think that luxury car drivers tend to have higher average speeds. Considering the driver's driving tendency of the other car, a driver might think that drivers with a small relative distance between two cars when following a car tend to drive aggressively. Then, it is expected that drivers make a decision about their own driving behaviors based on their impressions. In other words, driving decisions and behaviors can be strongly influenced not only by external information but also by the impression of others.

For instance, the appearance of a particular car type may affect the impression formation toward others. Doob & Gross (1968) found that low-priced light cars tend to give a low impression of the owner's social status, and the interaction partner's driving behaviors would become more aggressive [12]. Yazawa (2004) investigated that the three factors, "with or without learner plate," "high or low social status" and "driver's gender" affect the number time participants used their horns. In that study, the car type was used to control the social status of the owner (high status: luxury car; low status: light car) [13]. Consequently, it was shown that the number instances a horn was used was significantly higher when the other car had a learner plate or was considered of low status.

The above discussion suggests that the attributes of the other car and the driver may affect driving behaviors, and investigation about the effects of these factors is an important issue to be looked into in the future.

In this study, the car's exteriors were of the same type. Therefore, if the other car had a sports car's exterior, participants may have a more aggressive impression toward the driver of the other car. Conversely, if the other car had a light car's exterior, participants may have a more cautious impression toward the driver of the other car. Specifically, if the other car were a sports car, the lead probability of participants is expected to be lower than in this experiment, and if the other car had a light car, the lead probability is expected to be higher. Following on from this, it is necessary to further examine whether the driving behavior of participants is influenced by manipulating the impression of the other car and so the driver's behavior of the other car.

Acknowledgement. This work was supported by JSPS KAKENHI Grant Number JP18H05320.

References

1. Hiramatsu, M., Jang, H., Naemoto, H., Ito, Y., Yamazaki, M., Sunda, Y.: Jidou Soukou ni okeru Unten Sutairu Kojin Tekigou Shuhou no Teian (Method of driving style adaption for automated vehicle). Jidousha Gijutsu-kai Ronbunshu (Trans. Soc. Autom. Eng. Jpn.) **49**(4), 818–824 (2017). https://doi.org/10.11351/jsaeronbun.49.818
2. Ozawa, K., Ito, K., Takeda, K., Wakita, T., Itakura, F.: Unten-koudou-shingou ni hukumareru Kojinsei ni kansuru Kentou (Study of Individualities in Driver Behavioral Signals). Joho Kagaku Gijutsu Retazu (Inf. Technol. Lett.) **3**, 247–250 (2004). https://ci.nii.ac.jp/naid/110007634950/
3. Yoshikawa, S. Takagi, O.: Purotokoru-hou ni yoru Unten-koudou no Ishi-kettei no Katei no Kenkyu (A research of the decision making process of driving behavior by the protocol analysis). Shakai Shinrigaku Kenkyu (Jpn. J. Soc. Psychol.) **14**(1), 31–42 (1998). https://doi.org/10.14966/jssp.KJ00004622675
4. Premack, D.G., Woodruff, G.: Does the chimpanzee have a theory of mind? Behav. Brain Sci. **1**(4), 515–526 (1978). https://doi.org/10.1017/S0140525X00076512
5. Gallese, V., Goldman, A.: Mirror neurons and the simulation theory of mind-reading. Trends Cogn. Sci. **2**, 493–501 (1998). https://doi.org/10.1093/acprof:osobl/9780199874187.003.0003
6. Shimizu, T., Yai, T., Mimuro, T.: AHS heno Taiou-koudou wo Kouryo-shita Toshi Kousoku-douro Gouryu-bu no Unyou Hyouka Bunseki Shisutemu no Kaihatsu to sono Tekiyou (Microscopic traffic simulation system at merging section of urban expressway considering user's behavior under information services). Doboku Gakkai Ronbunshu 11–21 (2004). https://doi.org/10.2208/jscej.2004.758_11
7. Apperly, I.A.: Beyond simulation-theory and theory-theory: why social cognitive neuroscience should use its own concepts to study "theory of mind". Cognition **107**(1), 266–283 (2008). https://doi.org/10.1016/j.cognition.2007.07.019
8. Astington, J.W.: What is theoretical about the child's theory of mind: a Vygotskian view of its development. In: Carruthers, P., Smith, P.K. (eds.), Theories of Theories of Mind. Cambridge University Press, Cambridge, pp. 184–199 (1996). https://doi.org/10.1017/CBO9780511597985.013
9. Asch, S.E.: Forming impression of personality. J. Abnorm. Soc. Psychol. **41**, 258–290 (1946). https://dx.doi.org/info:doi/10.18999/bulfep.22.103
10. Shinotsuka, H., Hama, Y.: Koui Jouhou ni motoduku Taijin-inshou-keisei-katei no Jikkenteki-kenkyu (An experimental study of person impression formed on the basis of the informed behavior pattern of the target person). Shakai Shinrigaku Kenkyu (Jpn. J. Soc. Psychol.) **5**(1), 12–21 (1990). https://dx.doi.org/info:doi/10.14966/jssp.KJ00003725087
11. Hosokawa, T., Shino, M., Kamata, M., Kanamori, H., Fuwamoto, Y., Umemura, Y.: Koureiuntensha no Nichijo Unten-kodo Kiroku wo moto ni shita Usetsu-ji Fuanzen-kodo no Haaku to sono Hyoka (Classification of aged drivers based on their daily driving and prediction method for the classification using right turn driving parameters). Jidousha Gijutsu-kai Ronbunshu (Trans. Soc. Autom. Eng. Jpn.) **39**(4), 141–146 (2008). https://dx.doi.org/info:doi/10.11351/jsaeronbun.39.4_141

12. Doob, A.N., Gross, A.E.: Status of frustrator as an inhibitor of horn-honking responses. J. Soc. Psychol. **25**, 213–218 (1968). https://psycnet.apa.org/doi/10.1080/00224545.1968.9933615
13. Yazawa, H.: Effects of inferred social status and a beginning driver's sticker upon aggression of drivers in Japan. Psychol. Rep. **94**, 1215–1220 (2004). https://doi.org/10.2466/pr0.94.3c.1215-1220

Influencing Driver's Behavior on an Expressway with Intrinsic Motivation

Toshiki Takeuchi(✉)🆔, Ryosuke Mita, Naoya Okada, Tomohiro Tanikawa, Takuji Narumi, and Michitaka Hirose

The University of Tokyo, 7-3-1 Hongo, Bunkyo, Tokyo, Japan
{take,mita,m_okada,tani,narumi,hirose}@cyber.t.u-tokyo.ac.jp

Abstract. Uncontrollability of automobiles is one of the factors of traffic congestion on an expressway. The incentive to intrinsic motivation is particularly effective in influencing behavior decisions with continuity and low cost. Gamification is a method of providing motivation that introduces game design elements into non-game contexts. In this study, we proposed a method for inducing a driver on an expressway to take a rest using gamification based on the intrinsic motivation. We created a smartphone application for implementing the proposed method, thereby influencing each driver's behavior on an expressway in real-time. With our application, a driver takes a break at a rest area to complete quests provided by the application and automatically avoids a traffic congestion. The result of an experiment showed that our method induced significantly long rest compared to a conventional congestion presentation. The results of a user study on a real expressway indicated that the application achieved the rest events with 70% possibility in the best scenario.

Keywords: Gamification · Intrinsic motivation · Behavior induction · Traffic congestion · Automobile · Expressway

1 Introduction

Traffic congestion is a major problem in society. The estimated time loss resulting from traffic congestion in Japan reaches five billion hours per year, which is equivalent to the labor-hour of 2.8 million people [1]. However, automobiles are not directly controllable because each vehicle driver has a different intention and drives independently, and the road administrators can only control the traffic signals or electric bulletin boards. Therefore, traffic can become concentrated at a particular region or time, which can increase traffic congestion and accidents.

To disperse transportation demand by changing personal behavior, Okada et al. proposed a method of presenting the future congestion on a driver's smartphone and inducing them to take a rest at a rest area [2]. With this method, however, the chosen behavior by the driver is not always effective in reducing congestion because the location and time of the rest depends on the driver.

© Springer Nature Switzerland AG 2020
H. Krömker (Ed.): HCII 2020, LNCS 12213, pp. 114–124, 2020.
https://doi.org/10.1007/978-3-030-50537-0_10

A system needs to suggest an ideal driving behavior based on the congestion prediction and provide the driver with an appropriate incentive to choose the behavior.

Gamification is a method for providing the incentive to an individual using game factors and game principles in non-game contexts. Although there were some attempts to change people movements with gamification [3], most of them used monetary rewards and defrayed the cost. Furthermore, as the monetary reward is a type of extrinsic motivation that requires reason other than behavior, the efficacy of the monetary reward is temporary. In contrast, intrinsic motivation from pleasure and satisfaction of the behavior continues to be effective for a long time.

In this study, we aim to construct a system that induces a driver to take an ideal resting action on an expressway using gamification based on intrinsic motivation. Transportation demand may disperse and traffic congestion may ease by trying to take a rest and minimize the time involved in the congestion.

2 Related Work

2.1 Motivation

Many researches were conducted in the fields of psychology and social science to increase the motivation of people during activities. Various theories of motivation were developed with unique approaches. Although there are several psychological theories, motivation can be generally classified into "intrinsic motivation" and "extrinsic motivation" [4].

Extrinsic motivation comes from a reason unrelated to the action, i.e. external factors such as evaluation, reward, punishment, and coercion. Going to a theater to watch an uninteresting movie for free popcorn or being appointed as an organizer of a corporate drinking party despite being uninterested are examples of extrinsic motivation. Such reasons (the free popcorn and corporate coercion) are not directly related to the actions (going to the theater and organizing the party).

Intrinsic motivation, on the other hand, is motivation that generates interest and willingness in a person, thereby inducing them to complete the action because they want to do it. For example, actions such as interacting with friends, doing a favorite job, reading a book, and strolling on the beach at sunset are induced by pleasure from the actions themselves.

The Pleasure obtained from extrinsic motivation does not last long because people develop a resistance to the reward resulting in the requirement of greater and higher rewards for equivalent satisfaction or pleasure [5]. Kohn clarified that a student gradually stops doing an activity because of the big reward for promoting the activity in the field of education [6]. Additionally, extrinsic motivation keeps out the intrinsic motivation at the same time of the hedonic adaptation [7].

On the contrary, the actions induced by intrinsic motivation have continuity. The person feels positive emotions, such as happiness and pleasure, from the activity performed with an internal desire. Csikszentmihalyi named the satisfied and bracing feeling achieved with high creativity, energized focus, and full involvement as "flow" [8]. The flow state occurs frequently in games and game-like activities rather than in daily life.

2.2 Gamification

In the past several years, a technique called "gamification" has been introduced in various fields. The term "gamification" is used in its current meaning since a study by Deterding et al. in 2011 [9]. They defined gamification as using game factors and game principles in non-game contexts.

There are some examples of applying gamification to movements in the real world. In a metropolitan area, serious congestion that occur on a railway station when significant number of commuters gather is a problem. Uehara et al. conducted a stated preference survey for commuters' intention about four incentives: points for train ride, discount on station access cost, lounge service at the terminal station, and reduction in train congestion [10]. As a result of that, they found that seat availability and points for train ride had potential to change departure time.

Tokyo Metro Co., Ltd., in Japan has been conducting Tokyo Metro's early-rising campaign for Tozai Line since 2007 [11]. A participant of the campaign who takes an early train can obtain a gift such as a coupon. Half of the participants in the 2008 campaign answered that they had changed their departure time because of the incentive.

In Singapore, Pluntke and Prabhakar launched a project named INSINC that provided a monetary reward to a passenger who used a train by avoiding rush hours [12], which succeeded in reducing the congestion during those hours.

As we can see, these were successful instances of motivating people with gamification and changing people's movements in the real world. However, the designs used in these studies focused on monetary rewards, which is a type of extrinsic motivation that required high cost and was not suitable for long-term operation.

3   Gamification for Influencing Driving Behavior

As mentioned in the preceding section, intrinsic motivation is generally continuous and more cost effective than extrinsic motivation. In this study, we designed gamification based on intrinsic motivation and developed a system for inducing a driver to take a break at a rest area on an expressway.

3.1   Design of Gamification

We chose a smartphone application as the platform for influencing the behavior of individual drivers, which was accepted by the users en masse. We introduced the following three gamification factors to our system:

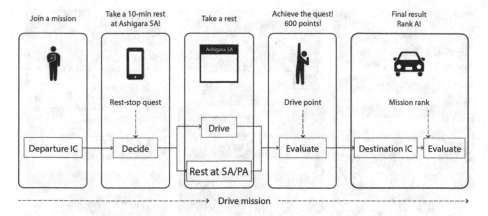

Fig. 1. Gamification design of proposed system.

- Mission and quest,
- Drive point, and
- Mission rank.

Figure 1 illustrates the flow of our gamification. First, a drive mission is offered to the driver, which is the journey from a departure interchange (IC) to a destination IC. The drive mission consists of some rest-stop quests that require the driver to take a break at a specific time and a specific rest area. If the driver takes the rest as instructed, the driver obtains drive points. A feedback that the rest-stop quest is achieved by the driver's own action provides an intrinsic motivation such as a pleasant sense of accomplishment.

When the driver arrives at the destination IC, he/she achieves the drive mission and obtains the mission rank. The mission rank is selected from four-grade ranks based on the drive points obtained during the drive mission. This is also an intrinsic motivation for the driver to aim for a higher rank in the mission.

McGonigal mentions that all games share four traits: a goal, rules, a feedback system, and voluntary participation [13]. In our gamification design, the drive mission corresponds to the goal, the quests prescribe the rules, and the drive point and mission rank are performed as the feedback system.

These gamification factors connect tasks in the real world with goals in the game world. Our gamification system changes the user's driving behavior and helps to reduce the traffic congestion.

3.2 Implementation

We constructed a gamification system, TomeiQuest, where a user enacts driving on the Tomei Expressway in Japan, which is a national expressway operated by Central Nippon Expressway Company Limited. The system was developed

118   T. Takeuchi et al.

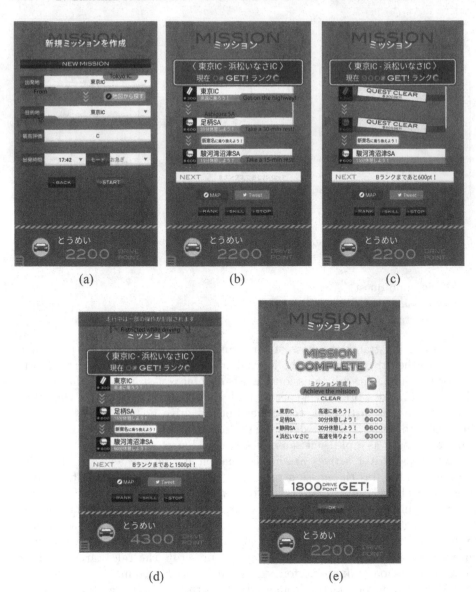

Fig. 2. Screenshots of TomeiQuest: (a) mission creation, (b) quest list at the start, (c) quest list in progress, (d) warning during driving, and (e) mission achievement. (Color figure online)

natively for an Android smartphone using Python on the server side. With the widespread use of smartphones today, we can use real-time location information and high-speed network connection to give feedback to individual drivers.

Figure 2 shows screenshots of the five main stages of TomeiQuest. TomeiQuest is a domestic application and mainly uses Japanese for its user interface. The red round rectangles in Fig. 2 are English translations of important parts for understanding.

First, a user needs to create a mission (Fig. 2 (a)), which is a journey from a departure IC to a destination IC. There are some rest-stop quests in the mission such as "Take a 30 min rest at Ashigara SA" (Fig. 2 (b)). If the user rests according to the quest, they obtain driving points (Fig. 2 (c)). There are two types of rest areas on the Tomei Expressway: service area (SA) and parking area (PA). SA has a larger area and richer facilities than PA. TomeiQuest creates two types of rest-stop quests for SA and PA. Crossing IC and resting at SA/PA are automatically detected with the smartphone's location information; thus the driver does not have to manipulate the smartphone during driving. The driver can view the application only at a rest area and when the car is parked for the safety of the driver (Fig. 2 (d)). Finally, when the user arrives at the destination, the mission is completed and the drive is evaluated using a four-grade achievement rank (Fig. 2 (e)).

The frequency of rest-stop quests depends on the driving distance and it increases every 80 km driving. The locations of the rest-stop quests are selected such that they are equally spaced in the total distance. The time of the rest-stop quest is decided based on the congestion forecast calendar distributed by Central Nippon Expressway Company Limited [14] and real-time congestion information. If there is no congestion in the route, the rest time is randomly selected from 5 min, 10 min, 15 min, 30 min, and 60 min. If there is congestion somewhere along the route or is predicted by the calendar, the rest time is decided to minimize the total driving time.

4 Comparison with Conventional Congestion Presentations

4.1 Experimental Design

We conducted a paper-based questionnaire to verify the superiority of our method over conventional congestion presentations. We recruited 11 people to participate in the experiment. All participants had a driver license. Their driving frequency varied from once a week to once a half year.

The participants answered what action to take in case the location and time of congestion were provided (conventional method) and in case of using TomeiQuest (proposed method) on the assumption that they were driving on an express way. In the conventional method, we showed a map of the Tomei Expressway including all the IC, SA, and PA. There was congestion information shown in the map that occurred in part of the sections during a period. We asked the participants to imagine a concrete driving situation including specific departure/destination IC and answer where and how long would they rest. In the proposed method, we asked them to answer whether they would achieve a rest-stop quest (i.e. rest at the SA/PA) displayed on TomeiQuest or not, imagining the same driving situation.

There were two driving situations: a private travel without time constraints and a business travel with time constraints. The rest time of a rest-stop quest just before congestion (target quest) was selected from 5 min, 10 min, 15 min, 30 min, 45 min, and 60 min.

4.2   Results

Figure 3 shows the maximum rest time on a target quest that the participants tolerated using our method. A rest-stop quest of up to 42 min could induce a real rest behavior in a private/SA condition. However, in the condition of business travel, the maximum rest time was approximately 20 min on SA and about 14 min on PA, which was nearly half of the time in private travel condition.

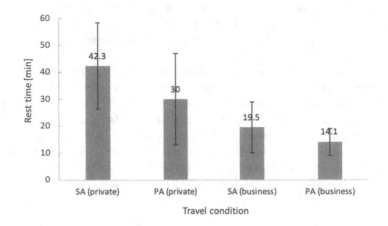

Fig. 3. Maximum rest time that participants tolerated.

Figure 4 shows the comparison of average rest time before congestion between our method and the traditional method. In the condition of private travel, the t-test showed a significant difference ($p = 0.01$) between the two methods. However, there is no significant difference in the condition of business travel.

From the results obtained in this experiment, we determined the acceptable rest time by condition of a driving purpose and a rest location. Except for time-sensitive drivers, the rest time can be up to 40 min on an SA quest and 30 min on a PA quest. The proposed method will be useful in a real environment because private travels especially increase in a vacation season and cause serious traffic congestion.

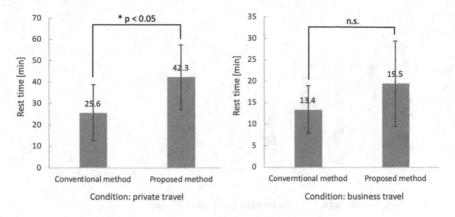

Fig. 4. Average rest time before congestion.

5 Feasibility of Our System

5.1 Investigation Summary

We conducted a real-world user study to evaluate the effect of our proposed method on users' driving behavior. We distributed our Android application through Google Play Store and requested the participants to freely use our system and drive on the Tomei Expressway. We conducted this user study for 15 weeks.

5.2 Results

In this study, 44 users installed TomeiQuest. Among them, the number of users who made and joined at least one mission was 25. During the study period, 31 valid missions with 45 quests were created except for missions that the users did not actually drove.

Figure 5 left shows the percentage of achievement of each rest-stop quest, which indicates that 70% of the 5-min quests are passed, whereas only 37.5% of the 60-min quests are passed. Figure 5 right is a chart that a single regression line based on a least-squares method is plotted. The achievement percentage linearly decreases based on the rest time in this study condition. If this model can apply to a longer rest than 60 min, the maximum rest time seems to be 120 min. In actually, however, there is probably a non-linear relation in the region more than 60 min.

Figure 6 represents the real rest time of each rest-stop quest. As seen in this figure, rest-stop quests within 30 min induce participants to take a longer rest. The 5-min quests, particularly, have the most efficient influence. In contrast, the average rest time of the 60-min quests is shorter than 60 min. It seems that the maximum time for a user to be patient is 50 min.

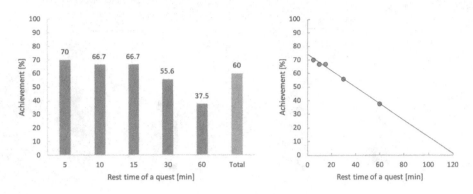

Fig. 5. Percentage of successful induction of each rest-stop quest.

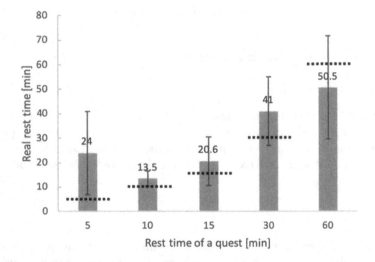

Fig. 6. Real rest time of each rest-stop quest. The red dotted lines indicate the same time as each quest for easy comparison. (Color figure online)

Figure 7 illustrates the real rest time comparison between in a quest and out of a quest. The number of rests in the quest is 27 and the number of rests out of the quest is 10. From the Welch's t-test, the driver takes a significantly longer rest in a quest. This result shows that the gamification effect exists compared to the no gamification condition.

As a result, our system can influence and change the driver's driving behavior in a real environment. We found that locating more small quests is a good design in this gamification.

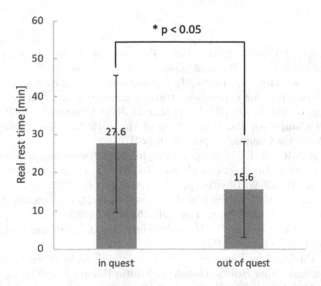

Fig. 7. Achievement percentage of rest-stop quests in a drive mission.

6   Conclusions and Future Research

In this study, we proposed a gamification method focused on intrinsic motivation for changing a driver's behavior on an expressway. We implemented the proposed method as a smartphone application named TomeiQuest and conducted two user studies to evaluate the effect of our method. Our paper-based questionnaire showed that our method could induce a user to take a significantly longer rest compared to the conventional congestion presentation. Through a real-world user study, we confirmed the efficacy of behavior change with our system and found the limitation that the time of the rest-stop quest should be up to 50 min. From the result that rest time was increased in the condition of private travel, our method will be especially useful for dispersion of traffic congestion in a vacation period.

In this user study, a participant used our system only a few times at the most, and therefore, we could not say that we revealed sustained effectiveness of intrinsic motivation. For future work, we wish to conduct a long-term experiment for investigating continuous behavior changes using our method. Additionally, other gamification mechanisms based on intrinsic motivation, such as badges on Swarm [15] and trophies on PlayStation [16], would be useful for enhancing behavior induction.

Acknowledgements. This research is conducted under a joint research by the University of Tokyo and Central Nippon Expressway Company Limited. We thank the project members and research participants.

References

1. The Ministry of Land, Infrastructure, Transport and Tourism (2016). https://www.mlit.go.jp/common/001132350.pdf. Accessed 25 Jan 2020
2. Okada, N., Nakazato, N., Takeuchi, T., Narumi, T., Tanikawa, T., Hirose, M.: Interactive interface for expressway travel planning with traffic predictions. In: Proceedings of the 2015 ACM International Joint Conference on Pervasive and Ubiquitous Computing and Proceedings of the 2015 ACM International Symposium on Wearable Computers, pp. 13–16 (2015)
3. Kracheel, M., McCall, R., Koenig, V.: Playing with traffic: an emerging methodology for developing. In: Emerging Perspectives on the Design, Use, and Evaluation of Mobile and Handheld Devices, pp. 105 (2015)
4. Ryan, R.M., Deci, E.L.: Intrinsic and extrinsic motivations: classic definitions and new directions. Contemp. Educ. Psychol. **25**, 54–67 (2000)
5. Brickman, P., Campbell, D. T.: Hedonic relativism and planning the good society. Adaptation-Level Theory (1971)
6. Kohn, A.: Punished by Rewards: The Trouble with Gold Stars, Incentive Plans, A's, Praise, and Other Bribes. Houghton Mifflin Harcourt (1999)
7. Frey, B.S., Oberholzer-Gee, F.: The cost of price incentives: an empirical analysis of motivation crowding-out. Am. Econ. Rev. **87**, 746–755 (1997)
8. Csikszentmihalyi, M.: Flow: The Psychology of Optimal Experience. Harper Perennial, New York (1991)
9. Deterding, S., Dixon, D., Khaled, R., Nacke, L.: From game design elements to gamefulness: defining gamification. In: Proceedings of the 15th International Academic MindTrek Conference: Envisioning Future Media Environments, pp. 9–15 (2011)
10. Uehara, K., Nakamura, F., Okamura, T.: Intention of railway commuters to departure time shifting by incentive measures. Transp. Policy Stud. Rev. **11**(4), 2–9 (2009). (in Japanese)
11. Tokyo Metro Co., Ltd.: Tozai Line off-peak project. https://t.metro-point-club.jp/form/pub/metro/tozailine. Accessed 25 Jan 2020
12. Pluntke, C., Prabhakar, B.: INSINC: a platform for managing peak demand in public transit. Journeys, Land Transport Authority Academy of Singapore, pp. 31–39 (2013)
13. McGonigal, J.: Reality is Broken: Why Games Make Us Better and How They Can Change the World. Penguin Publishing Group (2011)
14. Central Nippon Expressway Company Limited: Congestion forecast calendar. https://dc.c-nexco.co.jp/jam/cal/. Accessed 25 Jan 2020
15. Swarm—Remember everywhere. https://www.swarmapp.com/. Accessed 29 Jan 2020
16. PlayStation Official Site. https://www.playstation.com/en-us/. Accessed 29 Jan 2020

The Relationship Between Drowsiness Level and Takeover Performance in Automated Driving

Yanbin Wu[✉] ⓘ, Ken Kihara ⓘ, Yuji Takeda ⓘ, Toshihisa Sato,
Motoyuki Akamatsu, and Satoshi Kitazaki ⓘ

Human-Centered Mobility Research Center, National Institute of Advanced
Industrial Science and Technology (AIST), Tsukuba, Japan
wu.yanbin@aist.go.jp

Abstract. Although automated driving systems can perform dynamic driving tasks, at its lower levels, human drivers must still take over control of the vehicle whenever a Take Over Request (TOR) is issued. Human factors such as drowsiness may affect driver performance in responding to TOR. Results of research on the effects of different drowsiness levels on driver takeover performance appear inconsistent, for two possible reasons: 1. Some studies triggered TORs after a predefined duration of driving, while others triggered TORs based on a certain level of drowsiness. 2. Differences in drowsiness levels, which may have nonlinear effects on takeover performance, were not adequately considered. To investigate these inconsistencies, this experimental study recruited 40 participants and adopted a repeated-measures design. A total of 1436 available datasets, including a series of takeover performance measures and drowsiness measures, were collected. Analyses based on driving duration as well as drowsiness levels produced the following results: 1. Driving duration had a significant effect on drowsiness level but not on takeover performance. 2. Although mediumly drowsy drivers tended to sacrifice takeover quality for a fast reaction to ensure safety, highly drowsy drivers reacted significantly more slowly to TOR and were not able to maintain a safety margin comparable to that of drivers who were not highly drowsy. These findings have important implications for both researchers who are developing experimental studies to examine the effects of drowsiness on takeover performance and designers who use this information to design driver assistance systems.

Keywords: Automated driving · Takeover request · Takeover performance · Driver drowsiness

1 Introduction

Human factors have been reported to be connected to 65% of all automobile accidents and 80% of all heavy truck accidents (Dhillon 2007). Eliminating traffic accidents caused by human error is one of the most promising benefits that Automated Driving Systems (ADSs) can provide (Fagnant and Kockelman 2015). However, current ADSs are still not autonomous systems that operate in all situations and without any need for

© Springer Nature Switzerland AG 2020
H. Krömker (Ed.): HCII 2020, LNCS 12213, pp. 125–142, 2020.
https://doi.org/10.1007/978-3-030-50537-0_11

human intervention. The Society of Automotive Engineers (SAE) defined 5 levels of automated driving (SAE 2016). According to the SAE definition, ADS at its lower levels (SAE levels 2–3) is able to perform dynamic driving tasks, but the human driver still has to take over control of the vehicle whenever a takeover request (TOR) is issued. In other words, at these levels, the driver's role changes from an active operator to a fallback-ready driver.

For current vehicle drivers, the requirements of being a "fallback-ready driver" can be very confusing (Victor et al. 2018), and factors such as high workload (Radlmayr et al. 2019); Wu et al. (2020) are known to impair their correct use of driving automation. In addition, driver drowsiness level has been shown to increase faster in automated driving than in manual driving (Schömig et al. 2015). However, research on the effects of different drowsiness level on drivers' takeover performance has produced conflicting results. Saxby et al. (2008) reported slower response time to an emergency in a drowsiness (or "passive fatigue") condition evoked by a prolonged interval of automated driving (10 and 30 min). In a follow-up study, Saxby et al. (2013) replicated these results and observed a higher probability of crash. Similarly, Jarosch et al. (2019) found impaired reaction times and takeover qualities in a task-induced drowsiness condition. By contrast, a series of studies by Feldhütter and her colleagues produced a different pattern of results. Feldhütter et al. (2017) triggered a TOR after 5 min or 20 min of automated driving, and reported no difference in takeover performance between those two conditions. Feldhütter et al. (2018) triggered a TOR after drivers reached a certain level of drowsiness, and deterioration in takeover quality was observed, but slower reaction times were not found for the drowsy drivers. Feldhütter et al. (2019) adopted the same design in a more urgent condition but still found no evidence of slower reactions to TOR. Goncalves et al. (2016) triggered a TOR event when self-reported drowsiness reached a specific level, rather than after a constant duration, and reported poorer performance, but not reduced reaction time, in responding to TOR when drivers were drowsy compared to those who were not. Further, Schmidt et al. (2017) and Wu et al. (2019a) even reported no significant effects of driver's drowsy state on either takeover time or takeover quality. As suggested by Feldhütter et al. (2018), these inconsistent findings may be attributable to differences in study design.

Saxby et al. (2008, 2013); Feldhütter et al. (2017); Wu et al. (2019a) and Jarosch et al. (2019) triggered a TOR after a predefined duration of automated driving. This duration-based design overlooks the possibility that different drivers can develop drowsiness very differently, thus producing the inconsistent results. All drivers may not become drowsy after a constant and predefined duration. For example, Wu et al. (2020) recognized age as a significant factor that affects drivers' development of drowsiness during automated driving. The presence of both drowsy and alert drivers in a "drowsy" group defined by driving duration may have obscured significant differences in performance that produced unclear results.

To ensure that all drivers in the "drowsy" group are actually drowsy, TOR should be triggered when a certain level of drowsiness is self-reported or observed by a trained rater. However, such a design has also produced inconsistent results (Goncalves et al. 2016); Schmidt et al. (2017); and Feldhütter et al. (2018, 2019). In these studies, drowsiness was generally treated as a dichotomous state, with a threshold (usually the appearance of some drowsiness markers) predefined and a TOR triggered as soon as this threshold was reached. In this design, data from driver performance under extremely drowsy conditions were not available, because the TOR was already triggered (and the experiment completed) at the time the predefined threshold was reached. Considering that drowsiness is a transitional state between wakefulness and sleepiness (Johns 1998) and the interaction between drowsiness level and task performance seems to be non-linear (Kaida et al. 2006), the effect of drowsiness on takeover performance may also be nonlinear. In other words, it is possible that the driver could react appropriately when moderately drowsy but react inappropriately when highly drowsy. If so, the inconsistency in existing studies could be partially explained because their definitions and manipulations of the "drowsy" group were different.

The goal of this experimental study is to investigate the following two research questions regarding the effects of drowsiness on driver's performance in responding to TOR:

- Does an increase in drowsiness level after a long driving duration negatively affect drivers' takeover performance?
- If a moderate drowsiness level does not affect reaction times to TOR, will a high drowsiness level do this?

2   Methods

2.1   Participants and Apparatus

A total of 40 drivers (21–56 years old, mean = 39.1, SD = 11.9) were recruited from the local community. To avoid age and gender bias, five females and five males in their 20s, 30s, 40s, and 50s were recruited, respectively. In the recruiting advertisement, potential participants were required to have a legal driver license, drive frequently in his/her daily life, be in normal health, have no experience of motion sickness, and have normal vision. On average, the participants had 20.2 ± 11.6 years of driving experience and drove 14.8 ± 14.2 thousand kilometers per year. The National Institute of Advanced Industrial Science and Technology (AIST) Safety and Ethics committee approved the study, and each participant gave written informed consent before the experiment. Before the experiment, participants were required to complete a basic driver profile questionnaire, a driving style questionnaire, and a workload sensitivity questionnaire. All participants were paid for their participation.

All experiments were conducted in a fix-based middle-fidelity driving simulator (Mitsubishi Precision Co.). As shown in Fig. 1, the front and side views were projected on three liquid crystal displays. The driving simulator provided real-time and high-fidelity visual and auditory feedback to participants. A steering control loading system modulated reactive force against the input steering action (Moog Inc.). All data (e.g., speed of the ego vehicle) were sampled at a frequency of 60 Hz. In present study, the driving simulator was programmed to mimic the behavior of a SAE level 3 automated vehicle. In the automated mode, both longitudinal and lateral driving tasks were controlled by the system.

Fig. 1. The fix-based driving simulator and scene of the experiment.

2.2 Drowsiness Measures

Karolinska Sleepiness Scale (KSS). Subjective drowsiness was measured by the Karolinska Sleepiness Scale (KSS; Åkerstedt and Gillberg 1990), which is a widely used 9-step self-report scale of drowsiness (1 = extremely alert, 3 = alert, 5 = neither alert nor sleepy, 7 = sleepy but no difficulty remaining awake, 9 = extremely sleepy and fighting sleep). Kaida et al. (2006) translated KSS into Japanese and verified its association with objective drowsiness measures. In the present experiment, a printed Japanese version of the scale was pasted on the left side of the vehicle dashboard. After each takeover event, drivers were asked to reflect and rate their drowsiness level at the exact moment before the TOR was issued. The auditory instruction for the driver to orally report subjective drowsiness level was programmed into the simulator scenario and prompted automatically.

Rated Drowsiness by Trained Experts. Each driver's drowsiness level was also measured by two trained experts who observed driver eye movements, facial expressions, and behavioral indicators online on two separate screens during automated driving. Each observer independently estimated the driver's drowsiness level at 1 min intervals. The scale used to record observations was originally developed by Kitajima et al. (1997) and expanded by Homma (2016). As shown in Table 1, this scale has 6 levels (0 to 5), where 0 indicates that the driver is awake and motivated and 5 indicates that the driver is fully asleep. Following the definition of Homma (2016), scale levels 0 and 1 indicate low drowsiness, levels 2 and 3 indicate medium drowsiness, and levels 4 and scale 5 indicate high drowsiness. The rated level of drowsiness during the last 1 min epoch before the occurrence of TOR was used.

Table 1. Drowsiness and scale levels and respective behavioral markers used by trained experts. The descriptors have been translated from an original Japanese version developed by Kitajima et al. (1997) and adopted by Homma (2016).

Drowsiness level	Scale level	Behavioral markers
Low	0	Both body and eye movements are overall fast and agile; Gaze shifts quickly; Looking around restlessly
	1	Gaze focuses on specific points; Eye movement becomes small and stable; Slightly open lips
Medium	2	Eyelid-opening decreases; Frequent blinking; Occurrence of yawning; Changed sitting posture
	3	Eyelids almost close; Occurrence of blinks with duration > 1 s; Body and eye movements are overall slow
High	4	Frequent occurrence of eyelid closure with duration of 2–3 s; A few facial expressions of resisting falling asleep
	5	Without resistance to sleep; Fully asleep

2.3 Experiment Design

To evoke multiple levels of drowsiness, the duration of automated driving before a TOR was set to four levels: 2 min, 4 min, 8 min, and 16 min. To reduce the effect of surprise and to examine effects on an individual level, the experiment adopted a within-subjects repeated-measures design.

Procedures. Each participant came to the laboratory during one of four 2 h sessions (08:30–10:30, 10:30–12:30, 13:30–15:30, or 15:30–17:30) on 3 different days to complete the same experimental procedure 3 times. The participant was asked to sleep for the same duration on the night before each participation day and to maintain the same caffeine intake and amount of exercise across the 3 participation days. Participants signed informed consent forms only on their first day of participation; otherwise, the experimental procedure was consistent across the 3 participation days.

After participants arrived at the laboratory, the experimenters presented an overview of the experimental procedure and the operation of the driving simulator. Before starting the experiment, all participants were required to leave their cell phones, watches, and any other personal items that may interfere with the experimental process in a locker. The experimenter then instructed the participant to put on electrooculography (EOG) electrodes and electrocardiography (ECG) electrodes[1]. On each participation day, the participant first practiced driving the simulated automated vehicle for about 4 min to get familiar with the simulator and the ADS. During the practice drive, the participant experienced a takeover event that was exactly the same as that in the formal experiment.

---

[1] Analysis of EOG and ECG data is beyond the scope of this paper, and the results will be published separately.

As shown in Fig. 2 (top), the formal experiment on each of the 3 participation days consisted of 3 experimental drives, so each participant experienced $3 \times 3 = 9$ drives. A 2–3 min break was inserted between two consecutive experimental drives, during which the participant was allowed to drink water and use the restroom if needed. Within each of the 30 min experimental drives (Fig. 2, middle), the same 4 TOR events (Fig. 2, bottom) were presented after 2 min, 4 min, 8 min, or 16 min of automated driving. The occurrence order of the 4 durations was shuffled across different experimental drives, and the order was counterbalanced between different participants and different participation days, so that the participant would not be able to predict the occurrence of the TOR.

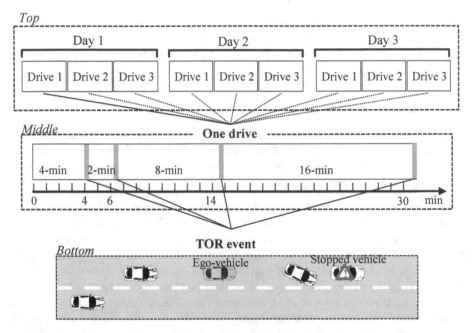

Fig. 2. Illustrative sketch of the experimental design. *Top*: Each participant experienced a total of 9 experimental drives on 3 participation days. *Middle*: Within each of the experimental drives, 4 TOR events were arranged after four drive durations: 2-min, 4-min, 8-min, and 16-min. The grey strips show the TOR events and the subsequent manual driving epochs. Note that the pattern of TORs shown in the figure (i.e., the ordering of 4-min, 2-min, 8-min, and 16-min) differed in other experimental drives. *Bottom*: traffic scenario of the TOR event.

Instructions to Participants. Participants were briefed about the capability of the ADS. They were told that the ADS can perform both longitudinal and lateral control of the vehicle, but a TOR may be issued when the ADS reaches its limit. During automated driving mode, they were instructed to relax and to keep their hands off the steering wheel and their feet off the pedals. Drivers were also instructed not to completely fall asleep, since they would be responsible to manually drive the vehicle when a TOR was issued.

Scenario of TOR Event. The scenario was built on a simulated Japanese dual two-lane motorway. To promote the development of driver drowsiness, the motorway was designed to be consistently straight, and the landscapes were very monotonous. The simulated ego-vehicle in the automated mode generally ran in the left lane with a cruising speed of 60 km/h. The takeover event was designed as a scenario involving 5 vehicles (Fig. 2, bottom). The vehicle running in front of the ego-vehicle changed to the right lane when a malfunctioning vehicle appeared in the left lane. At a speed of 60 km/h and a headway of 100 m, the time budget for takeover was 6 s. There were another two vehicles running behind the ego-vehicle, one in the left lane and the other in the right lane. After TOR was issued, the drivers had to confirm the position of the two vehicles and manually change to the right lane to avoid a collision with the malfunctioning vehicle. TOR was issued in two modalities, both auditory (a voice saying "Take over control" in Japanese) and visual (the color of the ADS symbol on the dashboard changing from green to orange).

2.4 Driving Performance Measures

Following the analysis in Wu et al. (2019b), this study also characterized driver's performance in response to TOR by fast reaction, smooth maneuvers, and maintenance of an adequate safety margin. The following measures were defined.

Reaction Time to Take Over Request (RTtor). RTtor was defined as the point in time when the driver began a conscious maneuver. A conscious maneuver was distinguished when a steering or braking input was greater than a threshold. Following Petermeijer et al. (2017) and Gold et al. (2013), two components were defined: RTsteer was the time consumed until the steering wheel was turned 2°, and RTbrake was the time consumed until the brake pedal was pressed at 10 percent of the full braking range. The smaller of the two values RTsteer and RTbrake was considered the reaction time to TOR (RTtor).

Minimum Speed (MinSpeed). MinSpeed was defined as the minimum vehicle speed before changing to the right lane. In the simulator, the vehicle dynamics were exactly same across all the drives. A smaller MinSpeed indicates a harder braking input. This measure reflects takeover smoothness in the longitudinal direction.

Standard Deviation of Steering Wheel Position (SDsteer). SDsteer was defined as the standard deviation of the steering wheel position within the first one-second epoch after changing to the right lane. A larger SDsteer value indicates more and/or larger steering operation in the running lane. This measure reflects takeover smoothness in the lateral direction.

Minimum TTC (MinTTC). MinTTC is defined as the time to collision at the moment of the lane change; it is the quotient of headway divided by vehicle velocity. TTC at the moment of the lane change reflects the safety margin in response to TOR.

3   Results

The dataset includes the data of drowsiness level measures and the data of takeover performance measures. Drowsiness levels were measured by drivers' subjective reports (KSS) as well as the rating scales of the two trained experts. Takeover performance was measured by reaction time measures (RTtor), smoothness measures (MinSpeed, SDsteer), and safety measures (MinTTC). Among the 40 participants, 39 participants completed all 9 experimental drives, and one participant completed 8 experimental drives. Each experimental drive had four TOR events after four different durations of automated driving (2 min, 4 min, 8 min, or 16 min), resulting in a total of $(39 \times 9) + (1 \times 8) = 359$ sets of data under each of the 4 driving duration conditions. The total number of datasets was 1436.

Shapiro-Wilk tests Razali and Wah (2011) were conducted to test the normality of the data. If the sample of one measure did not differ significantly from a normal distribution, a parametric analysis of variance (ANOVA) was conducted for this measure (F-distribution statistics). Otherwise, a non-parametric Kruskal-Wallis H Test was conducted (chi-squared distribution statistics). Post-hoc multiple comparisons were conducted using the Tukey-Kramer method. The significance level was set at 0.05.

3.1   Driving-Duration Based Analysis

KSS scores were considered as a subjective measure of drowsiness and the experts' rated scores were considered as an objective measure. As shown in Fig. 3, average KSS scores increased with an increase in driving duration. Figure 4 shows the results for drowsiness level evaluated by the two experts. The colored bars show percentages of expert estimations of drowsiness level (low, medium, high) under each condition. For example, in the 2-min condition, expert 1 (Fig. 4a) rated 69.8% of driver's drowsiness levels as low, 20.1% as medium, and 10.1% as high. As driving duration increased, both experts rated more data as high in drowsiness and less data as low in drowsiness.

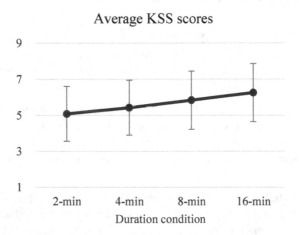

Fig. 3. Average KSS scores under four driving-duration conditions. Error bars indicate standard deviations.

To statistically examine the effects of driving duration on both drowsiness measures and driving performance measures, parametric or non-parametric analysis of variance were conducted. Because the data of both KSS scores and expert-rated drowsiness levels did not pass the normality test, non-parametric tests were conducted. A significant main effect ($\chi 2 = 13.6$, $p = 0.004$) of driving duration on KSS scores indicated that drivers reported a higher level of drowsiness after a longer duration of automated driving (Fig. 3). Significant main effects were also found for drowsiness levels rated by both expert 1 (Fig. 4(a), $\chi 2 = 17.7$, $p = 0.005$) and expert 2 (Fig. 4(b), $\chi 2 = 25.6$, $p < 0.001$).

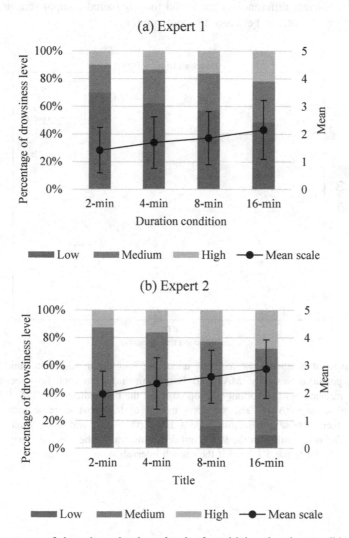

Fig. 4. Percentages of drowsiness levels under the four driving duration conditions rated by (a) expert 1 and (b) expert 2. The colored bars show the percentages of expert estimations of drowsiness level as low (0–1), medium (2–3), and high (4–5) under each condition. The solid line shows the mean of the drowsiness scale (Table 1, 0–5 scale), and error bars indicate standard deviations.

For driving performance measures, the data of MinTTC passed the normality test, while the other data did not. Parametric ANOVA was conducted for MinTTC and non-parametric Kruskal-Wallis H tests were conducted for the other measures. Although significant effects of driving duration on driver drowsiness levels were confirmed, effects of driving duration could not be found for RTtor (Fig. 5, $\chi^2 = 2.52$, $p = 0.47$), MinSpeed (Fig. 6, $\chi^2 = 1.28$, $p = 0.73$), SDsteer (Fig. 6, $\chi^2 = 3.35$, $p = 0.34$), or MinTTC (Fig. 7, F = 0.1, $p = 0.96$). As shown in Figs. 5, 6 and 7, obvious changes in driving performance measures with an increase in driving duration could not be confirmed. No significant differences were found for any paired comparison, as indicated by the overlapping notches between different conditions.

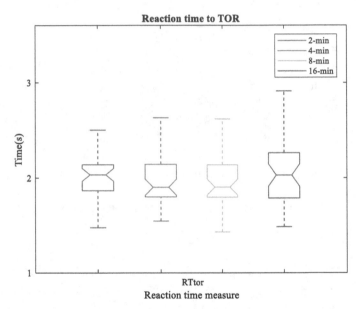

Fig. 5. Measure of reaction time to TOR under the four driving-duration conditions. This boxplot was produced using the MATLAB *boxplot* function. On each box, the central line indicates the median, and the bottom and top edges of the box indicate the 25th (Q1) and 75th (Q3) percentiles, respectively. The whiskers extend to the most extreme data points not considered outliers, whose upper limit is Q3 + 1.5 × (Q3 − Q1) and lower limit is Q1 − 1.5 (Q3 − Q1). The notch indicates the 95% confidence interval of the median. Two medians are significantly different at the 5% level if their notch intervals do not overlap.

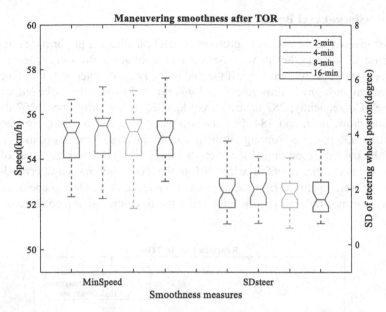

Fig. 6. Measures of maneuvering smoothness after TOR under the four driving-duration conditions. This boxplot was produced using the MATLAB *boxplot* function. Explanation of boxplots can be found in the caption of Fig. 5.

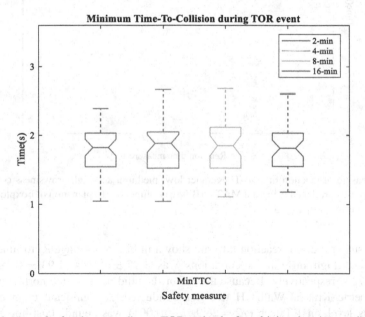

Fig. 7. Measure of safety in responding to TOR under the four driving-duration conditions. This boxplot was produced using the MATLAB *boxplot* function. Explanation of boxplots can be found in the caption of Fig. 5.

3.2   Drowsiness-Level Based Analysis

To further investigate the effects of drowsiness level on takeover performance, datasets for specific drowsiness levels were selected to conduct a drowsiness-level based analysis. A set of data was included if the two trained experts rated it at the same level (low, medium, or high) of drowsiness, and this set of data was then labelled with the rated level. Consequently, 287 datasets were labelled as low drowsiness, 262 datasets as medium drowsiness, and 184 datasets as high drowsiness, representing a total of 51% of the 1436 datasets. Among all the available datasets, 3 crash events occurred. Both of the trained experts rated the 3 crash events with a drowsiness level of 5; in other words, the participants fell asleep before the TOR and crashed afterwards. The crash events were excluded from an analysis of variance conducted to check whether driver performance differed significantly under the three drowsiness conditions.

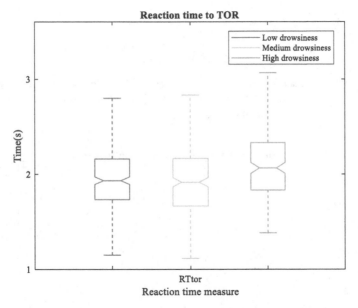

Fig. 8. Measure of reaction time to TOR under low, medium, and high drowsiness conditions. This boxplot was produced using the MATLAB *boxplot* function. Explanation of boxplots can be found in the caption of Fig. 5.

The results for driver reaction time are shown in Fig. 8. Average RTtor under low, medium, and high drowsiness conditions was 1.97 ± 0.40 s, 1.93 ± 0.38 s, and 2.18 ± 0.72 s, respectively. Because the RTtor data did not pass the normality test, a non-parametric Kruskal-Wallis H test was conducted. A significant main effect of drowsiness level on RTsteer ($\chi2 = 26.9$, $p < 0.001$) was found. Post-hoc multiple comparisons revealed that drivers reacted significantly more slowly under high drowsiness condition than under medium or low drowsiness conditions, and there was no significant difference between low and medium drowsiness conditions.

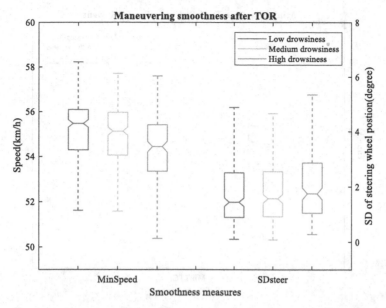

Fig. 9. Measures of maneuvering smoothness after TOR under low, medium, and high drowsiness conditions. This boxplot was produced using the MATLAB *boxplot* function. Explanation of boxplots can be found in the caption of Fig. 5.

The results for driver maneuvering smoothness are shown in Fig. 9. Average MinSpeed under low, medium, and high drowsiness conditions was 55.1 ± 2.2 km/h, 54.6 ± 3.1 km/h, and 53.2 ± 5.5 km/h, respectively. Because the MinSpeed data did not pass the normality test, a non-parametric Kruskal-Wallis H test was conducted. A significant main effect of drowsiness level on MinSpeed ($\chi 2 = 38.8$, $p < 0.001$) was found. Post-hoc multiple comparisons revealed that drivers drove significantly more slowly under high drowsiness conditions than under medium or low drowsiness conditions, and a significant difference between low and medium drowsiness conditions was also found. Average SDsteer under low, medium, and high drowsiness conditions was $1.88 \pm 1.36°$, $1.90 \pm 1.35°$, and $2.28 \pm 2.63°$, respectively. Because SDsteer data did not pass the normality test, a non-parametric Kruskal-Wallis H test was conducted. The main effect of drowsiness level on SDsteer was marginally significant ($\chi 2 = 4.47$, $p = 0.10$).

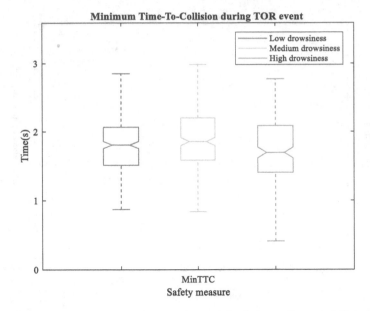

Fig. 10. Measure of safety in responding to TOR under low, medium, and high drowsiness conditions. This boxplot was produced using the MATLAB *boxplot* function. Explanation of boxplots can be found in the caption of Fig. 5.

The results for minimum Time-To-Collision during TOR events are shown in Fig. 10. Average MinTTC under low, medium, and high drowsiness conditions was 1.80 ± 0.44 s, 1.88 ± 0.45 s, and 1.71 ± 0.52 s, respectively. Because MinTTC data did not pass the normality test, a non-parametric Kruskal-Wallis H test was conducted. A significant main effect of drowsiness level on MinTTC ($\chi^2 = 11.8$, $p = 0.002$) was found. Post-hoc multiple comparisons revealed only a significant difference between high and medium drowsiness conditions.

4 Discussion

We address first the two research questions, followed by general discussion and limitations.

4.1 After a Longer Driving Duration, a Higher Level of Drowsiness Was Confirmed, but Takeover Performance Did not Deteriorate

As shown in Fig. 3 and Fig. 4, both self-report KSS scores and percentages of expert estimations of drowsiness level increased with an increase in driving duration, indicating that a simple manipulation of driving duration was able to induce statistically different levels of drowsiness. We assumed that a longer driving duration would not always lead to a high drowsiness level for all participants. This assumption was supported by the levels of drowsiness rated by the two trained experts. The percentages of driver states rated as highly drowsy after 16 min of automated driving were 22% for expert 1 and 28% for expert 2. The opposite pattern was also true: Expert 1 and expert 2 rated 10% and 12% of driver states as highly drowsy after only 2 min of automated driving. The presence of both highly and slightly drowsy drivers in the same driving-duration condition was therefore confirmed. Thus, averaging takeover performance data from different driving-duration conditions neutralizes the difference between highly drowsy drivers and slightly drowsy drivers, leading to the erroneous conclusion that drowsiness does not affect drivers' takeover performance.

4.2 Effects of Drowsiness on Takeover Performance are Multi-staged

By selecting and analyzing datasets assigned to specific drowsiness levels, we were also able to examine the effects of different drowsiness levels on takeover performance. Under medium drowsiness conditions, driver takeover quality deteriorated compared to low drowsiness conditions (Fig. 9), while driver reaction times to TOR did not significantly change (Fig. 8). This result is in line with the findings of Gonçalves et al. (2016) and Feldhütter et al. (2018, 2019). If we look only at average values, drivers seemed to react faster and were able to maintain a larger TTC when mediumly drowsy than when alert. We concur with Feldhütter et al. (2018, 2019) that drivers tend to sacrifice takeover quality (i.e., by harsh braking or unstable steering) to achieve a fast reaction and maintain an adequate safety margin.

In this study, a high drowsiness level was defined as the state of being fully asleep or in micro-sleep (frequent occurrence of eye closure for 2–3 s). Under high drowsiness conditions, evidence of significantly slower reactions as well as worse takeover qualities was found. Further, three crash events occurred when both of the trained experts rated the driver as fully asleep (scale level 5). These findings support the idea that the effects of drowsiness on takeover performance may progress in multiple stages: when mediumly drowsy, drivers were still able to sacrifice takeover quality to achieve faster reactions and to ensure safety; when drowsiness evolved to a high level, drivers reacted more slowly and more roughly to TORs; and when fully asleep, drivers' crash probability also increased.

This multi-staged effect may also partially explain the non-significance of reaction time after predefined durations of automated driving (Figs. 5, 6 and 7). Mediumly drowsy drivers tended to have similar reaction times, compared to slightly drowsy drivers. When TORs were triggered after 2 min or 16 min of automated driving, high drowsiness data and medium drowsiness data may have coexisted, and when reaction times were averaged, significant differences between these driving duration conditions may have disappeared.

4.3 Practical Applications and Limitations

The current findings provide the following evidence for researchers investigating the effects of drowsiness level on takeover performance:

- Following Feldhütter et al. (2018), we endorse an experimental design in which TOR is triggered by drowsy state rather than driving duration.

- Drowsiness should be treated as a multi-level state instead of a dichotomous (drowsy or not drowsy) state.
- Extremely drowsy conditions should also be considered.

The current findings also have the following implications for designers who intend to support fallback-ready drivers in automated driving:

- Driver assistance systems should support mediumly drowsy drivers in improving takeover quality.
- Safety countermeasures are necessary for TORs that happen when the driver is highly drowsy.

Finally, it is important to acknowledge the limitations of this study. First, the longest driving duration adopted in this experiment was 16 min. In the study of Gonçalves et al. (2016), a majority of the subjects reported a high level of drowsiness before 15 min of automated driving. It is possible that more drivers would become highly drowsy after a longer duration. Although this was not evaluated in the current study, individual differences can be significant in the development of drowsiness, and drowsiness can fluctuate over time Karrer et al. (2004). For example, while a duration of 60 min may be the minimum duration to elicit drowsiness for one driver, another driver may already have taken a nap and become alert again within 60 min. Second, the datasets used in the drowsiness-level based analysis were those with clear drowsiness labels and represented 51% of all the datasets. Since those two trained experts estimated driver drowsiness online, we did not require them to make repeated estimates of drowsiness levels until they provided consistent ratings. The authors hope to overcome these limitations in future research.

Acknowledgement. This work was supported by Council for Science, Technology and Innovation (CSTI), Cross-ministerial Strategic Innovation Promotion Program (SIP), "Large-scale Field Operational Test for Automated Driving Systems" (funding agency: NEDO).

References

Dhillon, B.: Human Reliability and Error in Transportation Systems. Springer Series - Reliability Engineering. Springer, Ottawa (2007). https://doi.org/10.1007/978-1-84628-812-8

Fagnant, D.J., Kockelman, K.: Preparing a nation for autonomous vehicles: opportunities, barriers and policy recommendations. Transp. Res. Part A: Policy Pract. **77**, 167–181 (2015)

SAE International: Surface vehicle recommended practice, taxonomy and definitions for terms related to driving automation systems for on-road motor vehicles, J3016. Revised September (2016)

Victor, T.W., Tivesten, E., Gustavsson, P., Johansson, J., Sangberg, F., Ljung Aust, M.: Automation expectation mismatch: Incorrect prediction despite eyes on threat and hands on wheel. Hum. Fact. **60**(8), 1095–1116 (2018)

Radlmayr, J., Fischer, F.M., Bengler, K.: The influence of non-driving related tasks on driver availability in the context of conditionally automated driving. In: Bagnara, S., Tartaglia, R., Albolino, S., Alexander, T., Fujita, Y. (eds.) IEA 2018. AISC, vol. 823, pp. 295–304. Springer, Cham (2019). https://doi.org/10.1007/978-3-319-96074-6_32

Wu, Y., Kihara, K., Takeda, Y., Sato, T., Akamatsu, M., Kitazaki, S.: Age-related differences in effects of non-driving related tasks on takeover performance in automated driving. J. Saf. Res. **72**, 231–238 (2020)

Schömig, N., Hargutt, V., Neukum, A., Petermann-Stock, I., Othersen, I.: The interaction between highly automated driving and the development of drowsiness. Procedia Manufact. **3**, 6652–6659 (2015)

Saxby, D.J., Matthews, G., Hitchcock, E.M., Warm, J.S., Funke, G.J., Gantzer, T.: Effect of active and passive fatigue on performance using a driving simulator. In: Proceedings of the Human Factors and Ergonomics Society Annual Meeting, vol. 52, no. 21, pp. 1751–1755. Sage Publications, Los Angeles (2008)

Saxby, D.J., Matthews, G., Warm, J.S., Hitchcock, E.M., Neubauer, C.: Active and passive fatigue in simulated driving: discriminating styles of workload regulation and their safety impacts. J. Exp. Psychol.: Appl. **19**(4), 287 (2013)

Gonçalves, J., Happee, R., Bengler, K.: Drowsiness in conditional automation: proneness, diagnosis and driving performance effects. In: 2016 IEEE 19th International Conference on Intelligent Transportation Systems (ITSC), pp. 873–878. IEEE, Rio de Janeiro (2016)

Jarosch, O., Bellem, H., Bengler, K.: Effects of task-induced fatigue in prolonged conditional automated driving. Hum. Fact. **61**(7), 1186–1199 (2019)

Feldhütter, A., Gold, C., Schneider, S., Bengler, K.: How the duration of automated driving influences take-over performance and gaze behavior. In: Schlick, C., et al. (eds.) Advances in Ergonomic Design of Systems, Products and Processes, pp. 309–318. Springer, Heidelberg (2017). https://doi.org/10.1007/978-3-662-53305-5_22

Feldhütter, A., Kroll, D., Bengler, K.: Wake up and take over! the effect of fatigue on the take-over performance in conditionally automated driving. In: 2018 21st International Conference on Intelligent Transportation Systems (ITSC), pp. 2080–2085. IEEE, Maui (2018)

Feldhütter, A., Ruhl, A., Feierle, A., Bengler, K.: The effect of fatigue on take-over performance in urgent situations in conditionally automated driving. In: 2019 IEEE Intelligent Transportation Systems Conference (ITSC), pp. 1889–1894. IEEE, New Zealand (2019)

Johns, M.: Rethinking the assessment of sleepiness. Sleep Med. Rev. **2**(1), 3–15 (1998)

Wu, Y., Kihara, K., Takeda, Y., Sato, T., Akamatsu, M., Kitazaki, S.: Assessing the mental states of fallback-ready drivers in automated driving by electrooculography. In: 2019 IEEE Intelligent Transportation Systems Conference (ITSC), pp. 4018–4023. IEEE, New Zealand (2019a)

Åkerstedt, T., Gillberg, M.: Subjective and objective sleepiness in the active individual. Int. J. Neurosci. **52**(1–2), 29–37 (1990)

Kaida, K., et al.: Validation of the Karolinska sleepiness scale against performance and EEG variables. Clin. Neurophysiol. **117**(7), 1574–1581 (2006)

Kitajima, H., Numata, N., Yamamoto, K., Goi, Y.: Prediction of automobile driver sleepiness (1st report, rating of sleepiness based on facial expression and examination of effective predictor indexes of sleepiness). Trans. Jpn. Soc. Mech. Eng. Part C **63**(613), 3059–3066 (1997)

Homma, R.: Evaluation of systems and strategies for driver assistance based on human characteristics. Doctoral thesis. The Graduate School of Human Sciences, Waseda University (2016)

Wu, Y., Kihara, K., Takeda, Y., Sato, T., Akamatsu, M., Kitazaki, S.: Effects of scheduled manual driving on drowsiness and response to take over request: a simulator study towards understanding drivers in automated driving. Accid. Anal. Prev. **124**, 202–209 (2019b)

Razali, N.M., Wah, Y.B.: Power comparisons of Shapiro-Wilk, Kolmogorov-Smirnov, Lilliefors and Anderson-Darling tests. J. Stat. Model. Anal. **2**(1), 21–33 (2011)

Karrer, K., Vöhringer-Kuhnt, T., Baumgarten, T., Briest, S.: The role of individual differences in driver fatigue prediction. In: Third International Conference on Traffic and Transport Psychology, Nottingham, UK, pp. 5–9 (2004)

Schmidt, J., Dreißig, M., Stolzmann, W., Rötting, M.: The influence of prolonged conditionally automated driving on the take-over ability of the driver. In: Proceedings of the Human Factors and Ergonomics Society Annual Meeting, vol. 61, no. 1, pp. 1974–1978. Sage Publications, Los Angeles (2017)

Petermeijer, S., Bazilinskyy, P., Bengler, K., De Winter, J.: Take-over again: investigating multimodal and directional TORs to get the driver back into the loop. Appl. Ergon. **62**, 204–215 (2017)

Gold, C., Damböck, D., Lorenz, L., Bengler, K.: "Take over!" How long does it take to get the driver back into the loop?. In: Proceedings of the Human Factors and Ergonomics Society Annual Meeting, vol. 57, no. 1, pp. 1938–1942. Sage Publications, Los Angeles (2013)

Urban and Smart Mobility

Toolbox for Analysis and Evaluation of Low-Emission Urban Mobility

Felix Böhm$^{(\boxtimes)}$, Christine Keller, Waldemar Titov, Mathias Trefzger,
Jakub Kuspiel, Swenja Sawilla, and Thomas Schlegel

Institute of Ubiquitous Mobility Systems, Karlsruhe University
of Applied Sciences, Moltkestr. 30, 76133 Karlsruhe, Germany
{Felix.Boehm, iums}@hs-karlsruhe.de

Abstract. The evaluation of people's mobility is crucial for understanding traffic, traffic security and the effects of traffic planning. In this paper, we present our toolbox for analyzing and evaluating aspects of different mobility modes. Some of these tools support the participation of road users in the analysis. The tools either can be applied to implement analyses for planning purposes or for the evaluation of implemented measures. Our goal is to improve the understanding of mobility in all its facets and ultimately to increase user comfort, safety and the overall user acceptance in urban mobility.

Keywords: Urban mobility · Transportation planning · Eye tracking

1 Introduction

Urban mobility is currently facing big changes. For a century, cars were the center of attention and dominated our perspective of urban mobility. The road and urban design was optimized for cars, which contributed to their explosive growth. Many cities currently face similar problems that are related to the rising number of cars. The most severe being congestions, pollution and space usage. Nowadays, urban planners realize that to effectively combat those problems they have to change their perspective on traffic and urban mobility as a whole. All road users have to be equally represented in the planning and design processes. This will ensure that we create sustainable, efficient and safe urban mobility. This, in turn, requires us to understand the needs, wishes, behavior and the physical and psychical limitations of every road user. During our research on the evaluation and analysis of different modes of transport, we found that the availability and quality of the tools needed to collect this information greatly vary from one mode of transport to another. In some cases, necessary tools simply did not exist. We therefore utilized or specifically developed several tools to fill these gaps. Our work resulted in a toolbox for mobility analysis, which we will present in this paper. We want to raise awareness for these tools as well as report on problems we encountered, limits to the tools and our solutions to these problems. To structure this paper, we have divided the tools into two categories. The first category consists of tools that can be used across transport modes and those are described in Sect. 2. The second category, which we cover in Sect. 3, consists of tools that are more bound to specific

© Springer Nature Switzerland AG 2020
H. Krömker (Ed.): HCII 2020, LNCS 12213, pp. 145–160, 2020.
https://doi.org/10.1007/978-3-030-50537-0_12

transport modes (walking and cycling). We describe several studies and case studies for the tools and discuss the lessons learned in using the tools.

2 Comprehensive Tools

2.1 MobiDiary - an Instrument for Multi-modal Mobility Analysis

For understanding and simulating people's mobility, traffic planners need precise data. This includes information on preferred routes, on transport modes, trip purposes, covered distances and travel duration. Unfortunately, gathering this kind of data usually requires time-consuming surveys. The conventional method, called 'mobility diaries' is realized on paper and must be filled out by hand by participants. Those diaries are usually recorded after a taken trip, for example each evening during the survey period. A challenge of this method is the difficulty of participants remembering the exact mobility procedure of a given day. This often results in incorrect or incomplete statements and thus inaccurate conclusions. Another problem is the digitization of this information for easier further analysis is a very time-consuming act.

We think that apps on mobile devices are the best solution to eliminate some of the above-mentioned limitations. Mobile devices, like smartphones, are in widespread use and are highly available. Another advantage is that they are equipped with most of the sensors required for automated data acquisition, which assists the participants in data recording. Other researchers also saw the possibilities of using Smartphones for data acquisition and developed several solutions. Schelewsky et al. [1] developed such an application and examined the chances and challenges of using mobile devices for automated acquisition of intermodal routes. They report that the feasibility of smartphone tracking is very high, because most routes can be determined more precisely than with classic methods. Safi et al. [2] employed their system SITSS (Smartphone-based Individual Travel Survey System) as a pilot in the national household survey of New Zealand. They conclude that the application reduced the required human and financial resources for the data collection. Patterson et al. [3] created 'Itinerum' a smartphone travel survey platform that allows researchers to customize the Itinerum app with their own questions and prompts, distribute these surveys, monitor, visualize and process the collected data without a background in programming. All these projects show the demand and benefit of those apps.

Nonetheless, we were missing some key features in the developed solutions. That is why we created our own android application for mobile devices called 'MobiDiary'. What differentiates our App from the Apps mentioned above is the possibility to track additional context information. The App queries the current weather and links the Information to the recorded routes. The App also checks the audio jack, and records, if headphones are connected. That enables to verify multiple hypotheses like:

- When its sunny people are more likely to ride to work by bike
- People listen to their headphones while riding a train rather than while riding a bike

The app also captures factors like start time, travel path, the chosen mode of transport, the transfer time, the time of arrival and finally, yet importantly, the purpose of the trip. The data acquisition is based on existing and time-proven methods like the 'Mobilitätspanel', which is a standardized questionnaire for accessing travelers mobility designed and conducted by the German ministry of traffic and infrastructure [4]. The app asks the user to state their mode of transport after each stage of their way. Simultaneously, the Google Awareness API automatically identifies the modes. This makes it possible to verify the provided data from the users and vice versa, the identification Google provides. Similar we handle the purpose of the route. As stated above the user can give us the purpose of the route. Additionally, we also use the Google nearby Places API to identify the purpose automatically. Furthermore, the App enables the users to add a description of the situations in which they made a transport change decision. We use this information to validate the specifications on the purpose of a trip given by the app users. If for example, a user indicated that the route purpose is free time when being located in a university building, the given information might be wrong. To record a trip, the user has to actively start the recording. After finishing the trip, the data is stored on the local device and can be shared by the user. The device uploads the data to our server as soon as a Wi-Fi connection is established.

Lessons Learned

We evaluated the application in a study in which 38 student subjects participated. They used the app to record their mobility behavior over two weeks in October 2018. The study is part of their program in transportation planning. In the years before, students had to record their trips by hand. In comparison to this method, we conclude that the app does facilitate and speed up the procedure of collecting the information. Additionally, the acquisition of context-based information enabled us to evaluate more scenarios than it would have been possible using only traditional surveys. However, we also found some flaws in our approach. Using the different Google APIs it was easy for us to implement the required features. However, Google changed the pricing of those APIs in the middle of 2019. We are now forced to pay more money for large-scale surveys. Because of this, we plan to design our own algorithm to classify the trips purpose and transportation mode. We are still too early in the developing process to provide concrete details. Another problem we encountered was that user often forgot to mark in the app that they finished a trip. This resulted in the app still recording the trip while the users were already at their destinations. For this reason, we want to implement a feature that detects inactivity and asks the user by means of a notification if they have finished their journey. Despite these initial difficulties, the advantages of electronic route recording using apps clearly outweighed the disadvantages of conventional methods.

2.2   Eye Tracking

Eye tracking technology is becoming increasingly precise and affordable. Eye tracking devices make it possible to record eye movements. The gaze points show exactly where study participants looked at. In situations where there is a high cognitive load, it can be assumed that the viewpoints also reflect the actual focus. This is the case in many traffic situations, e.g. when crossing a street or driving a vehicle.

Head mounted eye tracking glasses enable us to record the eye movements in multiple situations in real life traffic. Current eye-tracking glasses, like the 'Tobii Glasses 2' we use, are lightweight and allow complete freedom of movement. After a few minutes of adaptation, the glasses are also hardly perceived as disturbing by test persons [5].

Various statistical and visual analysis methods can be used to evaluate the gaze data, allowing the following basic questions to be answered:

- When was an object or area (Area of Interest – AOI) focused on first
- In which order were different objects looked at
- How long are objects focused on
- How often are objects focused

Road users continuously analyze and process visual information. Traffic situations and how they are perceived are constantly changing, e.g. due to other road users, the mood of the user or weather conditions. In general, the majority of the necessary information in traffic is observed visually. Depending on the situation, drivers, pedestrians and cyclists receive up to 83% of the information visually [6]. While researchers have analyzed the visual perception of car drivers in many eye tracking experiments [7], the number of eye-tracking experiments with other and more vulnerable road users is still comparably low. Though in recent years, researchers have published a growing number of eye tracking experiments analyzing pedestrians and cyclists (e.g., Davoudian and Raynham [8]; Kiefer et al. [9]; Mantuano et al. [10]; Vansteenkiste et al. [11]; Wenczel et al. [12]). To our knowledge, eye tracking studies on the gaze behavior of e-scooter drivers are not yet published.

We have conducted several eye tracking studies in which we evaluated the behavior of different road users in real traffic. First, we shortly summarize how and why we used eye tracking in the different studies. Second, we present our combined lessons learned in conducting those studies.

Eye Tracking with Pedestrians and Cyclists
To investigate the behavior of pedestrians and cyclists, we conducted an experiment, in which participants had to follow a predefined route in an urban area of a town in Germany. Participants had to cycle and walk the same route, subsequently. With the collected data, we wanted to answer multiple questions:

- Which objects as well as obstacles are focused on most and how distracting are advertisements at the side of the road?
- What common gaze sequences do pedestrians and cyclists perform, and how does their behavior differ?
- How long and when do pedestrians and cyclists perform shoulder checks, especially if entering a potentially dangerous traffic situation?

We analyzed a part (160 m) of a longer route (1.5 km) participants had to cycle and walk. We sub-divided this part of the route into three sub-segments.

For investigating eye movement data in real world traffic situations, we had to define a high number of AOIs. Therefore, we used an abstract reference image for mapping eye movement data onto AOIs. We used a visual approach for a detailed analysis and comparison of the recorded multi-modal data. The visual analysis showed that the explorative analysis with multiple data streams leads to an efficient overview of the collected data. Our main results of the analysis are that participants paid most attention to the path itself; advertisements do not distract pedestrians and cyclists much; and participants focus more on pedestrians than on cyclists. We were also able to extract common gaze sequences, and we analyzed how often pedestrians and cyclists perform shoulder checks.

Eye Tracking with Cyclists and e-Scooter Drivers

E-scooters are a new means of transport that has been permitted on German roads since summer 2019. Just at the beginning, there were many reports in the media about an apparently large number of accidents involving e-scooter drivers. Due to these starting difficulties, the e-scooter currently has the image of being a dangerous means of transport. We wanted to revise this image, analyze the driving behavior of e-scooter drivers and, if necessary, develop proposals to improve the infrastructure.

We therefore conducted three studies that deal with the gaze behavior of e-scooter drivers. We conducted the studies with both cyclists and e-scooter drivers, to identify different or similar gaze patterns in comparison. Two of the studies were based on cycling studies of Vansteenkiste et al. [11], and evaluate the influence of the surface condition and how driving speed and lane width influence the gaze behavior. We did find that on a road in good condition e-scooter drivers focused longer on the road in front of them then cyclists. The narrower the lane becomes, the further the view wanders from outer, non-relevant areas towards the near area. A comparison of the average duration of the different AOIs shows no significant differences between e-scooter drivers and cyclists. In the third study, we analyzed the performance of shoulder checks in typical accident situations. We did find that e-scooter drivers executed the same number of shoulder checks as the cyclists. Also the average shoulder check duration differed minimally.

Eye Tracking with Cars

Previous eye tracking studies with car drivers have mostly been carried out under laboratory conditions or with simulators. Therefore, it makes sense to investigate the gaze behavior in real traffic with cars. This allows the real traffic context to be taken into account. Participants have the chance to act free and it is possible to evaluate unexpected traffic situation and the participants' behavior in these situations. It is possible to evaluate situations and circumstances that are not represented by simulators. The results help to understand human behavior in real situations.

Before carrying out an eye tracking study in real traffic, it is helpful to carefully consider the goal of the study. Depending on the study's focus, it is useful to gather and analyze further data, like stress by pulse measurement, for example. We did a real life study with car drivers, using mobile eye tracking glasses and pulse measurements. Our goal was to get a better sense of driver's stress and behavior. In our study, the participants drove their own car, which reduced possible distraction. We gave all participants a destination, which they then had to reach independently. During the study, we gave the participants two tasks. One task was parking at a car park. We let the participants choose where and how to park. The second task was turning. According to the

German Federal Statistical Office (2018) turning is one of the main causes of accidents caused by driver misconduct [13].

Eye Tracking with Public Transit

Providing good passenger information is considered a key factor for a well-functioning public transportation system. Eye tracking as a classical tool for the evaluation of usability is a very good instrument for the evaluation of the visual part of the passenger information. To include the context factors of real public transport use, it is possible and advisable to conduct field studies in real world environments. So far, we have conducted eye tracking studies in the field of public transport only in our laboratory. However, we are already planning field studies in this area. We conducted most of the eye tracking studies for a project dealing with the display of passenger information on semi-transparent train windows. For this purpose, we have tried to create a laboratory environment that is as realistic as possible. For example, we used real train seats in our experimental setup (Fig. 1). We have taken care to maintain the correct seat spacing so that we can transfer the results to the real passenger cabin. We already presented the detail results of this approach in Keller et al. 2019 [14].

Fig. 1. Participants with eye tracking glasses in our experimental setup

Lessons Learned

Eye tracking is one of the best tools to analyze and understand visual focus and perception of road users. The captured gaze points allow relatively reliable statements about when, for how long, and in which order different elements were perceived in road traffic.

Nevertheless, some points have to be considered when conducting eye tracking studies in real traffic situations to guarantee meaningful results. In real traffic studies, many parameters are difficult or impossible to determine. It is therefore all the more important that the procedure, objective and general conditions of the study are clearly defined in advance. Especially weather can have a strong influence on the study. From a technical point of view, strong sunlight is problematic, as it prevents reliable recording of gaze points. Glare from sunlight, snow and rain can influence driving behavior and visual perception. Another challenge lies in the comparison of the collected data. No situation is similar due to weather conditions, time, other road users, vehicles etc. It helps to consider which parts are important for the analysis, like gaze behavior while turning or parking the car and to focus on the conditions for these parts. It is advisable to include alternative appointments in planning due to bad weather conditions or technical issues with vehicles.

In addition to capturing eye tracking data, it may be helpful to collect more data using other sensors, like pulse measurement or GPS tracking. Additional data can be used to enrich obtained results or simplify the comparison of different data sets.

The evaluation of eye tracking images in real traffic situations is currently still relatively time-consuming. For the evaluation, it is necessary to transfer the gaze points to reference images, which currently has to be done manually. Furthermore, it is required to define AOIs. When mapping the gaze points, it can often happen that they lie between two AOIs and then it must be decided which AOI the gaze point is assigned to. All those steps are labor-intensive and time consuming.

2.3   360° Camera System- Analyzing Traffic Situations as a Whole

Road users are often confronted with complex and confusing traffic situations, especially in an urban environment. It is correspondingly difficult to examine these situations and to understand how the various parties involved interact with each other. For this reason, we have developed and manufactured a 360-degree camera system that enables us to capture complex situations in traffic. The system consists of a circular bracket on which up to 32 cameras can be mounted (Fig. 2). We aligned the cameras horizontally and vertically in a way that the videos of the individual cameras intersect at a distance of 10 m. This corresponds to the typical distance between the roadway and the facade in inner-city areas. The circular arrangement of the cameras allows a continuous observation of the traffic area in 360°. The bracket itself can be mounted on a car roof rack. This allows us to easily use the system and record the traffic situation either statically, when the car is parked or dynamically, while driving.

As a possibility to review the recorded footage, we have access to multiple virtual and augmented reality glasses. We also have a video display wall consisting out of eight 4K curved 78" displays. We arranged them in a 2 × 4 pattern, which results in a seven meters wide display area with a resolution of more than 130 million pixels. The video data of our camera system can be stitched together and viewed on our display wall, in order to identify interesting situations in traffic and to examine the interplay of several road users. On our display wall, we can achieve a highly immersive viewing experience, which introduces another use case for the recorded images. Using the video footage of our camera systems, we can develop realistic driving simulations. These would allow us to observe how participants react in complex or dangerous situations without really putting them in these situations and it guarantees comparability, since all participants can be put in the exact same situations. However, we still see the biggest benefit of the system in analyzing and evaluating the public space and the interaction between cars, pedestrians, cyclists and other road users. Conceivable scenarios are the observation of overtaking maneuvers, the traffic flow at highly frequented intersections or the behavior of road users at traffic lights.

Fig. 2. The 360° camera system mounted on top of a car

Lessons Learned
We have not yet been able to use and evaluate the system in a complex scenario. Therefore, the lessons we have learned in using the systems are limited. In our initial approach, we found that mounting the system on a car was quite simple. On the other hand, stitching the many individual videos together correctly was a challenge. Primarily because the process requires a lot of computing capacity and computing time. Currently we are looking for suitable projects to utilize the system to its full extent.

3 Tools for Specific Modes of Transport

3.1 Walking

Pedestrians are directly affected by most traffic and urban planning decisions, yet very often this mode of transport and pedestrian's requirements are not taken into consideration. Even if they are, it is rare that walking is investigated in the field rather than by surveys or in workshops. Common participation processes for urban planning in Germany rely on directly addressing residents in areas that are selected for restructuring.

Residents are reached via postal letters and are invited to participate in public meetings and discussions about possible improvements in the infrastructure in their home district. Very often, their experience as pedestrians is not considered as important as their requirements as car drivers, for example. This approach has three major shortcomings. These shortcomings concern the homogeneous structure of the audience at participative events, the lack of addressing pedestrians, especially those that are no residents in the area and the missing context and location reference at workshops and discussion panels.

The conventional instruments of urban planning are not suitable for the evaluation of the requirements and problems of pedestrians in urban spaces. A report of the International Transport Forum (ITF) of 2012 calls walking the "neglected transport mode" and describes, that sufficient data on pedestrians is missing [15]. It also reports, that specific methods to collect such data are missing as well. For example, walked distances are often not reported when people give an account of their mobility in studies. Because the distances are short and people do not remember them. They are also not measured, when the main mode of transport of a distance is another, for example using the train. However, many people walk to reach the train station, but this part of the trip is not reported on separately.

We therefore designed and developed an approach, which provides an investigative approach for pedestrian related problems and improves the participation of pedestrians, specifically in the planning process. We came up with combination of two tools: online feedback and feedback pillars.

In order to locate problems on footpaths, we developed both a mobile app and a web application, which any citizen can use to mark specific locations on a map and give statements describing the conflict or potential for improvement related to pedestrians in these places. Figure 3 gives an overview of the developed mobile App. Depending on the quantity of problem submissions for an area, this problem can then be further investigated and validated using our feedback pillars.

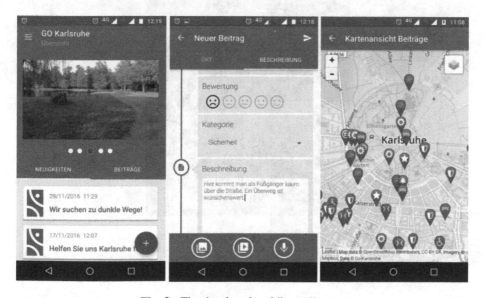

Fig. 3. The developed mobile application

Since most conflicts for pedestrians do affect them solely in the moment, it is important to get their feedback right away and without many barriers. We therefore wanted to develop an instrument that is easily set up, can be placed in most environments and does not need a power outlet. Our solution, the feedback pillars, consist of acrylic posters containing embedded buttons, a metal box filled with the electronics necessary to register participation and a bracket to mount it onto a pillar. A feedback pillar is shown in Fig. 4. The poster serves both as an eye catcher for the feedback pillar itself as well as to communicate the question being evaluated through this pillar. The question asked might be, for example: "I feel safe crossing the street at this traffic light" followed by four mechanical buttons mapped to smileys ranging from a happy smiley to a sad one.

After locating a potential site that may need to be adapted in terms of its footpaths, we used one or more feedback pillars in this area to get an overall view of how this problem is perceived by the broader public using the footpath. Due to the simplicity of the design of the feedback pillar and its direct placement at the location that should be evaluated, it is possible for any pedestrian passing by the feedback pillar to give short feedback on their current situation. This results in a lot of responses from pedestrians. On some days of our field test, for example, we were able to collect more than 1000 inputs. The application of these feedback pillars has two main objectives: first, the collected data can give a quantitative validation for whether or not further action might actually be needed in this location. Second, it serves as a baseline for further evaluations if there are actions being taken to adjust the problem.

Fig. 4. The developed feedback pillar

Lessons Learned

A lesson learned from using the presented online tools is the observation of the same challenge every app or web page has: the acquirement of users. Attempts to spread awareness for the app throughout the population were made through advertisements in local radio broadcasting stations and trams. Yet, even the usage of lotteries for users of the app did not result in significantly high download rates. This is also due to the frequency of how often one might use the app. Not many people would download an app just to place a complaint once. Furthermore, the potential of misuse of the app by making non-productive statements should also be taken into account, as it is the case with most messaging boards.

Some of the lessons learned from our deployment of feedback pillars are the need for longer lasting batteries, since the ones used for now had to be exchanged every two days, which is an unnecessary additional workload. Another aspect that could be improved is the acrylic poster containing the question. Up to now, the whole poster has to be replaced in order to ask a different question, which is linked to cumbersome disassembling and reassembling and therefore additional effort. The possibility to monitor the data gathered by the pillar instantly is also something that could be implemented in the future. Right now, the data is being saved on a SD-Card inside the feedback pillar and can only be retrieved by removing it, which causes the risk of data loss in the event of damage or theft. In our current configuration, it is impossible to ask the user follow-up questions. For example, why they do not feel safe crossing the street. This could be solved by switching from static displays to electronic displays. Another difficulty that we encountered while analyzing the collected data, is how we can determine that a person is not falsifying the result by pressing the button more than once. In our first approach, we applied a threshold of five seconds between recording inputs. Every input under this threshold is dismissed. Of course, this approach is not ideal and we are currently researching better methods to solve this problem.

3.2   Cycling

Monitoring road conditions is a very relevant task in mobility research and has received a significant amount of attention. However, most monitoring systems focus on roads for cars. We introduce an analysis tool for monitoring various comfort factors of cycle paths by collecting crowdsourced data from specially designed sensor modules. The primary goal is the consistent, automated acquisition of sensor data with regard to their statement about the comfort of partial bicycle routes. Unlike another, previously presented approach, presented in Titov et al. 2019 [16] that monitors road roughness by collecting crowdsourced data from a mobile application, we have additionally developed a tool for collecting data on comfort factors using sophisticated sensor modules.

Without local knowledge, it is difficult for a cyclist to identify a comfortable cycle path. Navigation tools, like GoogleMaps, often focus on computing the fastest routes. However, it is not clear whether these routes also offer the expected comfort for a cyclist. If the cyclist does not want to ride on narrow streets with heavy car traffic or avoid signaled intersections, additional information on the respective road sections is required. Also to efficiently promote cycling, cities and communities must know about the quality of their bicycle infrastructure.

A study at the University of Maryland confirms these assumptions [17]. The researchers conducted an online survey of students at the University of Maryland who live within five miles of the campus. While the greatest motivation for bicycle owners to ride a bike was to have their own bike path and a separate guide to motor vehicle traffic. Better lighting conditions and a good map of local cycle paths also contributed significantly to motivating test subjects to use the bicycle. When asked what prevents the participants from using the bicycle, the majority responded with an insufficient feeling of safety in road traffic and the poor condition of the road spaces.

We determined which factors could hinder, disturb or possibly stimulate a cyclist. Previous studies have focused primarily on weather conditions and seasons in connection with comfort and cycling [18]. These rather temporal influencing factors are not the focus of our approach. Instead, we considered spatially differentiated influencing factors. A study from 2008 put the influence of "local factors" on the proportion of bicycle traffic at around 70% [19]. Subsequently, we investigated to what extent these factors can be measured by small-scale, low-cost sensors. We developed a sensor module that collected data on these factors. Utilizing our sensor module, data will be collected automatically to measure and localize these factors. In this way, mass data can be collected quickly and efficiently, when the sensor module is adequately distributed.

The developed sensor module shown in Fig. 5 consists mainly of microcontroller components and sensors from the manufacturer Tinkerforge. The Tinkerforge modules have the advantage that they can be linked together very easily and offer a uniform, well documented API. In this way, the use and application range of the sensors can be adapted without having to use additional hardware and different programming languages. The developed sensor module is build up with the listed Tinkerforge components: IMU (Inertial Measurement Unit) Brick 2.0, temperature bricklet, GPS (Global Positioning System) Bricklet, dust detector bricklet, sound intensity bricklet, color bricklet and the RED (Rapid Embedded Development) bricklet.

Fig. 5. Sensor module to collect bicycle-related data

The IMU includes an acceleration sensor, a gyroscope and a magnetic field sensor. A British study dealt with the suitability of microelectromechanical systems sensor technology on bicycles [20]. It found out that neither acceleration sensors nor gyroscopes alone provide useful results. Both measuring methods are independently subject to high error influences. By combining the data of both systems with the help of Kalman filters, accurate results can be obtained, where error influences have been eliminated as far as possible. The IMU Brick 2.0 implements the Kalman filter, calibrates itself and calculates Quaternions, linear acceleration, gravity vector and independent yaw, roll and pitch angles. Strong deflections in the Z-acceleration indicate fast, jerky changes in height. These can be caused by potholes and road bumps and are an indicator of poor road quality. Increased negative Y-accelerations, on the other hand, indicate heavy braking that could be caused by dangerous situations. The value of the pitch angle gives us information about the slope of the road. While enthusiastic cyclists sometimes even want steep gradients, they tend to disturb the normal cyclist during his ride. Already from a gradient of 3%, the effort required to cope with the gradient is as high as the effort required to cope with the air resistance [21].

The gradient can also be measured with the temperature bricklet. The bricklet contains not only a temperature sensor, but also a humidity sensor and a barometer. In order to detect changes in altitude while driving, the difference values between the successive individual values must also be determined. If the difference is negative, the air pressure has increased. Increasing air pressure means (under the same conditions) decreasing altitude. If the difference is positive, this means an increase in altitude. Temperature and humidity serve more for the classification or correction calculation of other sensor values and less as a direct indicator for spatially differentiated comfort conditions of bicycle traffic.

The GPS bricklet allows us to georeference the collected data points. Furthermore, information on the current speed of the object on the earth's surface can be obtained from the RMC (Recommended Minimum Navigation Information) and VTG (Course and speed information relative to ground) data supplied. In addition to speed detection, it is also possible to determine downtimes and delays. However, the extent to which waiting times at intersections affect cycling behavior has not yet been empirically investigated.

High concentrations of air pollutants occur mainly along main traffic routes and in street spaces with little air exchange. Unlike car drivers, cyclists are directly exposed to these air pollutants. The dust collector bricklet can detect particles from 1 μm. According to this the fine dust PM10 and PM2.5 can be detected. If a high level of particulate matter is measured in the road space, it can be assumed that other combustion-related air pollutants are increasingly present. In another study, a bicycle sensor was developed and tested in the Chinese city of Changzhou, which was intended to measure air pollution in street spaces [22]. An aerosol gas sensor and a sensor for $CO2$, $N2$, NOx, were used. The sensors and a GPS receiver were connected to a microprocessor. The investigation showed that significantly increased amounts of particulate matter were found on roads with large amounts of traffic. The particulate matter values turned out to be a particularly good indicator of traffic-related air pollution at road spaces.

The color bricklet measures the light intensity and can help to identify poorly lit cycle paths or areas. 40% of the participants in the study in Maryland stated that better street lighting would encourage them to use bicycles more often [17]. Reporting these areas to the competent authorities could therefore help to promote cycling.

We tested the sensor module in and outside the city of Karlsruhe to obtain information about the suitability of the individual measurement groups as well as their statement and sensitivity. In the field test, we collected data from four test drives. During these test drives, 2,756 valid periodic data sets and 21,941 event data were collected. Each of the 2,756 data sets contains up to eight event data listed as next. Velocity data, dust pollution data, temperature and humidity data, altitude data, deceleration data, road surface data, curviness data and lightning conditions. The collected data were subsequently formatted and exported, so that first analyzes and visualizations can be created with a Geographic Information System (GIS). Using the GIS software, the respective sensor data are mapped and differentiated. The result can then be checked for logical correctness, expected value and significance. An exemplary result of the analysis is shown in Fig. 6. It displays the light condition as a result of the light intensity measured with the sensor during a night bicycle ride on a street in Karlsruhe.

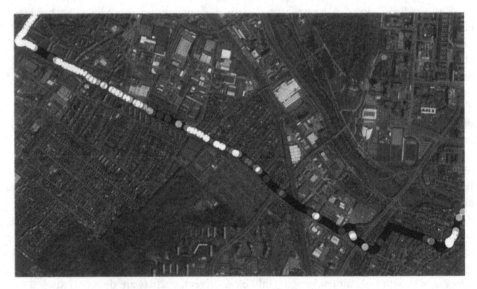

Fig. 6. Measured light intensity on a road in Karlsruhe (white = bright, black = dark)

Lessons Learned

In our approach previously presented in Titov et al. 2019 [16] we used only mobile phones as sensor platform. Compared to this approach, the sensor module enables a much greater variety in data acquisition and evaluation and offers many applications. The sensor module can be used to collect data on bike usage in a city and to classify routes for cyclists in order to improve route guidance based on comfort factors. The data is also very relevant for urban planning, since it can indicate locations with comfort or safety problems for cyclists. These indicators can be used to decide on measures to increase the safety and comfort of cyclists and to improve the overall situation of cyclists. A benefit of the approach with smartphones is that practically everyone has access to them. This enables crowdsensing, which is not possible with the sensor module due to its design and cost.

Our research shows that in individual case studies most of the acquisition methods used by the sensor module have proven to be resilient. From the data obtained, conclusions can be drawn about the comfort of the bicycle routes travelled, even with little interpretation effort. But due to the small extent of the field test, final statements about the individual functionalities are not yet possible. Comparisons of our test tracks, however, have shown that it is possible to determine whether the sensor evaluation provides useful results. Especially in the area of height determination and light or color value determination, conclusions that are more precise could be drawn with the help of better calibration and subsequent smoothing of the results.

We are continuing the analysis of the collected data and will report our full findings in a separate article. For further use of the sensor module, we also want to develop a sensor housing that protects the sensor from weathering and other damage, but does not negatively affect the sensor functions.

4 Conclusion

In this paper, we have presented a comprehensive toolbox for the analysis of mobility. We have described our application of existing, state of the art technology such as smartphones and eye tracking systems for the analysis of traffic situations and driving behavior of several modes of transport. In addition, we presented tools we have developed ourselves for general mobility analyses or for the analysis of specific modes of transport. We have tested our tools in several studies and applied them in mobility analyses for different periods of time. We reflected on the use and development of our tools and present the lessons learned. In this way, we want to contribute to the continuous improvement of these instruments and the resulting research on our urban mobility.

References

1. Schelewsky, M., Bock, B., Jonuschat, H., Stephan, K.: Das elektronische Wegetagebuch: Chancen und Herausforderungen einer automatisierten Wegeerfassung inter-modaler Wege (2014). https://www.researchgate.net/publication/259897501_Das_elektronische_Wegetageb-uch_Chancen_und_Herausforderungen_einer_automatisierten_Wegeerfassung_intermodaler_Wege
2. Safi, H., Assemi, B., Mesbah, M., Ferreira, L., Hickman, M.: Design and implementation of a smartphone-based travel survey. Transp. Res. Rec. **2526**(1), 99–107 (2015)
3. Patterson, Z., Fitzsimmons, K., Jackson, S., Mukai, T.: Itinerum: the open smartphone travel survey platform. SoftwareX **10**, 100230 (2019)
4. BMVI: Article regarding the inquiry of mobility data by the German ministry of traffic and infrastructure (2019). https://www.bmvi.de/SharedDocs/DE/Artikel/G/deutsches-mobilitaetsp-anel.html

5. Kunze, U.: Evaluation einer Focus+Context Visualisierungstechnik mit Eye-Tracking. Institut für Visualisierung und Interaktive Systeme, Stuttgart (2018)
6. McEvoy, S., Stevenson, S., Woodward, M.: The impact of driver distraction on road safety: results from a representative survey in two Australian states. Inj. Prev. 12(4), 242–247 (2006)
7. Kapitaniak, B., Walczak, M., Kosobudzki, M., Jozwiak, Z., Bortkiewicz, A.: Application of eye-tracking in the testing of drivers: a review of research. Int. J. Occup. Med. Environ. Health 28(6), 941–954 (2015)
8. Davoudian, N., Raynham, P.: What do pedestrians look at night? Light. Res. Technol. 44(4), 438–448 (2012)
9. Kiefer, P., Straub, F., Raubal, M.: Towards location-aware mobile eye tracking. In: Proceedings of the Symposium of Eye Tracking Research & Applications, pp. 313–316. ACM (2012)
10. Mantuano, A., Bernardi, S., Rupi, F.: Cyclist gaze behavior in urban space: an eye-tracking experiment on the bicycle network of Bologna. Case Stud. Transp. Policy 5(2), 408–416 (2016)
11. Vansteenkiste, P., Cardon, G., Philippaerts, R., Lenoir, M.: The implications of low quality bicycle paths on gaze behavior of cyclists: a field test. Transp. Res. Part F: Traffic Psychol. Behav. 23, 81–87 (2014)
12. Wenczel, F., Hepperle, L., von Stülpennagel, R.: Gaze behavior during incidental and intentional navigation in an outdoor environment. Spat. Cogn. Comput. 17(1–2), 121–142 (2017)
13. Bundesamt, S.: Unfallentwicklung auf deutschen Straßen, Wiesbaden (2018)
14. Keller, C., Titov, W., Sawilla, S., Schlegel, T.: Evaluation of a smart public display in public transport. In: Mensch und Computer 2019-Workshopband (2019)
15. ITF: Pedestrian Safety, Urban Space and Health. OECD Publishing, Paris (2012)
16. Titov, W., Schlegel, T.: Monitoring road surface conditions for bicycles – using mobile device sensor data from crowd sourcing. In: Krömker, H. (ed.) HCII 2019. LNCS, vol. 11596, pp. 340–356. Springer, Cham (2019). https://doi.org/10.1007/978-3-030-22666-4_25
17. Akar, G., Clifton, K.J.: The influence if individual perceptions and bicycle infrastructure on the decision to bike. In: 88th Annual Meeting of the Transportation Research Board, Washington, DC (2009)
18. Bosselmann, P., et al.: Sun, Wind, and Comfort A Study of Open Spaces and Sidewalks in Four Downtown Areas. University of California, Institute if Urban and Regional Development, Berkeley (1984)
19. Thomas, T., Jaarsma, R., Tutert, B.: Temporal variations of bicycle demand in the Netherlands: the influence of weather on cycling. Transp. Res. Board (2008)
20. Miah, S., Kaparias, I., Liatsis, P.: Evaluation of MEMS sensors accuracy for bicycle tracking and positioning. IEEE, London (2015)
21. Suhr, W., Schlichting, H.-J.: Mit Pedalkraft gegen Berge und Wind. In: Physik unserer Zeit, Weinheim, pp. 294–298. Wiley-VCH Verlag (2007)
22. Xiaofeng, L., Bin, L., Aimin, J., Shixin, Q., Chaosheng, X., Ning, X.: A bicycle-borne sensor for monitoring air pollution near roadways. IEEE, Changzhou (2015)

Training Pedestrian Safety Skills in Youth with Intellectual Disabilities Using Fully Immersive Virtual Reality - A Feasibility Study

Robin Cherix, Francesco Carrino[✉], Geneviève Piérart,
Omar Abou Khaled, Elena Mugellini, and Dominique Wunderle

University of Applied Sciences and Arts Western Switzerland,
Fribourg, Switzerland
{Robin.Cherix,Francesco.Carrino,Genevieve.Pierart,
Omar.AbouKhaled,Elena.Mugellini,
Dominique.Wunderle}@hefr.ch

Abstract. The possibility to move independently outdoor has a huge impact on the quality of life. However, it requires complex skills, difficult to acquire for youth with intellectual disabilities (ID). They need an engaging and varied environment in which they can safely train these skills for all the time they may need. We present an exploratory study that aims to evaluate the usability of virtual reality (immersive headset) as learning tool for youth with ID. We developed a simulator of a pedestrian crossing able to reproduce different environmental conditions (i.e., weather, day-time/night-time, and drivers' kindness). We tested our simulator with 15 people (9–18 years old) with ID. The tests showed good acceptability and a learning effect was visible after only four consecutive sessions, for a total of sixteen simulated crossings. However, additional studies are required (i) to assess in which measure this effect is imputable to actually learned crossing road skills or to a better control over the tool, (ii) to measure the transfer of the learning from virtual reality to real word conditions.

Keywords: Virtual reality · Intellectual disability · Pedestrian crossing

1 Introduction

For Wehmeyer, self-determined behaviour refers to "the attitudes and abilities required to act as the primary causal agent in one's life and to make choices regarding one's actions free from undue external influence or interference" [1]. Youth with intellectual disabilities (ID) are able to have personal control on their life, but this ability is often underestimated. Because of the ID, they need more time and opportunities to develop self-determined behaviours [2, 3]. ID can lead to lower autonomy (ability to fulfil complex tasks without excessively depending of the environment) and self-regulation (ability to make decisions about what skills to use in a situation and evaluate a plan of action, with revisions when necessary) [1]. Regarding these difficulties, family circle tends to be overprotecting, reducing the youth learning opportunities [2].

Teachers usually develop individualized strategies and supports in learning programs. Heterogeneous levels of difficulties and resources characterize the ID

H. Krömker (Ed.): HCII 2020, LNCS 12213, pp. 161–175, 2020.
https://doi.org/10.1007/978-3-030-50537-0_13

population. Individualized programs are very important for youth with ID, because this kind of disabilities can affect their ability to identify undue external influence. Research on motivation [3] shows that people with ID tend to be more externally directed than people without disabilities (they use external indications instead of personal cognitive resources to resolve problems); they also have a greater dependence on social reinforcement. Thus, the risk of abuse is more important for them than for the population without disabilities. However, this is not an inner characteristic: it results from internalization of environmental attitudes, such as overprotection, that reduce the young child exploratory behaviours. People with ID can develop autonomy and self-regulation, two components of self-determined behaviours, on a long-life way [4].

To move independently in public spaces requires complex skills, such as to use the sidewalk and the pedestrian crossing (with and without traffic lights), to take the bus, to interact with others, etc. [5]. Shows et al. demonstrated that people with ID living in a semi-independent way are twice more at risk to be involved in a pedestrian accident than person living in residential contexts [6]. Thus, youth with ID need to train several times the learning behaviours with a teacher to get sufficiently self-confident and skilled. They also need to practice the behaviours in various conditions (e.g., by sunny or rainy weather, by day or by night, etc.).

For these reasons, it seems necessary to provide youth with ID the possibility to safely experiment situations related to mobility in public spaces. It is also important to take into account their specific difficulties in acting the behaviours (e.g., to cross the road, to manage an unexpected situation in the bus, to speak with strangers, etc.). It is complicated to recreate this kind of situations in real life because of the dangers they imply and the important level of human support they need. New technologies such as virtual reality (VR) could allow simulating various conditions in a secure way to train youth with ID before testing real situations.

Fully immersive VR is rarely used to support youth with disabilities in learning tasks. Yet it allows testing complex scenarios that are difficult to be tested in real life (because of danger, costs, setup time, etc.). This paper is a translation of the paper [7], previously published in French. It presents the results of an exploratory and interdisciplinary study that aims to evaluate the usability of a VR (immersive headset) as learning tool for youth with ID and it. It focuses on a pedestrian crossing scenario without traffic light. The headset has been chosen in the aim to exploit its immersive possibilities, even more so since it is very accessible in terms of costs and portability and it could be integrated.

This article is structured the following way: Sect. 2 presents the state of the art, with a special focus on the use of VR by people with ID. Section 3 reports the methodology used in the study. Section 4 shows the results in terms of feasibility and Sect. 5 proposes a discussion of the results. Section 6 concludes the article and proposes some perspectives for future research.

2 State of the Art

For many years, researchers have studied the use of VR for the rehabilitation of people with ID. In 2005, Standen and Brown [8] presented a systematic review of the literature showing how VR could be used as an intervention and/or assessment tool. In their analysis, they pointed three fields of application: promote the skills for an autonomous life, improve cognitive performances and improve social skills. With regard to autonomy, they had analysed some work that focused on very concrete cases such as: shopping, cooking, following directions and wayfinding, developing crafting skills and safety related skills like crossing the road. The authors stated that, besides studies concerning people with autism spectrum disorder (ASD), these experiments showed the possibility of a transfer of learned skills from virtual reality to real world.

Yannick Courbois and his colleagues published many works between 2010 and 2015 addressing the learning of new routes, shortcuts and navigation skills for children and adults with a Down syndrome and/or a Williams syndrome in a virtual environment [9–12]. In their works, VR was used mainly as an assessment tool for skills learning, without analysing a transfer of this learning to reality. Most of their studies still showed that people with ID could learn to find their way or learn a route in VR, even though they need more time if compared to people the same age without ID.

In 2015, Freina and Ott [13] presented a new review of the literature about the use of VR in education in general. In their analysis, they focused on the works published in the 2014–2015 period that use VR in an immersive manner, thus using a Cave Automatic Virtual Environment (CAVE) system or an immersive headset like an Oculus Rift[1] or HTC VIVE[2]. The authors assert that, since 2005, very few works studied the VR as a supportive, learning tool for the people with ID, referencing only two works focused on children with ASD.

About learning crossing the road on a pedestrian crossing, in [14], the authors investigated the differences between teenagers with an attention-deficit hyperactivity disorder (ADHD) and a control group. This study showed that VR was able to find significant differences between the two groups. Thus, authors stated that VR could be used to identify and educate people that are most likely to be exposed to risk and be involved in dangerous situations.

In [15], the authors used VR to specifically try to determine if children with an ASD can learn to cross the road on a pedestrian crossing in safe manner and assess to what extent these new skills transfer to real life. For their tests, six children used a non-immersive system based on 3D simulation displayed on a screen. The results showed that children with ASD can learn in VR and highlighted a substantial improvement of their capacity to cross the virtual road safely. Half the participants also significantly improved their behaviour as pedestrian in real life after just one intervention in VR. Keeping in mind the small number of participants and the brevity of the intervention, these results seem to show that VR can be used as a learning tool for children with ASD.

A more recent work (2015) investigated a similar scenario with another group of six children with ASD. However, in this case, they used an immersive environment (i.e., a CAVE system) [16]. The results showed that most children (four out of six) reached the desired learning goal. This was verified at the end of a four-day period by asking the children to cross the road on a pedestrian crossing in the real world. The authors noticed that these results are motivating for the children themselves, as they feel they

---

[1] https://www.oculus.com/.
[2] https://www.vive.com/.

succeeded, and for their parents, as they feel proud of their children's capabilities, even though they were sceptical at first.

Another study in 2015 conducted with seven adults with ASD showed different results [17]. The protocol included ten sessions, 45 min each, one session a week. These sessions were in non-fully-immersive VR as the participants stood in front of a two-by-two meters screen. Six participants fulfilled the protocol (one was excluded due to issue with depth perception). They easily learned the body gesture necessary to move and interact with the virtual environment. However, even if the participants may have considerably improved their navigation performances (i.e. control system comprehension, needed time to accomplish the task, etc.), they did not significantly reduce the number of mistakes neither in the road crossing simulation or the test questionnaire.

To summarize, existing studies about the use of VR for teaching new skills to people with ID show encouraging conclusions. However, most of these studies focus on a very specific group of participants (most of them being affected by ASD). In addition, the limited number of participants to these works and the sometimes contradictory results highlight the need for more research. Furthermore, at the best of our knowledge, no study has yet investigated the acceptability of an immersive VR headset for youth with ID.

3   Methodology

In this section, we will explain the VR scenario that our interdisciplinary team composed of researchers in Human-Computer Interaction and Social workers developed. The scenario focused on the needs of youth (10–18) with an ID going from mild to moderate. In the second part, we describe more in details the test protocol we used as well as the objective and subjective measures considered for this experiment. The exact description of the profile of the subjects that took part to the tests is provided in Sect. 4.

3.1   Scenario

The scene is composed of a sidewalk and a two-way street surrounded by buildings. Vehicles travel the road in both directions. Vehicles vary in terms of colour and type (i.e., bicycles, cars, bus, and trucks). The road has four pedestrian crossings that the user must cross. In order to move around, the user has to use a controller to point to the wanted direction and press and keep pressed a button to move forward. To facilitate the interaction, we show a 3D model of the controller in VR that mimic the position and the orientation of the real one. To make clearer the pointed direction, a "virtual stick" comes out from the top of the virtual controller (see Fig. 1).

The vehicles stop when the users show their "intention" to cross the road. To do this, they have to stand in an invisible waiting area close to the road while being in front of the pedestrian crossing and facing the opposite side of the road (with a certain tolerance angle). The position and size of the waiting area, as well as the range of the angle have been determined during a pre-test phase.

Fig. 1. The participant's point of view in the simulation. A "virtual stick" comes out from the top of the controller to highlight the pointed direction and, consequently, the direction of movement.

We developed this scene using Unity. The users wear an HTC VIVE headset equipped with a Tobii eye-tracker[3].

Given the importance of training the skills in different conditions, to improve the generalisation of the learned skills, five parameters can be modified in real-time while the experiment runs to bring diversity to the scenario.

First of all, the weather can be changed among three choices: sunny, cloudy or rainy. In addition, it is possible to play the scenario by day or night.

The second parameter is about the "kindness" of the drivers. This parameter determines the probability that a vehicle stops for users that showed their intention to cross (as described above). For instance, a kindness of 75% means that each vehicle has a 25% chance to not let the user cross the road.

The third parameter determines how long the vehicle waits for the user to start crossing once stopped. It consists in a minimal and a maximal waiting time. The actual waiting time is picked randomly in the interval defined by these two values.

The fourth parameter concerns the travel speed of the player that can be modified to fit users' preferences and reduce motion sickness. Motion sickness can occur when the brain receives contradictory information from the visual perception (which perceive a movement) and the vestibular system (which does not perceive this movement). Dropping the speed decreases this contradiction thus diminishing the motion sickness symptoms. For the same reason, according to guidelines provided by headsets manufacturers [18, 19], the up and down movements of the head when walking (head bobbing) is not recreated in VR as it is usually done in first-person video games. Instead, the player "slides" on the ground.

---

[3] https://www.tobii.com/.

The last and fifth parameter enables or disables a visual and audio feedback in case of collision between the user and a vehicle. The visual feedback is a red filter that covers the entire user's field of view. The audio feedback consists in an approaching sound of an ambulance siren. The feedback is meant to send a strong signal to the users that something went wrong. This was added to the system after the suggestion of a child psychiatrist. The psychiatrist was particularly concerned that one of the young people she follows may be hit by a "virtual" vehicle in our scenario and, without a strong feedback, the participant may think that being hit by a car in reality is as harmless as it is in our simulation. It turned out that, during our tests, a vehicle hit this person, thus triggering the feedback. This child got so scared that he/she immediately removed the VR headset from his head.

All the parameters described above cannot be seen in the VR headset. They are only visible on the control screen, i.e., the monitor of the computer that runs the scene (see Fig. 2). Thus, a person in charge of the technical supervision can access and modify these parameters when needed, for instance to test new conditions or to change the difficulty level.

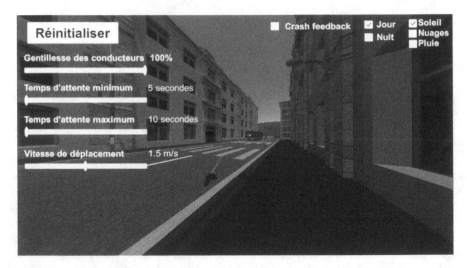

Fig. 2. The control screen shows the parameters of the scene. On the left side (button): Reset. On the left side (sliders): Drivers' kindness, Minimum and maximum waiting time and the travel speed. On the right side (checkboxes): Crash feedback, Day/Night, Sunny/Cloudy/Rainy. The blue circle represents the participant's gaze (in this pictures, it is positioned on the approaching bus). (Color figure online)

During the tests presented in this paper, in order to decrease the variability of the test conditions, only the visibility (day/night) and the driver's kindness parameters were modulated. The weather was set as "sunny". All the other parameters were set beforehand, based on pre-tests made with six teenagers without ID (for more details, see the next section).

The eye-tracker embedded in the VR headset allows to monitor the users' gaze direction. This data is used to analyse the users' attitude toward the traffic and, in particular, their attention level concerning the vehicles before they start to cross the road ("whether they correctly look on both sides or not"). The gaze is displayed as a blue circle. Just like the other parameters, this circle is only visible on the control screen (see Fig. 2). Moreover, for every step of the simulation, the system records what object was looked at (for example car_01 or bicycle_03) as well as the gaze position on this object (relative to the object origin). The system also records the users' head position and orientation as well as the positions of all the vehicles. This set of data can be used to replay the experiment at a later time for analysis purpose.

In order to gather qualitative data about the experiment, the participants were asked to fill a questionnaire at three moments of the experiment. This questionnaire was written in easy-to-read[4] (easy-to-read is a method of presenting written information to make it easier to understand for people with difficulty reading as people with intellectual disabilities). This questionnaire was presented with pictograms for the participants who have difficulty or the impossibility to express themselves verbally. This support was elaborated with the help of social workers, partners of the project.

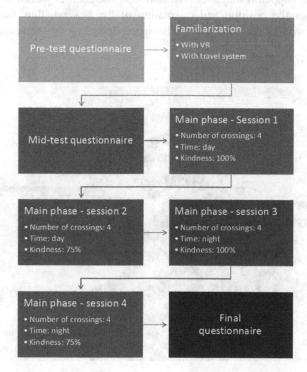

Fig. 3. Progress of the test protocol for participants susceptible to be particularly annoyed by the change between the day-light scene and the night-light scene. For the other participants, the order of the test sessions 2 to 4 were randomized.

---

[4] https://www.inclusion-europe.eu/easy-to-read/.

3.2   Test Protocol

The research team on the field was composed of three people: an IT technician and two social workers. One of the social workers was in charge of the accompaniment of the participants during the experiment (answering their questions, guiding them, encouraging them, etc.) while the IT technician and the remaining social worker observed the participants' verbal or non-verbal actions and reactions in the VR scene and in reality.

The test protocol was divided in multiple phases in order to gather information before, during and after the experiment but also put the participants at ease with VR. The scheme in Fig. 3 shows the detailed progress of the protocol.

Before the experiment begins, the participants can ask to be accompanied by an educator if they do not feel at ease with idea of being alone with the research team.

During the experiment, the participants sit on a chair that can be adjusted in height and rotate (see Fig. 4). This chair has no wheels to provide more stability and to prevent the users from moving from the centre of the experiment, as some elements of the simulation rely on the centre to work correctly. It was decided to have the participants sitting to prevent any loss of balance, causing them to hurt themselves in their fall, which could happen if they were standing up. The chair is placed in between two sensors that track the position and orientation of the headset, thus the head of the participants, as well as the controllers. The space delimited by the sensors is at least 4 m^2 wide.

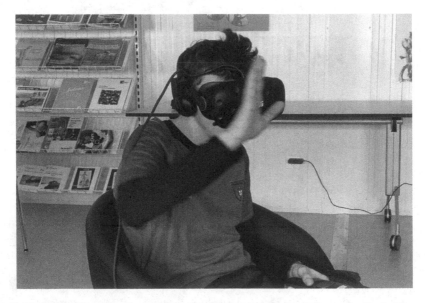

Fig. 4. A participant during the test while thanking a virtual driver who stopped to let him cross.

Before the proper experiment begins, a first questionnaire is filled to assess the capacity of the participants to cross a road with a pedestrian crossing autonomously as well as their experience with technology in general (smartphone, video games, computer, etc.) but also with virtual reality precisely. This first questionnaire is filled with the participant but also the educator responsible of the participant as well as the parents or curator of the participant through a consent form they have to fill before the experiment takes place (this form is also the occasion to inform all the people involved about the goal of the research and the details of the experiment).

During the experiment, notes concerning the attitude of the participants are taken. A structured form was created and used to help and guide observers (the team of three described above) to take notes alongside the experiment. These notes concern the following aspects:

- Technical: putting on the headset was difficult; there was a problem with the simulation; the wires of the headset disturbed the participant.
- Comprehension: the participant had difficulties understanding what he/she had to do; the participant looks at the traffic before crossing like he/she was taught to.
- Emotions: the participant had to be reassured; what kind of help/encouragement the participant had; how frequently; the participant easily accepts wearing the headset; which reactions were observed, verbally and non-verbally.

After the familiarization phase and at the end of the experiment, questions are asked to the participants, orally and with the help of the pictograms about how they felt:

- Physiologically: the participant had a stomach ache, pain in the eyes, or a headache; the headset was felt as annoying.
- Related to the learning: the participant thinks he/she learned something new.
- Emotionally: the participant felt fear, surprise or joy; the participant found that it was easy to cross the road; the participant found the simulation was realistic.

The experiment starts off with a familiarization phase so that the participants can get used to the equipment and to the interaction with a virtual world in VR. This phase is composed of two steps. During the first step, the participants can choose a game between two options: an active air hockey game or a more relaxing experience with musical bubbles. The air hockey is a game played on a table similar to ice hockey where the user must push a puck in the opponent's goal. In the musical bubbles, the user has to burst bubbles touching them with the controllers. When exploding, the bubbles play a note of music. The pitch of the note is based on the height of the bubble when burst.

This duration of this step differs from participant to participant with an upper limit of 15 min. Once this step is done, the participants learn how to move around in the VR scene. To do this, the participants are placed in a scene similar to the one that they will meet during the main phase of the test but without vehicles around. The participants must then move up to one of the pedestrian crossings and cross the road. This phase continues until the participants are able to move around the VR scene. The ability to do so is a requirement to take part in the main test. Once this phase is done, the participants can start the main phase.

The main phase is composed of a series of four sequential sessions. The participants are asked if they want to take a break at the end of each session. During a session, the participants have to cross the road on a pedestrian crossing four times, for a total of sixteen crossings. The sessions differ from one another by their environmental conditions, more precisely the kindness of the drivers and the time of the day the scene takes place (i.e., day or night). The possible percentages for the kindness are: 100% (drivers always stop if the participants are in the right waiting position) and 75% (drivers stop three times out of four). These conditions are tested once in daylight and once during the night, for a total of four sessions. The series always starts with the "easy" session characterized by 100% kindness and daylight conditions. The order of the three remaining sessions depends on the estimated capacity of the participants to handle multiple day to night and vice versa changes. This capacity is estimated by educators who know the participants. If it is estimated that this would not disturb the participants at the expense of the experiment, the order of these Sessions 2–4 is selected randomly. On the contrary, in order to reduce the discomfort, the test was performed as follows:

- Session 2: 75% kindness; day;
- Session 3: 100% kindness; night;
- Session 4: 75% kindness; night.

These are all the phases that compose the experiment.

4   Results

The experiment was conducted 18 times with 15 participants. Three of them took part twice in a five-month interval. The goal of this second participation was to observe the impact of the first experiment on the participants and what they reminded of this first time. The experiment could only be reconducted with three of the participants due to the scholar calendar constraints.

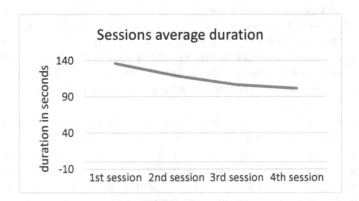

Fig. 5. The average duration for each session.

The youngest participant was nine years old and the oldest was eighteen years old, the average age was 13.8 years old. The group was composed of four girls and eleven boys. All the participants have ID, some of them have associated troubles, five of them having an ASD. In Switzerland, where the experiment took place, it is not possible to provide precise information about the severity of ID since there is no diagnosis before the age of 18. We relied on the experience of the educators to recruit participants with mild to moderate ID levels.

About the capacity of each participant to cross the road, their educator and parents were asked to estimate a score going from 1 (the participant knows how to cross autonomously) to 5 (the participant does not know how to cross at all). Our participants had a mean score of 4.12.

Among these 18 experiments, 10 of them could be completed. One had to be interrupted because of motion sickness and one had to be shortened due to weak understanding of the exercise. Four participants could not perform the experiment due to a total lack of understanding of the experiment and two did not want to do it. To estimate the performances of the participants, we measured the needed time to complete a session (four crossings in a given condition) and the mean time the participants waited before starting to cross the road as well as how many times a collision with a vehicle occurred.

Concerning time needed to complete four crossings, it decreased continuously along the experiment and this, no matter the environmental conditions (i.e., day, night, drivers' kindness). The average time to complete a session decreased from 136 s to 101 s (see Fig. 5).

The mean waiting time before starting to cross globally decreased from session to session and decreased or stayed the same within a session, at the notable exception of one participant who waited a lot before the third crossing of the fourth session (see Fig. 6).

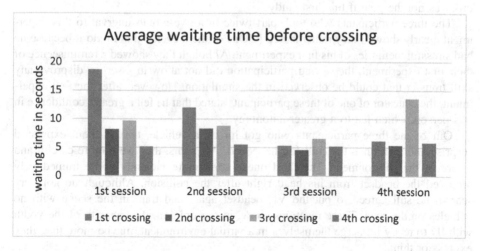

Fig. 6. The average waiting time before crossing.

Out of the 175 crossings that were done during this experiment, 3 led to a collision between a participant and a vehicle. No participant was hit more than one time.

5   Discussion

These primary results are encouraging and show the feasibility of our approach. Most participants (10 out of 15) rapidly understood the goal of the experiment, how to move around in the VR scene using the proposed approach and what they had to do to reach the goal. In one case, we had to go over the basics of crossing the road multiple times. In another case, we had to slightly adjust the words we used, i.e., speaking of "zebra crossing" instead of "pedestrian crossing" (please notice that "zebra crossing" in not a term that exists in French, which was the language of the experiment). Participants did not get bothered by the gear used (wires and headset), at the exception of one who found the headset heavy but he/she agreed to complete the experiment nevertheless. The simulation was realistic enough for the participants to quickly immerge themselves in the virtual environment. Three participants described this environment as realistic to very realistic; six of them claimed it more looked like a game to them; the remaining participants could not express what they felt about it. Spontaneous reactions were observed in the case of three participants, like thanking the drive who let them cross with a hand sign or getting angry at the ones who would not let them cross.

As for the eye-tracking, it turned out very useful to have a better understanding of the attitude of these youth toward the traffic. In one particular case, the tracking could highlight that the participant had a good understanding of the traffic around him and knew what to do before crossing, which reassured his/her parents and educator and led them to let him/her have more autonomy.

As stated in the previous section, data about the simulation and the participants' actions were recorded during these experiments in order to proceed to deeper analysis. These data will be used to compare the participants to teenagers without ID, although this was not the goal if this first study.

The three participants who took part twice in a five-month interval to this experiment clearly showed they remembered it. One of them even did not redo it because he had stressful memories of his first experiment. Although they showed a reminiscence of their first experiment, the second participation did not allow to assert or disprove any skill transfer that could be observed in the simulation. However, after the first experiment, the educator of one of these participants stated that he felt a greater confidence in this person, which led to a greater autonomy.

Out of the three participants who got hit by a vehicle, two to them expressed surprise and fear when the collision happened. After this, they were more careful and aware of the environment. The third one reacted more violently as he immediately removed the headset from his head right after the collision. Although he got very scared, he still agreed to put the VR headset again and train in the scene with no vehicles on the road. These reactions show, in our opinion, the capacity of the youth with ID to really immerge themselves in a virtual environment, maybe more than they express or think.

Giving the heterogeneity of this population, it will be necessary to have more flexible scenarios, adapted to the specific needs of each one. Ideally every individual would have their own scenario. This could be done by developing new scenarios and functionalities with the help of the institutions and specialized schools that would benefit from this tool. For instance, the collision feedback described in Sect. 3 was introduced at the behest of one of the institutions that took part to this feasibility study.

6 Conclusion and Future Work

In this paper, we presented a preliminary study which goal was to determine the feasibility of using VR, head-mounted display headsets in particular, to teach youth with ID. The results are encouraging as they show the acceptability of this technology and signs of a learning curve within VR.

Although this experiment showed a good acceptability of the VR among youth with ID (11 out of 15 were interested in taking part again or to a similar experiment), further, more in depth studies are necessary. To acknowledge, or not, a capacity to transfer experience from a virtual world to the real one, it will be necessary to repeat this experiment on a regular basis, on a larger group and on a long enough period of time. Our goal is to integrate this VR learning tool in a scholar schedule. In addition, this will allow distinguishing in which measure the performance improvement (e.g., the reduced time to complete the crosses) comes from real learned crossing road skills or from a better control over the tool. The initial, continuous and final assessment of this learning, including the transfer and generalization of the learned skills, will be evaluated using the single-case experimental design [20, 21].

From a usability perspective, the current system needs the support of an IT technician during the experiments to set up the gear and make sure it runs smoothly. This could be improved by using a simpler VR system, like the standalone versions that do not require any wire or sensors. The control interface could also be improved by simplifying the simulation settings. The goal is to develop this new interface adapted to educators and specialized teachers by including them in the conception-development loop. In addition, given the wide heterogeneity of the aimed population in terms of skills and need, it will be important to develop more scenarios by involving since the conception phase people with ID through co-design protocols [22].

Lastly, in this paper, we speak about "simulation" and not "serious game". In addition, we avoided to introduce gamification elements. Educators fear that game elements (e.g., scores, badges, etc.) that do not exist in real world would introduce an additional difficulty to transfer skills from the VR to reality. In other terms, this choice was made to prevent the participant from thinking this simulation is "just" a game with no consequences on their life. However, we are well aware that longer and repeated learning sessions will probably require some form of motivation supports. Educators suggested to keep these supports outside the simulation. In a similar manner of what is usually done today in which the motivation support comes after the actual work: if the children work well in class, the educators allow them to go play. Nevertheless, in the future, it could be very interesting to study the impact of serious game and gamification in terms of motivation and transfer of learned skills from virtual reality to the real world for people with ID.

7   Special Thanks

The authors would like to thank the directors, teachers and educators of the six partner structures for their support. The authors would also like to thank Sara Daniela Leite Alves de Sousa, Alice Masozera, Amandine Meystre, Ludovic Patoureau, and Marion Ziegenhagen, students at the Haute Ecole de Travail Social Fribourg who contributed to the development and implementation of the tools of this project.

References

1. Wehmeyer, M.L.: Self-determination and the education of students with mental retardation. Educ. Train. Ment. Retard. Dev. Disabil. **27**, 302–3014 (1992)
2. Normand-Guérette, D.: Stimuler le potentiel d'apprentissage des enfants et adolescents ayant besoin de soutien. PUQ (2012)
3. Juhel, J.C.: La personne ayant une déficience intellectuelle: découvrir, comprendre, intervenir. Presses de l'Université Laval (2012)
4. Tassé, M.J., Morin, D. (eds.): La déficience intellectuelle (2003)
5. Farran, E.K., Courbois, Y., Van Herwegen, J., Blades, M.: How useful are landmarks when learning a route in a virtual environment? Evidence from typical development and Williams syndrome. J. Exp. Child Psychol. **111**(4), 571–586 (2012)
6. Strauss, D., Shavelle, R., Anderson, T.W., Baumeister, A.: External causes of death among persons with developmental disability: the effect of residential placement. Am. J. Epidemiol. **147**(9), 855–862 (1998)
7. Cherix, R., Carrino, F., Piérart, G., Khaled, O.A., Mugellini, E., Wunderle, D.: Training road crossing skills for young people with intellectual disabilities using virtual reality: a feasibility study. In: Proceedings of the 31st Conference on l'Interaction Homme-Machine, pp. 1–9. ACM (2019)
8. Standen, P.J., Brown, D.J.: Virtual reality in the rehabilitation of people with intellectual disabilities. Cyberpsychol. Behav. **8**(3), 272–282 (2005)
9. Mengue-Topio, H., Courbois, Y., Farran, E.K., Sockeel, P.: Route learning and shortcut performance in adults with intellectual disability: a study with virtual environments. Res. Dev. Disabil. **32**(1), 345–352 (2011)
10. Farran, E.K., Courbois, Y., Van Herwegen, J., Blades, M.: How useful are landmarks when learning a route in a virtual environment? Evidence from typical development and Williams syndrome. J. Exp. Child Psychol. **111**(4), 571–586 (2012)
11. Courbois, Y., Farran, E.K., Lemahieu, A., Blades, M., Mengue-Topio, H., Sockeel, P.: Wayfinding behaviour in Down syndrome: a study with virtual environments. Res. Dev. Disabil. **34**(5), 1825–1831 (2013)
12. Purser, H.R., et al.: The development of route learning in Down syndrome, Williams syndrome and typical development: investigations with virtual environments. Dev. Sci. **18**(4), 599–613 (2015)
13. Freina, L., Ott, M.: A literature review on immersive virtual reality in education: state of the art and perspectives. In: The International Scientific Conference eLearning and Software for Education, vol. 1, no. 133, pp. 10–1007 (2015)
14. Clancy, T.A., Rucklidge, J.J., Owen, D.: Road-crossing safety in virtual reality: a comparison of adolescents with and without ADHD. J. Clin. Child Adolesc. Psychol. **35**(2), 203–215 (2006)

15. Josman, N., Ben-Chaim, H.M., Friedrich, S., Weiss, P.L.: Effectiveness of virtual reality for teaching street-crossing skills to children and adolescents with autism. Int. J. Disabil. Hum. Dev. **7**(1), 49–56 (2008)

16. Tzanavari, A., Charalambous-Darden, N., Herakleous, K., Poullis, C.: Effectiveness of an immersive virtual environment (CAVE) for teaching pedestrian crossing to children with PDD-NOS. In: 2015 IEEE 15th International Conference on Advanced Learning Technologies, pp. 423–427. IEEE, July 2015

17. Saiano, M., et al.: Natural interfaces and virtual environments for the acquisition of street crossing and path following skills in adults with Autism Spectrum Disorders: a feasibility study. J. Neuroeng. Rehabil. **12**(1), 17 (2015)

18. Shupak, A., Gordon, C.R.: Motion sickness: advances in pathogenesis, prediction, prevention, and treatment. Aviat. Space Environ. Med. **77**(12), 1213–1223 (2006)

19. Yao, R., Heath, T., Davies, A., Forsyth, T., Mitchell, N., Hoberman, P.: Oculus vr best practices guide. Oculus VR **4**, 27–35 (2014)

20. Petitpierre, G., Lambert, J.L.: Les protocoles expérimentaux à cas unique dans le champ des déficiences intellectuelles. Méthodes de recherche dans le champ de la déficience intellectuelle. Nouvelles postures et nouvelles modalités **57**, 102 (2012)

21. Tate, R.L., et al.: Normes de présentation de recherche utilisant les protocoles à cas unique en interventions comportementales (SCRIBE-2016). Pratiques Psychologiques **25**(2), 103–117 (2019)

22. Arfaoui, A., Edwards, G., Morales, E., Fougeyrollas, P.: Designing interactive and immersive multimodal installations for people with disability. In: Virtual Reality. IntechOpen (2020)

A Decision Support System for Terminal Express Delivery Route Planning

Jiazhuo Fu[1] and Wenzhu Liao[2(✉)]

[1] Department of Industrial Engineering, Chongqing University,
Chongqing 400044, China
jiazhuo_fu@cqu.edu.cn
[2] Department of Engineering Management, Chongqing University,
Chongqing 400044, China
liaowz@cqu.edu.cn

Abstract. The transportation decision support system (DSS) is contributing to improving transportation efficiency and reducing costs. It will play an important role in resource scheduling of logistics enterprise. However, the existing literatures pay more attention to the resource scheduling problem of large hubs such as airports, ports and distribution centers, and rarely studies the express delivery route planning. Besides, the traditional way of terminal express delivery is that the courier uses his own intuition and experience to complete the daily express delivery tasks [1], which seriously affects the efficiency of terminal express delivery as the volume of express delivery increases. This paper proposes a decision support system (DSS) for terminal express delivery route planning due to the lack of route planning assistive systems at the terminal express delivery in reality. The results of benchmark test show excellent performance of Clarke & Wright algorithm with an adaptive large neighborhood search algorithm (CW-ALNS) in terms of computational results and time compared with other algorithms. And the real case at Chongqing city shows that the DSS can effectively plan the delivery route for the courier and analyzes the effect of the maximum load limit of the vehicle used by couriers on route planning. The DSS provides a theoretical support on the express delivery enterprise developing a terminal express delivery route planning tool in reality.

Keywords: Decision support system · Terminal express delivery · Route planning

1 Introduction

With the rapid development of e-commerce, Chinese express delivery market has ushered in explosive growth. As of 2018, Chinese express delivery business volume was 50.71 billion pieces, an increase of 26.6% year-on-year. Since 2014, it has ranked first in the world for five consecutive years, and has become a driving force and stabilizer for the development of the global express parcel market [2]. However, the huge express delivery business has brought opportunities and challenges to Chinese express delivery enterprise. On the one hand, the average daily delivery volume of the courier is 80 to 100 pieces, and more than 80% of the courier work on average more

H. Krömker (Ed.): HCII 2020, LNCS 12213, pp. 176–189, 2020.
https://doi.org/10.1007/978-3-030-50537-0_14

than 8 h due to the lack of express delivery enterprises' attention to planning terminal express delivery route. In particular, Chinese unique online shopping festival activities such as Taobao Double Eleven, the number of parcels generated on that day may be five times the daily, which will put the courier under tremendous delivery pressure. In addition, unreasonable route planning will also affect the core competitiveness of express delivery enterprises. On the one hand, unreasonable delivery routes will increase delivery costs such as fixed costs, transportation costs, and energy costs. On the other hand, if the parcel is not delivered to the customer in a timely manner, it will affect customer satisfaction with the express delivery enterprise and reduce the enterprise market share.

Therefore, how to optimize the terminal express delivery route has become the key to improve the delivery efficiency and reduce the delivery cost of the enterprises. Previous studies have focused on DSS for a single courier personal, but lack of considerations for the overall delivery of terminal express delivery center. This article attempts to build a theoretical model for the DSS of terminal express delivery route planning, and provides a reference for developing terminal express delivery auxiliary tools.

2 Literature Review

The terminal express delivery DSS has only been developed in recent years, so the relevant literatures are relatively lacking. The two typical ones are shown below. Ye et al. (2017) used Baidu Map API to design a complete and application-oriented express delivery routing optimization system based on the actual road distance between nodes and the actual delivery time of real-time traffic conditions, based on JavaScript, C# programming language, and genetic algorithms [3]. Li et al. (2019) used the Gaode Map API to obtain the route information between distribution points, and combined the C-W saving algorithm and Shiny-R technology to develop a fast delivery path optimization tool based on Internet [4]. The delivery route planning of the couriers in the above literatures is essentially a traveling salesman problem, which can help the courier to deliver the express to a certain extent. However, there is a lack of consideration of the overall delivery route planning of the terminal express delivery center. The traveling salesman problem is considered to be the shortest path for a single courier personal to deliver the express from the terminal express center. However, a terminal express center generally has multiple couriers, so how to assign a batch of express to couriers and what route does the couriers use to deliver the express need to be considered. The route planning problem for the courier at the terminal express center is essentially vehicle routing problem.

The vehicle routing problem was proposed by Ramser and Dantzing in 1959 [5]. The problem can be described as: n vehicles in a depot want to deliver a batch of products to m customers, and finally return to the depot with the shortest total path.

Similarly, the terminal express center assigns a batch of express to the couriers, and the couriers deliver the express to each customer's address, and finally returns to the center. The research on vehicle routing problem is mainly focused on heuristic algorithms. The C-W saving algorithm proposed by Clarke and Wright in 1964 is a heuristic algorithm to solve the vehicle routing problem [6]. In view of the characteristics of order distribution in the e-commerce environment, Li et al. established a vehicle routing problem with time window mathematical model with the goal of minimizing driving costs and order penalty costs [7]. Feng et al. established a multi-objective vehicle scheduling model with time constraints based on actual problems, and used particle swarm optimization to solve vehicle scheduling problems [8]. Wang et al. built the lowest cost green low-carbon cold chain logistics distribution route optimization model, and further proposes a cycle evolutionary genetic algorithm to solve the model [9]. Li et al. developed an optimization model of green vehicle routing for cold chain logistics, and modified particle swarm optimization are applied in this study to solve the routing problem of a real case [10]. Liao et al. developed a hybrid genetic algorithm to traditional fuel vehicle routing model [11].

To sum up, the research on the vehicle routing problem in the existing literatures is mainly focused on the improvement of algorithms, and lack of research on DSS. The proposed DSS uses the CW-ALNS algorithm to calculate optimal route, and the route information between any two nodes is obtained through the Baidu Map. The terminal express center can plan the delivery route with the help of DSS. This DSS is expected to provide great convenience to the majority of courier and greatly improves the efficiency of terminal express delivery.

3   Constructing Decision Support System Structure

The extensive effects of DSS are used in logistics activities [12]. The proposed DSS for terminal express delivery route planning is shown in Fig. 1. First, customer information included address and expected delivery time is imported into enterprise database. Then, the dispatcher of terminal express center enters the customer information and the courier information into the DSS system. DSS next converts the customer address into coordinates on the Baidu map and creates a distance matrix by using Baidu map obtaining actual distance between any two nodes. Then a delivery scenario is generated by the route planning algorithm. Finally, the dispatcher sends each delivery route to the handheld terminal of the corresponding courier. The core of this decision support system is how to build a terminal express delivery model and which algorithm to use to solve it.

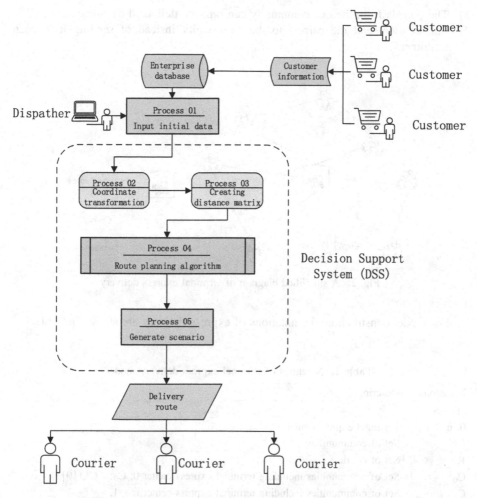

Fig. 1. DSS for terminal express delivery route planning

4 Model Construction

4.1 Problem Description

A simplified diagram of terminal express delivery is shown in Fig. 2. There is a terminal express center with a certain number of couriers, and a set of communities need to be served. The location of the terminal express center and communities is known, as well as the parcel quantity and time window of each community. All couriers must return to the terminal express center while completing their delivery tasks. The main purpose is to find an optimal delivery route considering total cost. Some assumptions of this study are listed as follows:

1) Traveling distance between two locations is provided by Baidu Map;
2) The delivery address will not be changed during the express delivery process;

3) The parcels from the one community can only be delivered by one courier;
4) The courier send the parcels to the community instead of sending it to each customer.

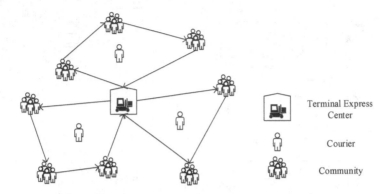

Fig. 2. A simplified diagram of terminal express delivery

For model construction, the notations of express delivery are shown in Table 1.

Table 1. Notations of terminal express delivery model

Notations	Description
(1) Sets	
0, n + 1	Terminal express center
C	Set of communities
K	Set of couriers
C_0	Set of communities including terminal express center 0, $C_0 = C \cup \{0\}$
C_{n+1}	Set of communities including terminal express center n + 1, $C_{n+1} = C \cup \{n+1\}$
(2) Input variables	
q_i	Parcel quantity of community i, $i \in C$
Q	The maximum load limit of the courier
v	Courier traveling speed
d_{ij}	Travel time between vertices i and j
u	The average speed of unloading a parcel
E_i	Earliest acceptable time window for community i
L_i	Latest acceptable time window for community i
F_1	Cost of courier operating
F_2	Unit cost of delivery
(3) Decision variables	
x_{ijk}	0–1 variable, $x_{ijk} = 1$ if the courier k travels from node i to j, otherwise $x_{ijk} = 0$
t_{ki}	Time when the courier k arrives the node i

4.2 Objective Function

The proposed model in this study consists of two costs, which are courier operating cost (C1) and delivery cost (C2).

Vehicle Operating Cost.

The operating cost of the courier includes the salary of the couriers and the cost of their vehicle maintenance:

$$C_1 = \sum_{k \in K} F_1 \sum_{\substack{j \in C_{n+1} \\ i=0}} x_{ijk} \tag{1}$$

Delivery Cost. The delivery cost mainly refers to the energy cost of the transportation tools during the delivery of the couriers.

$$C_2 = \sum_{k \in K} \sum_{i \in C_0} \sum_{j \in C_{n+1}} F_2 d_{ij} x_{ijk} \tag{2}$$

4.3 Modeling

Based on the above analysis, the express delivery model is given as followed:

$$\min TC = \sum_{k \in K} F_1 \sum_{\substack{j \in C_{n+1} \\ i=0}} x_{ijk} + \sum_{k \in K} \sum_{i \in C_0} \sum_{j \in C_{n+1}} F_2 d_{ij} x_{ijk} \tag{3}$$

Constraints:

$$\sum_{k \in K} \sum_{\substack{j \in C_{n+1} \\ i \neq j}} x_{ijk} = 1, \forall i \in C \tag{4}$$

$$\sum_{\substack{i \in C_0 \\ i \neq j}} x_{ijk} = \sum_{\substack{i \in C_{n+1} \\ i \neq j}} x_{jik}, \forall j \in C, k \in K \tag{5}$$

$$\sum_{\substack{j \in C \\ i \neq j}} q_j \sum_{i \in C_0} x_{ijk} \leq Q, \forall k \in K \tag{6}$$

$$E_i \leq t_{ki} \leq L_i, \forall i \in C, k \in K \tag{7}$$

$$t_{ki} + \frac{q_j}{u} x_{ijk} - l_0 \left(1 - x_{ijk}\right) \leq t_{kj}, \forall i \in C_0, j \in C_{n+1}, i \neq j, k \in K \tag{8}$$

$$x_{ijk} \in \{0, 1\}, \forall i \in C_0, j \in C_{n+1}, i \neq j, k \in K \tag{9}$$

Equation (3) denotes to minimize total cost to complete delivery tasks. Equation (4) denotes that each community j must be served by a courier once. Equation (5) denotes that the number of incoming arcs is equal to the number of outgoing arcs. Equation (6) denotes that the courier loading parcels cannot exceed its maximum load limit. Equation (7) denotes that the time for the courier to reach the community is within the time window. Equation (8) denotes the time constraint from node i to j when the courier departs from community i. Equation (9) denotes 0–1 variable constraint.

5   Route Planning Algorithm Design

A hybrid heuristic algorithm combined Clarke & Wright algorithm and adaptive large neighborhood search algorithm (CW-ALNS) is applied to DSS in order to optimize the courier delivery route. CW algorithm is used to obtain the initial solution of the model. ALNS algorithm is used to further optimize the initial solution.

5.1   The Initial Solution of Courier Delivery Model

Initially, an actual distance matrix between any two nodes is obtained by Baidu map, where the nodes include the terminal express center and all communities. Then, each community is served by a courier who connects the node of each community to the terminal express center. Next, the saving distance ΔV of connecting node p and node q is calculated by $V_{0p} + V_{0q} - V_{pq}$ and the saving table M is made by sorting ΔV. Eventually, if the route after merging meet the constraints, community p and community q are merge into the same route according to the M. Once no communities in the M can be merged into the same route due to the constraints, the initial solution of courier delivery model is output.

5.2   ALNS Improvement Scheme for the Initial Solution

An ALNS algorithm developed by the paper consists of two algorithms: Community Removal (CR) and Community Insertion (CI). An iterative approach in ALNS algorithm is used to optimize the solution until the stop-criterion is met. CR algorithm firstly removes the communities in each route. Then CI algorithm reinsert the communities removed by CR into existing or new routes. Based on the probability, the removal and insertion algorithms are applied by dynamically selecting them.

The details of the ALNS algorithm structure is described in Ropke et al. [13]. The program code of ALNS is shown below.

```
Input.
W: An initial solution
Output.
W^f: An improved solution
program ALNS()
    var W^b:= W;
    var p^-:= (1,1);
    var p^+:= (1,1);
    repeat
        Select remove and insert methods r ∈ Ω^-and i ∈ Ω^+ using
        p^- and p^+;
        (W^r, n^r):= = CR(W);
        while(n^r is not empty) do
            W^i:= = CI(W^r,n^r);
        end while
        if c(W^i) < c(W^b) then
            W^b:= = W^i;
        end if
        update p^- and p^+;
    until stop-criterion met
    const W^f:= = W^b;
```

Removal Algorithms. Two removal algorithms are applied in ALNS:

1) Random removal algorithm randomly removes m communities from the solution.
2) Shaw [14] removal algorithm removes communities based on relatedness to each other. The formula of relatedness between community i and j is shown in Eq. (10).

$$R_{ij} = \frac{1}{c'_{ij} + V_{ij}} \tag{10}$$

Where all c'_{ij} are normalized in the range [0, 1], and c'_{ij} is calculated as $c_{ij}/\text{max}c_{ij}$. c_{ij} is the distance of getting to j from i. V_{ij} evaluate to 1 if i and j are served by different couriers.

Insertion Algorithms. CI algorithm includes Greedy Insertion algorithm (GIA) and Furthest Insertion algorithm (FIA) [13]. In GIA, firstly the minimum value of inserting each community node in the set n_r into each route is obtained. Each community in the set n_r is inserted into the solution, and then whether the new route satisfies the constraints is determined. If the constants are not met, the next community is inserted into the current solution. Otherwise, the difference between the new solution and current solution is calculated. Therefore, the best inserting position and the minimum increment after community node inserted into routes is obtained. The community node with the smallest increment in set n_r is inserted into its optimal inserting position in GIA, and FIA is opposite. All communities in the set n_r is repeated to insert into the route in the process until set n_r is empty.

6 Computation Experiment

In Sect. 6.1, the performance of the proposed CW-ALNS algorithm is tested using a vehicle routing problem with small, middle and large size benchmark of Solomon [6], and the results are compared with the CW algorithm and the tabu search (TS) algorithm. And then a case study at Chongqing, China is presented in Sect. 6.2.

6.1 Benchmark Instances Analysis

There are a set of 6 small instances with 25 customers, a set of 8 middle instances with 50 customers and a set of 12 large instances with 100 customers. The parameter settings are main factors which affect the performance of these three algorithms. In TS algorithm, the number of iterations is 100 and tabu length is 20. In CW-ALNS algorithm, the number of iterations is 1000, the number of customers to be removed are set as 15% of total customers and Shaw removal determinism factor is 5. The experiments are implemented on Inter(R) Core(TM) i5-5200U with 4 GB of RAM. The computational distance and time of three algorithm under three-scale instances are shown in Table 2, Table 3 and Table 4.

Table 2. Results of small-scale instances

Instance	CW		TS		CW-ALNS	
	Distance	Time(s)	Distance	Time(s)	Distance	Time(s)
C101	224.58	0.26	224.58	1.11	191.81	1.29
C205	297.89	0.35	273.25	1.33	268.53	1.74
R105	585.63	0.93	568.97	1.17	560.67	1.16
R203	473.10	0.24	459.35	3.06	435.38	1.08
RC102	453.36	0.19	451.74	1.06	352.94	1.12
RC203	389.75	0.27	330.25	1.51	327.69	0.97

Table 3. Results of middle-scale instance

Instance	CW		TS		CW-ALNS	
	Distance	Time(s)	Distance	Time(s)	Distance	Time(s)
C101	224.58	0.26	224.58	1.11	191.81	1.29
C204	297.89	0.35	273.25	1.33	268.53	1.74
R104	585.63	0.93	568.97	1.17	560.67	1.16
R204	473.10	0.24	459.35	3.06	435.38	1.08
RC101	1273.84	1.78	1115.83	4.36	992.02	6.52
RC104	813.35	2.12	710.10	4.47	549.03	6.20
RC201	453.36	0.19	451.74	1.06	352.94	1.12
RC204	389.75	0.27	330.25	1.51	327.69	0.97

Table 4. Results of large-scale instances

Instance	CW		TS		CW-ALNS	
	Distance	Time(s)	Distance	Time(s)	Distance	Time(s)
C101	916.87	34.29	861.7	40.12	828.94	36.42
C103	1135.4	29.95	1084.7	62.04	887.87	48.18
C205	1167.0	40.84	1090.5	57.77	1003.0	57.60
C208	1207.5	40.62	1206.6	61.24	1019.7	56.88
R101	1930.9	24.22	1813.0	47.00	1719.9	44.25
R103	1364.4	27.89	1356.5	53.27	1279.8	46.07
R205	1179.8	27.99	1179.4	51.45	1078.7	45.25
R208	977.36	33.53	964.60	52.74	901.68	52.38
RC101	2063.9	24.80	2048.9	36.13	1810.7	45.40
RC103	1608.1	28.32	1474.8	39.34	1437.0	50.18
RC205	1573.8	31.00	1532.2	55.57	1399.1	43.18
RC208	1110.8	32.06	1109.0	56.34	1070.3	52.01

The CW-ALNS algorithm is a further optimization of the solution obtained by CW algorithm, so its solution is better than the CW algorithm. In small and middle instances, the actual difference of computation time between the two algorithms is not large. However, the computation time of CW-ALNS algorithm is 10–20 s longer than that of CW algorithm in large instances. Besides, the solution obtained by the CW-ALNS algorithm is better compared with the TS algorithm. In terms of computation time, there is no significant difference in calculation time between the two algorithms. Therefore, the CW-ALNS algorithm can effectively solve the express delivery problem.

6.2   Case Study

Case Description. There is a terminal express center (TEC) is located in Chongqing, China. Its main business is to deliver the express from the TEC to each community within the scope of services. The selected community data is 16 communities mainly served by the center. The map of the TEC and 16 major communities is shown in Fig. 3. The information included latitude and longitude coordinates, express quantity and delivery time of these nodes is shown in Table 5. In real life, because the route of any two node A and node B on urban roads is affected by traffic rules, the time of courier traveling from node A to node B and from node B to node A may be different, and the straight line distance between two nodes differs greatly from the actual distance. Therefore, Baidu map is used to create a distance matrix between any two nodes.

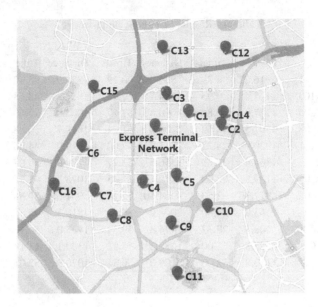

Fig. 3. The map of the express terminal network and 16 major communities

Table 5. The information of selected communities

Number	Coordinates		Express quantity	Delivery time (h)	
	Longitude	Latitude		E	L
TEC	106.506	29.604	0	0	16:00
C1	106.514	29.607	10	13:30	14:30
C2	106.522	29.605	8	13:00	14:30
C3	106.508	29.611	18	13:30	15:30
C4	106.502	29.592	4	14:30	15:30
C5	106.511	29.593	7	14:00	15:00
C6	106.487	29.600	12	15:00	15:30
C7	106.490	29.590	15	13:30	14:30
C8	106.495	29.585	11	14:00	15:30
C9	106.509	29.583	7	13:00	15:30
C10	106.518	29.587	9	13:30	15:30
C11	106.511	29.573	3	14:30	15:00
C12	106.523	29.621	12	13:00	14:00
C13	106.507	29.621	16	13:30	15:30
C14	106.522	29.607	5	14:30	15:00
C15	106.490	29.613	2	13:30	14:30
C16	106.480	29.592	14	13:00	14:30

The related parameters required in the express delivery model are shown in Table 6.

Table 6. The related parameter in express delivery model

Parameter	Value	Parameter	Value
Q	50	K	4
u	0.5 min	F_1	50 Yuan
F_2	10 Yuan/km	v	30 km/h

The courier delivery route of the case study is shown in Table 7 and the detailed results of the case study are shown in Table 8. The terminal express center can refer to the results of this route planning to provide a reference for the courier in the actual express delivery process.

Table 7. The courier delivery route of the case study

Objective function	Courier	Routes
Minimize the total cost	1	TEC→C3→C1→C6→TEC
	2	TEC→C7→C16→C8→TEC
	3	TEC→C5→C10→C9→C11→C4→TEC
	4	TEC→C15→C13→C12→C2→C14→TEC

Table 8. The detailed results of the case study

Objective function	Courier	Express quantity	Traveling distance(km)	Total cost (yuan)	Computation time (s)
Minimize the total traveling time	1	40	11.50	752	1.39
	2	40	8.40		
	3	30	19.60		
	4	43	15.70		

Sensitivity Analysis. In this section, the impact of the maximum load limit on route planning is analyzed by the DSS. The larger the maximum load limit, the higher the fixed cost. Therefore, assuming that the maximum load limit is increased by 5, the fixed cost is increased by 7.5 yuan. The results under different maximum load limit are shown in Table 9. From the results, when the maximum load limit is 55, the cost of the express delivery center is the lowest. When a terminal express center purchases vehicles for couriers, the sensitivity analysis obtained by DSS can provide a reference for determining the maximum load limit of the vehicle.

Table 9. The result under different maximum load limit

Q	Couriers	Total traveling distance(km)	Total cost (yuan)	Computation time (s)
40	4	58.1	721	4.13
45	4	56.6	736	4.01
50	4	55.2	752	3.38
55	3	53.9	711.5	3.45
60	3	52.6	721	4.12
65	3	52.6	743.5	3.83

7 Conclusions

This paper builds a DSS that serves terminal delivery of courier. The effectiveness of the CW-ALNS algorithm proposed in this paper is verified by small, middle and large size benchmarks. This DSS was applied to solve a case of express terminal delivery in Chongqing, China, and analyzed the impact of the maximum load limit on total cost. This will help reduce the delivery time of courier, reduce the cost of enterprise delivery and improve delivery efficiency.

Acknowledgements. This work is supported by project of science and technology research program of Chongqing Education Commission of China (No. KJQN201900107) and project of Chongqing Federation of Social Science Circles (No. 2019PY43).

References

1. Encheva, S., Kondratenko, Y., Solesvik, M.Z., Tumin, S.: Decision support systems in logistics. In: International Electronic Conference on Computer Science, pp. 254–256. AIP (2008)
2. China Express Development Index Report 2018. http://www.spb.gov.cn/xw/dtxx_15079/201904/t20190417_1814716.html. Accessed 16 Jan 2020
3. Ye, W.H., Zhang, F.Z.: Express distribution route optimization under real-time road condition. Comput. Eng. Sci. **39**(8), 1530–1537 (2017)
4. Li, L.Y., Zhang, K.: Express delivery route optimization and software design. Comput. Eng. Sci. **41**(08), 1406–1412 (2019)
5. Dantzig, G.B., Ramser, J.H.: The truck dispatching problem. Manag. Sci. **6**(1), 80–91 (1959)
6. Clarke, G., Wright, J.W.: Scheduling of vehicles from a central depot to a number of delivery points. Oper. Res. **12**(4), 568–581 (1964)
7. Li, L., Liu, S., Tang, J.: Optimal model and two-stage algorithm of order delivery problem in electronic commerce. J. Syst. Eng. **26**(02), 237–243 (2011)
8. Feng, W., Li, X.Q.: Solution of multi-objective vehicle scheduling model based on particle swarm optimization. Syst. Eng. **4**, 15–19 (2007)
9. Wang, S., Tao, F., Shi, Y., Wen, H.: Optimization of vehicle routing problem with time windows for cold chain logistics based on carbon tax. Sustainability **9**(5), 694 (2017)
10. Li, Y., Lim, M.K., Tseng, M.L.: A green vehicle routing model based on modified particle swarm optimization for cold chain logistics. Ind. Manag. Data Syst. **119**(3), 473–494 (2019)

11. Liao, W., Liu, L., Fu, J.: A comparative study on the routing problem of electric and fuel vehicles considering carbon trading. Int. J. Environ. Res. Public Health **16**(17), 3120 (2019)
12. Kuraksin, A., Shemyakin, A., Byshov, N.: Decision support system for transport corridors on the basis of a dynamic model of transport flow distribution. Transp. Res. Procedia **36**, 386–391 (2018)
13. Ropke, S., Pisinger, D.: An adaptive large neighborhood search heuristic for the pickup and delivery problem with time windows. Transp. Sci. **40**(4), 455–472 (2006)
14. Shaw, P.: Using constraint programming and local search methods to solve vehicle routing problems. In: Maher, M., Puget, J.-F. (eds.) CP 1998. LNCS, vol. 1520, pp. 417–431. Springer, Heidelberg (1998). https://doi.org/10.1007/3-540-49481-2_30

A Tactile Interface to Steer Power Wheelchairs for People Suffering from Neuromuscular Diseases

Youssef Guedira[1]([⊠]), Delphine Dervin[2]([⊠]), Pierre-Eric Brohm[2]([⊠]), René Farcy[3]([⊠]), and Yacine Bellik[1]([⊠])

[1] Üiversité Paris-Saclay, CNRS, LIMSI, 91400 Orsay, France
{youssef.guedira,yacine.bellik}@limsi.fr
[2] Centre de rééducation et de réadaptation fonctionnelle Le Brasset,
77100 Meaux, France
{Delphine.dervin,Pierre-Eric.Brohm}@croix-rouge.fr
[3] Université Paris-Saclay, CNRS, ENS Paris-Saclay,
Laboratoire Aimé Cotton, 91405 Orsay, France
rene.farcy@universite-pari-saclay.fr

Abstract. For many people around the world, electric wheelchairs remain a practical means of regaining mobility. Unfortunately, some are not able to use power wheelchairs because of difficulties using a standard joystick. People suffering from neuromuscular diseases who experience a loss in muscular strength can find it difficult to use a standard joystick. In this paper, we explore an alternative steering device in the form of a tactile interface on smartphones. We outline some design choices that are meant to alleviate some of the challenges that may rise with respect to users with neuromuscular diseases. We then detail a study that investigates the usability of this interface over two phases. In the first one, 11 users with neuromuscular diseases tried this type of steering in free learning sessions. 4 among them took part in the second phase where we tested their kinematic performance between the use of the tactile steering interface and the joystick. The paper presents data and observations from both phases and tries to detect tendencies and draw hypotheses that can guide further in-depth clinical testing of the tactile steering for wheelchair users suffering from neuromuscular diseases. Overall, the user performance with the tactile interface was close to or the same as their performance with the joystick. In addition, the users reported a lesser level of physical demand of the tactile steering over the joystick and some of them even preferred the former over the latter.

Keywords: Power wheelchair · Tactile interface · Joystick · Kinematic evaluation · Neuromuscular diseases

1 Introduction

Power wheelchairs can be a practical mobility aid for many people around the world that can allow them to regain independent mobility. They can be used indoors and outdoors and adapted to low and high speed scenarios. However, a problem arises when the person can no longer utilize the steering device that comes with the

H. Krömker (Ed.): HCII 2020, LNCS 12213, pp. 190–210, 2020.
https://doi.org/10.1007/978-3-030-50537-0_15

wheelchair (in many cases this steering device is a joystick). In particular, people suffering from neuromuscular diseases can experience, to varying degrees, a loss of muscular strength that renders the use of a standard joystick hardly achievable.

Many devices have been used as alternatives to a standard joystick when the latter is not compatible with the disability profile of the person. For example, eye-gaze [23], brain-computer interfaces [24] and voice commands [25] were used to issue movement instructions to wheelchairs by users suffering from severe types of disabilities. Unfortunately, many of these solutions do not fulfill the needs of people with neuromuscular diseases. In a previous publication [3], we have introduced a tactile steering interface on smartphones/tablets as a potential alternative to the joystick for people who have difficulties using it. The underlying assumption is that tactile steering is physically less demanding than a standard joystick. Also, the smartphone can be a hub for various applications so that the user can not only steer the power wheelchair but also interact with surrounding artifacts thanks to the emerging domotic technologies.

In a previous preliminary study [10], several wheelchair users not suffering from neuromuscular diseases tested the tactile interface. This previous study showed that the performances using a standard joystick or our tactile interface were similar or close. In this paper, we present the results of a study that gages the usability of the tactile interface by teenagers suffering from neuromuscular diseases. For each participant in the study, we first conducted a training session. Then, in two separate sessions, the participant could perform various tasks that are common in daily life, respectively with the joystick and the tactile interface, to compare their respective performance. The goal of this evaluation was to detect tendencies that can guide future clinical trials in this direction.

The paper is organized as follows: We first give a brief description of some of the challenges faced by people suffering from neuromuscular diseases. Second, we outline some alternatives that have been used in the past to provide power wheelchair users with means of steering that are easier to handle than a joystick. We then give a brief description of the tactile interface on smartphone to steer a power wheelchair and outline its main features that are relevant to the study in this paper. Then, we describe the evaluation of this interface in two steps. The first one is the training session. We outline the relevant observations from this session and how they can be useful to accompany users in their first encounter with the tactile interface. The second one is a kinematic evaluation where we compare user performance, in common daily life tasks between the use of the tactile interface and their own joysticks. We outline the results of each task then the results of the subjective evaluation given by the users. Finally we discuss these objective and subjective results under the light of user characteristics as well as results from previous studies.

2 Neuromuscular Diseases

2.1 Definition

In this section, we briefly present some of the challenges faced by people suffering from neuromuscular diseases. Then, we present some joystick alternatives that have been used as alternatives to a standard joystick for this category of users.

A neuromuscular disease designates an illness that causes an abnormality in the muscular function. It can result from genetic mutations that later have negative repercussions on the motor and cognitive abilities of the person [7]. One of the most known diseases in this spectrum is Duchenne Muscular Dystrophy (DMD) which affects approximately 1/3500 male births worldwide according to [6]. The disease can either touch the nervous system (the case of neuropathies), the muscle (the case of myopathies) or the chemical junction between the latter and the motor nerve ending (case of neuromuscular junction diseases). Regardless of the type of neuromuscular disease classification, there are repercussions that can be found across the spectrum. In this paper, we will restrict ourselves to the manifestations on the motor function as it is useful to understand the challenges that are faced by our target user population.

2.2   Repercussions on the Motor Function

The dominant characteristic of patients with neuromuscular diseases is muscle weakness [7]. When the weakness attains lower limb muscles, the person can lose locomotion. When the weakness attains upper limb muscles (whether it be at the level of gross or fine motor functions [4]), it can translate into a reduced range of motion in fine manipulation and reach. We differentiate weakness from fatigue (a second characteristic of neuromuscular diseases [12]) in the sense that with the latter, the person has an initial strength at the beginning of the movement but loses it abnormally.

Furthermore, the degree of user weakness can vary from one user to another [7] depending on the specific illness and its progression. The condition of the user can even fluctuate in time as a result of intrinsic or extrinsic factors. An example of the latter is the loss of muscular strength when the surrounding temperature gets colder.

Another motor repercussion of some neuropathies can be the presence of spasmodic movements [13]. These movements can sometimes be triggered by an external factor such as a stressful situation or a highly emotional reaction.

These manifestations can impede the manipulation of artifacts in the daily life of the person. When the latter needs to use a power wheelchair, manipulating a standard joystick can be very hard and sometimes impossible as it is a force transducer. In the next section, we will see some steering devices that have been used to replace a joystick, especially for people suffering from neuromuscular diseases.

2.3   Alternative Wheelchair Steering Devices for Neuromuscular Patients

A joystick is an efficient means of steering a power wheelchair and is the most used means for this task [1]. Unfortunately, many people suffering from neuromuscular diseases find it very difficult to use because of their muscle weakness. Sometimes, an adjustment can be made to the joystick software in order to decrease the lever travel distance required to attain full throttle, partially reducing the physical demand. However, this requires the intervention of a specialized professional. When such modification is still insufficient, a special type of joystick can be prescribed to the user, mainly a mini-joystick [15]. This type of joysticks is much more sensitive and requires considerably less force to be used. In fact, [16] investigated the use of such devices (Fig. 1) to steer power wheelchairs and found that they could help users with neuromuscular diseases to steer with less fatigue than a normal joystick. The biggest inconvenience for such systems however is the price as they can cost over $ 3000 for a single mini-joystick [15]. Another inconvenience with force transducers (the mini-joystick being one) is the need to maintain a certain force, even if it is small, for as long as the user needs to maintain the movement. An open interview with functional therapists from the French Association against Myopathies (AFM) revealed that it is important for a person with muscle weakness to maintain movement without having to maintain a hand effort.

Fig. 1. A collection of steering devices for neuromuscular people, used in [16]

A more subtle – but non negligible – inconvenience of using a mini-joystick, is the fact that it might signal that the person reached a critically low level of muscle strength. In a social environment, this can carry a negative stigma that would make the user uncomfortable or even feel ashamed of his/her steering device to be used in public.

Consequently, we needed to investigate novel steering interfaces that would be more suitable to the needs of users suffering from neuromuscular diseases. The steering interface should mainly accommodate for the lack of physical strength. It should also be adaptable enough to comprise as wide of a range of users across the spectrum of neuromuscular diseases. After discussions with healthcare professionals, the choice for a novel steering interface was set on tactile interaction. The main premise is that the person would need only to maintain touch in order to keep the movement of the wheelchair which may result in a lesser physical requirement than a standard joystick.

3 Tactile Steering Interface on a Smartphone

3.1 The Use of a Smartphone to Steer the Wheelchair

In previous studies [3], we introduced a novel tactile interface to steer power wheelchairs. In this section, we revisit some of its main characteristics with respect to the condition of users suffering from neuromuscular diseases.

This tactile interface is in the form of an android app on a smartphone or a tablet that connects to the wheelchair through USB or Bluetooth. Although a tactile steering pad already exists as a commercial steering device [17] (Fig. 2), not only is it expensive (over $3000) but it does not offer a wide enough range of personalization to the user profile. On the contrary, the use of a smartphone can have many added benefits in comparison to a touch-pad.

Fig. 2. Switch-it TouchDrive 2 steering pad

First, nowadays smartphones are ubiquitous [17] and their cost is cheaper than other state-of-the-art steering devices. This makes them more accessible to a wider number of users. A possible advantage of this is that the smartphone being commonly used device would carry less social stigma for people with neuromuscular diseases than a mini-joystick would.

Second, smartphones and tablets are becoming more and more powerful in terms of processing capabilities which can offer a fertile ground for adaptation as well as new interaction possibilities. Their storage capacities may also allow to store a large number of wheelchair user profiles. A profile is a combination of wheelchair driving and seating settings that can be stored in the wheelchair control module and accessed when needed. All the wheelchairs that we have seen so far can store between 5 and 8 profiles as a maximum. We can clearly imagine that in a tablet we can store much more profiles which allows for more flexibility in the daily use.

Third, more and more companies are providing domotic environments with connected artifacts that can be controlled via a smartphone application. Hence, the user's smartphone can serve not only as a steering device but also as a hub for environment control applications. Therefore, the user will have less interaction devices to care for, which in turn means a lower number of user interfaces to get accustomed to.

Finally, the advantage of using a smartphone can go beyond human-object interaction to human-human interaction. For wheelchair users who also suffer from difficulties articulating speech, a speech synthesis application could be installed on the same smartphone. An example would be Commob [9] that can offer multiple levels of interaction to synthesize speech using pictograms.

In [5], we have presented in detail the features of the tactile interface. In the next section, we provide a reminder of the main ones that are relevant for the rest of the paper.

3.2 The Steering Interface and Personalization Features

The interface offers a circular steering panel (Fig. 3). To steer the power wheelchair, the user simply needs to touch the screen inside this panel and displace the controlling finger. The center of the circle represents a neutral zone where the user can keep his/her finger without moving. The size of this neutral zone can be adjusted depending on the user's preferences. Inside the circle, the user can continuously vary the wheelchair movement direction by maintaining touch with the intended direction. The user can also control the speed of the wheelchair movement as it increases the further the finger is from the circle center. Meaning that the further the user's finger is from the center of the circle, the faster the wheelchair movement is. The direction of the movement is set by the vector from the center to the user's touch. With that, the user attains maximum speed right at the edge of the circle. However, the thicker the user's finger, the harder it becomes for the detected touch to reach the edge. This may penalize the movement speed. For this reason, we added a thick tripe at the edge of the steering circle. It allows the user to keep a maximum speed even if the detected touch is not exactly at the edge. We adopted this steering technique (similar to a joystick) to keep a certain level of consistency if a user decides to switch from the joystick to the tactile interface and vice versa. In later design iterations, we may explore other interaction techniques like bimanual steering or a trackball-like steering.

Fig. 3. The tactile steering panel

In order to avoid the interference of accidental touch activations, we can set the interface to consider only the first touch. Thus, the person can rest his/her hand on the screen if needed to alleviate fatigue without worrying that it might disturb the steering. The interface can also be configured to consider the centroid of all points of contact which can be useful for users having tense hand posture (Fig. 4).

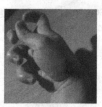

Fig. 4. An example of a tense hand posture

To handle the lack of precision that some users might have, we added the option for the circle to be relocated with the first touch of the user on the screen. This can also come handy for users having low gross motor abilities. If their movement amplitude cannot allow them to attain the steering circle positioned in a certain location, they can simply touch wherever they can on the screen and the circle follows. As for people with low fine motor space, we added the possibility to resize the steering circle diameter in a calibration interface to accommodate for the low movement range.

Another important personalization feature is the possibility to segment the steering circle into different regions for the directions and speeds. The range of speed control can be segmented to have discreet levels of speed. The number of levels can range from one level (representing an all-or-nothing logic for the speed) to 127 levels which is practically continuous given the size of the interface/tablet. The direction segmentation can go from 4 directions only (forward, backward, left and right) to finer levels of direction control with up to 256 segmentations (Fig. 5). On one hand, this offers an opportunity for users with lower dexterity to steer with more stability. On the other hand, this can present an opportunity for progressive learning for the interface use. At first, we can give the user the least amount of segmentations. As he/she get more comfortable using the interface, we can increase the number of segmentations gradually until we reach an optimal level for the specific user. In [10] we observed how this progression helped some users suffering from motor and cognitive deficiencies to learn the use of the tactile interface.

Fig. 5. Screenshots, from left to right, showing different widths of forward direction subdivision when the level of subdivision is 4, 8, 16 and 32 regions for directions and 3 speed levels

Lastly, we added a 3D printed cover (Fig. 6) that can be used when the steering circle is at a fixed position. It delimits the steering area so the user can steer the wheelchair without having to visually focus on the screen. In order to have a smoother contact surface, we added a thin, granulated film that covers the touch area. According to Hayward et al. [11] such contact surface would minimize the friction coefficient between the finger and the touch screen. Thus, giving less of a resistive force to finger displacement, would make the use over a longer period of time less tiring.

Fig. 6. 3D printed cover

4 Testing the Tactile Interface with Neuromuscular Users

4.1 Background of the Study

In a previous publication [10], we presented how several wheelchair users suffering from various backgrounds were able to use the tactile interface to steer a power wheelchair in common daily life steering tasks. We observed how their performance could vary between the use of a joystick and the tactile interface. We also saw that a person having more experience with the tactile interface could perform better with the latter. Finally, we noticed that user preference correlated with their performance. In other words, a person who performed better with the tactile interface had a preference towards it and vice versa.

In the study presented in this paper, we test the tactile steering interface with users suffering from neuromuscular diseases. We wanted to see if the results obtained previously were replicable with this population of users. We also wanted to observe the subjective level of fatigue using the tactile interface compared to a joystick. The tests were carried out inside the facilities of "Le Brasset" functional rehabilitation center in the Paris region. One of our priorities conducting the tests was that the users feel comfortable during the test sessions and avoid external straining factors. This meant that the joysticks that they had to use in the comparison were theirs. Furthermore, even the wheelchairs had to be the ones they daily use as they are more adapted to them and their morphology. This helped us avoid the negative impact that using a maladjusted wheelchair could have on the performance of the users.

Most of the different evaluation sessions were supervised by the chief occupational therapist. Furthermore, throughout the whole sessions where the participants used the tactile interface, one of the experimenters was holding a wireless activation switch which stops the wheelchair in the case of emergency. All the participants signed an informed consent if they were adults. If not, the form was given to their parents to be signed. In both cases, the informed consent was read to the participants. The experimentation protocol was approved by the ethics committee of Paris-Saclay University.

4.2 Demographic characteristics of the users.

Eleven teenagers ($age_{average}$ = 16.64, SD = 3.72), 10 males and 1 female, participated in the study, all suffering from a neuromuscular disease with a loss of lower limb mobility. Their level of education ranged from 4th grade to 2nd year after high school. They all use a power wheelchair indoor and outdoor with a joystick. The latter was either a standard one or a joystick with increased sensitivity to account for the loss of muscular strength that they experience. In this paper, we will not focus on the specific condition of the users but rather their general motor abilities. We asked their chief functional therapist to fill in a questionnaire to rate on a scale from 0 (nonexistent) to 5 (intact) the physical ability, cognitive and perceptive functions of the patients.

At the level of physical strength, for the proximal muscle (closer to the trunk), strength, it was avg = 2.364, SD = 0.809, this correlated with their gross motor impairment: avg = 2.364, SD 0.809. Concerning distal muscle (further from the trunk) strength: avg = 2.545, SD = 0.934 which in turn correlated with fine motor ability avg = 2.727, SD = 1.001.

At the level of cognition and perception, none of the participants had any notable deficiency. The exceptions were only that P4 tends to lose attention and get distracted easily and P9 has difficulties to stay focused for a prolonged period of time.

On the same scale as above, the participants' ability to take risk was slightly moderate with an average = 3.818 and SD = 0.874. The willingness to learn new things was slightly lower with an average = 3.727, SD = 0.647. The capability to integrate and socialize in a group was even slightly lower with an average = 3.455, SD = 0.934.

4.3   Training Session

Getting Used to the Tactile Interface

We conducted a first training session with the 11 participants in our evaluation. This session was conducted inside a wide room within the care center and in most cases under the close supervision of the chief occupational therapist of the center. The training session was also an opportunity to observe the first interaction of our participants with the tactile interface and to determine the best interface settings for each participant. The main goal was to get the participants acclimated to the use of the tactile interface before the kinematic comparison session with the joystick. On this matter, all 11 participants were able to successfully learn to use the tactile interface and steer the power wheelchair within the experimentation room. We also asked the participants to freely steer within the room while avoiding collisions with various artifacts in the room (tables, chairs…). Most of the participants could steer with ease within 5 to 10 min and reported that it was relatively easy to grasp and use. They also reported, as the training session was progressing, that they felt more and more confident using the tactile interface. A further discussion with the participants' chief occupational therapist revealed that even for their first encounter with a power wheelchair, they were quickly able to steer it, using a joystick, in a very natural manner. We can hypothesize that the consistency of the steering logic helped the participants grasp more quickly how to steer with the tactile interface.

Personalizing the Interface to User Needs

For a better use of the interface, it is important to tailor the different settings to the needs of the user. As previous informal tests suggested, the positioning of the tablet can play a crucial role in its usability. This can be more critical in the case of users with neuromuscular diseases as they already suffer from muscle weakness and/or a limited range of motion at their hand. So, for each participant, we tried to set up the positioning of the tablet to be as comfortable to use as possible (Fig. 7).

Fig. 7. Left: the articulated fixation of the tablet, center: P 10 at the training session using the tactile interface with a cover and, right: participant 4's wrist elevated with a block of foam

As a general procedure, we decided to place the tablet at the same place as the user's joystick, then work from there to find an optimal position. As for the inclination of the tablet, the specific angle was less important than the relative position of the wrist. For both the placement and the inclination, we proceeded by gradually tweaking both of them until the user felt the most ease. There is however one point to take into consideration. According to the functional therapist, it is best to have a position/inclination combination that puts the wrist in the most neutral position possible. At times, this may not be the most comfortable position for the user at that very moment, but we should also consider the soundest configuration from a therapy standpoint. For this reason, we would advise that, at least for the first time, the positioning of the tablet should be performed under the supervision of the user's functional therapist. In some cases, like for P4, a better wrist posture was achieved by simply elevating the wrist using thick foam (Fig. 7, right).

Before the training session, P5 frequently asked us to reposition his hand over the joystick whenever it slid away, as his proximal muscles are too weak to do it by himself. When he tried the tactile interface, he needed considerably less assistance in repositioning his hand as it required him to only slide it rather than bringing it up and over the joystick lever.

Steering Application Parameters: User Preferences
Configuring the settings as well as the position of the tactile interface was most challenging with participant P7. The progression of his condition makes it so that he can barely move a couple of fingers and has a very limited range of motion. In fact, at the time of the evaluation, he was struggling to use his joystick with which he could hardly steer in reverse or rotate to the left because of his limited force and range of motion. After several minutes of trial and error, we reached a configuration that allowed him to fully use the tactile steering interface. We made the steering circle as small as necessary for him to reach the whole circle surface (ø = 2 cm) with the thumb. The rest of the hand rested on the cover. Then, we set the positioning of the steering circle to move with every first contact of his controlling finger. Using this configuration, he was able to steer around obstacles and move freely in the experimentation room. He was also able to steer in reverse and rotate right and left. After only a few minutes of steering, he started to develop his own steering strategy: as he had a limited range of motion and the steering circle was repositioned with the first touch, he started

anticipating his direction changes in order to use the whole available motor space. For example, in order to roll forward, he first moved his thumb as far back as he could, then made the first contact with the screen. This positioned the steering circle a little lower but gave him more movement amplitude in the upper portion. This can be seen as an example of a co-adaptation [14] where the configuration of the interactive system was set as close as possible to the needs of the user. At the same time, the user was able to rethink and adapt his own behavior in order to utilize in a more efficient manner the functionalities of the system.

To summarize, all 11 participants were able, at the end of the training session to grasp the use of the tactile interface. Overall, we could also notice that their steering improved as the training session progressed.

4.4   Kinematic Evaluation

After observing that our participants were successfully able to steer their power wheelchair using the tactile interface, we conducted a kinematic evaluation to quantify the performance level that the tactile interface gives to our participants compared with their own joysticks. For this reason, we asked our participants to perform 5 different tasks with both the tactile interface and the joystick. The first three tasks (straight line in a hallway, 90° corner and doorway crossing) were inspired from the Wheelchair Skill Test [21] and were in conformity with the previous kinematic evaluations of the tactile interface. In the fourth task, the 90° corner was directly followed by a doorway crossing. Then, the fifth task was a slalom which is a task that alternates between left and right turns and requires a certain level of coordination. Even though the participants were already used to the joystick, we still alternated the use of the devices to avoid the task learning effect. Each task was performed a first time as a training then three trials were recorded. We asked the participants to perform the whole experiment at two different speed levels. First, the speed of the wheelchair was limited to 2.5 km/h. Then, the speed limit was increased to 4 km/h which demands a higher level of concentration and control especially in tight maneuvers. This whole process made the experiment quite long and tiring but this was an opportunity for us to observe how the users would react, in the case of each device, as they get more tired. The users were however informed that they could halt the experiment at any given time if they wished without any consequence. We held a separate testing session for each device as it was too tiring to do both devices during the same day. The time between the two sessions was within a week.

To assess user fatigue, we relied mostly on self-reported fatigue with the help of questionnaires and post-test informal interviews/questionnaires. In [16], experimenters judged the success in using a steering device destined to users with neuromuscular diseases by the users completing 30 min without having to stop because of muscle fatigue. For our tests, we relied on a similar assumption given that the tests lasted between 45 and 60 min on average for each session. Unfortunately, due to various personal and logistical issues, only 4 patients (P2, P4, P10 and P11) among the initial users were able to take part in the kinematic evaluation.

Apparatus

As mentioned earlier, the participants used their own wheelchairs during the entire evaluation process. More precisely, P2 and P10 used a Q6-Ultra with 6 wheels while P4 used a You-Q with 6 wheels and P11 used a You-Q with 4 wheels, all four wheelchairs by Quantum. We also made sure that all the wheelchairs were reprogrammed, using the OEM programming software to have a similar speed and acceleration profile. The participants also used their own joysticks which were standard with P11 having a U-shaped handle (Fig. 8).

Fig. 8. Left: standard joystick used in the tests [19], right: U-shaped joystick handle [20]

The tablet used in the study was the same one used in the training session: Sony Xperia Z Ultra. As for the configuration of the steering interface, each participant used the optimal configuration determined during the training session.

The experiment was filmed by 4 different cameras. The first one (an HD Sony camera) captured the global scene. The second one, a GoPro camera (60fps) was embedded on the wheelchair to capture the ground track. With the help of marked lines in the track of each task, we could calculate, through a frame by frame visual analysis, the time spent in each portion of the task. Then, two GoPro cameras (120 fps), were also embedded on the wheelchair and pointed towards the wheelchair's driving wheels. Each wheel was dotted with 11 equally spaced colored markers (Fig. 9). This allowed us later to calculate the instantaneous speed of each wheel trough a frame by frame visual analysis and deduce the instantaneous speed.

Fig. 9. The wheel speed capture setup with the colored markers (Color figure online)

The recorded data were then analyzed through statistics package in RStudio. The normality analysis of the data gave a non-normal distribution. Because of that, and the small sample size, we chose to run a non-parametric statistical analysis using Wilcoxon rank test with a significance value $\alpha = 0.05$. In the following sections, we detail the results of each task separately by looking at the differences in performance between the joystick and the tactile interface for the whole sample. Then, we report results from the subjective feedback. Finally, we discuss the implications of these results.

Task 1: Straight line in a Hallway
In this task, a hallway of 5 meters was simulated by small signaling cones. The width of the hallway was set to 90 cm which is the minimum allowed for hallways according to accessible building regulations [2]. Figure 10 shows participant 11 performing this task using a joystick.

Fig. 10. Participant 11 performing a straight line in a hallway

We recorded the traversal along the whole 5 m distance, but we only counted the middle 3 m. In this task, we wanted to see if the participants could hold a straight line using the steering device and whether they will attain the full potential of speed while on the straight line. The recorded performance criteria were: the average speed, the number of collisions and the total time to roll the middle 3 m. Table 1 gives a summary of the results of comparison between the joystick and the tablet.

Table 1. Performance criteria for the straight-line task

Median value of	Speed level 1			Speed level 2		
	Joystick	Tablet	P value	Joystick	Tablet	P value
Total time to traverse 3 meters (s)	4.478	4.503	>0.05	2.765	2.8	>0.05
Average speed (m/s)	0.681	0.687	>0.05	1.115	1.067	>0.05
Sum of the collisions for all 4 participants	0	0	–	0	3	–

The results suggest no significant difference between the time spent (as well as the average speed) with the joystick or the tablet. However, from the video recordings, and especially at speed level 2, they were a little less stable with the tablet which got reflected on the number of collisions and may have also penalized the linear speed due to successive trajectory corrections

Task 2: 90° Corner

In this task, a hallway with 90° corner was simulated using signaling cones. The corner entry portion of the hallway was the same as the one used in the first task (with width of 90 cm). The perpendicular portion was made in the same way but with a width of 110 cm according to wheelchair accessible building regulations [2]. The start of the corner, as well as the end, were counted 1 m away from the apex. Figure 11 shows participant 4 performing the task.

Fig. 11. Participant 4 performing a 90° corner

The recorded data were: the speed of entry which gives an idea about the participant's level of confidence before taking the corner, the total cornering time, the number of collisions as well as the speed at the exit of the corner. The latter gives an idea about how well the participant could pick up speed right after the corner. Table 2 gives performance figures for the cornering task.

Table 2. Performance criteria for the 90° corner task

Median value of	Speed level 1			Speed level 2		
	Joystick	Tablet	P value	Joystick	Tablet	P value
Total cornering time (s)	5.247	6.008	**0.02258**	3.225	3.962	**0.006099**
Speed at corner entry (m/s)	0.6655	0.68	>0.05	1.121	1.1295	>0.05
Speed at exit (m/s)	0.675	0.6415	>0.05	1.029	0.879	>0.05
Sum of the collisions for all 4 participants	1	1	–	1	2	–

The results suggest no significant difference in the speeds between the two devices, therefore the level of confidence entering the corner. The difference appears in the total time which can be explained by the longer planning time when using the tablet.

Task 3: Doorway Crossing

In this task, two cylindrical tubes were used to simulate a doorway (Fig. 12). The width of the latter was set to be 80 cm which is the minimum width according to wheelchair accessible building regulations [2].

Fig. 12. Participant 4 crossing the doorway

The recorded data (summarized in Table 3) concerned the instantaneous speed before entering the doorway (indicates the level of confidence of the user approaching the doorway), the average speed during the doorway crossing, the speed at the exit of the doorway, the number of collisions as well as the total amount of time spent in the task. The start and end of the task were taken at 1 m before and after the door.

Table 3. Performance criteria for the doorway crossing task

Median value of	Speed level 1			Speed level 2		
	Joystick	Tablet	P value	Joystick	Tablet	P value
Total time to traverse the doorway (s)	3.138	3.118	>0.05	1.945	2.96	>0.05
Average traversal speed (m/s)	0.6485	0.688	>0.05	1.02	0.4205	**0.01414**
Speed at entry	0.6565	0.6895	>0.05	0.9505	0.697	>0.05
Speed at exit	0.6585	0.6805	>0.05	1.1004	0.9125	>0.05
Sum of the collisions for all 4 participants	0	0	–	0	3	

The data in Table 3 do not show a significant difference at the level of the speed at entry even between the two devices even in speed level 2. This can indicate a similar level of confidence. The same observation can be made for the speed at the exit which indicates that participants were able to pick up speed with both devices after traversing the doorway. The difference shows however at the level of the average traversal speed where, in speed level 2, participants were slower with the tablet than with the joystick. This can also indicate that once in the middle of the traversal, the participants felt the need to be more careful with the tactile interface as they have less experience using it. The complexity of the task and the novelty of the tactile interface had also an impact on the number of collisions in speed level 2 (more collisions registered for with the tactile interface).

Task 4: 90° Corner Followed by a Doorway Crossing

In this task, we took the same corner as before and added a doorway passing right afterwards (Fig. 13). The start of the doorway crossing (1 m before the door) was set the same as the end of the corner (1 m after the apex). The goal of this combined task was to see if combining two consecutive tasks that both require planning could impact the user's performance in either of them.

Fig. 13. Participant 2 performing the corner followed by a doorway crossing

The recorded data (summarized in Table 4) concerned the amount of time spent separately in the cornering and the doorway passing as well as the total number of collisions in the whole task. Table 4 gives performance figures for this task as well as a comparison between the performance of in the corner/doorway tasks alone and the same task when put back to back with another task.

Table 4. Performance criteria for the 90° corner followed by a doorway crossing task

Median value of	Speed level 1			Speed level 2		
	Joystick	Tablet	P value	Joystick	Tablet	P value
Total traversal time (s)	8.562	9.418	>0.05	5.37	6.963	**0.03509**
Cornering time (s)	5.26	6.133	>0.05	3.308	3.853	**0.012**
Doorway crossing time	3.345	3.535	>0.05	1.988	3.185	>0.05
Sum of the collisions for all 4 participants	0	0	–	0	4	–
p-value against cornering time from task 2	0.388	0.6101	–	0.9687	0.367	–
p-value against doorway passing from task 3	0.666	0.07756	–	0.1698	0.8445	–

The difference between the two devices can be most seen in the traversal time with speed level 2. In fact, the complexity of the cornering coupled with the high speed render this task quite challenging in terms of planning giving a clear advantage to the joystick. This also reflects on the number of collisions which are higher with the tablet. However, we do not detect a significant difference between coupling the corner and doorway compared to when the tasks are separated. This may indicate that, with both devices, the participants were able to handle two consecutive tight maneuvers with the same performance as when they were separated, and that with both devices.

Task 5: Slalom

In this task, we asked the participant to perform a slalom (Fig. 14). This task requires a high level of coordination between rotation and straight movement while avoiding collision with cones. The slalom was comprised of 3 in-line cones at 2 m from one another. The distance between the first cone and the entrance was 2 m, the same distance from the last corner to the exit. Both the entrance and the exit were 90 cm wide.

Fig. 14. Participant 11 performing a slalom

The data recorded were the time to perform the slalom as well as the number of collisions. Table 5 summarizes user performances for both devices.

Table 5. Performance criteria for the slalom task

Median value of	Speed level 1			Speed level 2		
	Joystick	Tablet	P value	Joystick	Tablet	P value
Total traversal time (s)	15.2	19.83	**0.0007315**	10.148	12.07	**0.002939**
Sum of the collisions for all 4 participants	1	4	–	0	4	–

In this task, we can see that the participants performed better with the joystick than with the tactile interface. This task is the one that demands the more planning and coordination. In the other tasks, the difference between the two interfaces was less apparent. If we put together all this information, we can hypothesize that the difference in the performance during the slalom task is more due to the fact that the participants are more used to the joystick than the tactile interface. Thus, the planning load with the joystick would naturally be lighter than with the tactile interface. In a complex task like the slalom, this can make a noticeable difference.

Subjective Evaluation

After each testing session, we asked the participants to fill a System Usability Scale [8] form (summarized in the Table 6) and, asked them open questions for a more open feedback on the system, especially in the case of the tactile interface.

Table 6. SUS scores given by the participants

Participant	Tablet SUS score	Joystick SUS score
P2	95	100
P4	95	90
P10	62.5	90
P11	62.5	45

We also asked the participants to fill a post-test questionnaire inspired from the NASA TLX [22] concerning the physical and mental demand of the use of each steering interface during the test. Participants' responses on the questionnaire indicate that both devices were easy to learn and use during the tests. None of the participants reported to feel frustrated during the tests with neither device. The answers of P2 and P10 seemed to agree on the fact that neither device was physically demanding. However, P10 indicated that the tactile interface demanded more mental planning than the joystick during the experiment. The mental demand of the tactile interface reported by P2 was slightly higher than the joystick. P4 reported about the same level of mental demand for both devices contrary to P11 who reported a higher level of mental demand with the tactile interface. For P4 and P11, the tactile interface demanded considerably less physical effort than the joystick. This is further backed by the fact that P11 asked to pause the session with the joystick because of fatigue (to continue on the following day), which was not the case with the tactile interface. Overall, the tactile interface had a lower physical demand but a higher mental demand.

Finally, we asked each participant to choose one favorite device between the tactile interface and the joystick: P2 and P10 chose the joystick while P4 and P11 chose the tactile interface. This choice consolidates the respective scores the participants gave on the SUS questionnaire. The more important result was that the preference of the users tilted towards the interface that had the least physical demand for them. This was especially true for the two participants with the least motor strength.

Discussion of the Kinematic Evaluation Session Results

Overall, all 4 participants were able to complete the tests successfully with both devices. Performing the tasks required them also to move around the experimentation room. This actually required repetitive speeding and slowing, avoiding static obstacles and sometimes performing tight maneuvers. With the tactile interface, all 4 participants were able to complete the test sessions with over 30 min of wheelchair steering without having to stop because of muscle fatigue. However, they all felt tired at the end of the session, especially with the repetitive nature of the tests. This was also the case for the joystick except that for participant 11, who had to stop in the middle of the test because she felt tired and asked to finish the test the day after. This suggests that for her, the tactile interface was significantly less tiring from a physical standpoint and confirms her post-experiment feedback.

All 4 participants agreed that while the tablet was less tiring in terms of physical force, it was more demanding in terms of planning. This was clearly reflected by their performance in tight maneuvers where they made more collisions and spent more time planning their traversal. This can be attributed to the fact that they have been used to the joystick for years and that the tactile interface was a new steering device and they had yet to get more used to it as they expressed at the end of the tests. Post-test informal interviews also revealed that P2 and P10 still preferred the joystick while P4 and P11 preferred the tactile interface. In [10], users tended to prefer the device with which they performed better. In contrast, the tests with neuromuscular users suggest that their preference can correlate with both the level of physical strength and the age. P4 and P11 are at the same time the youngest and the ones with the lesser physical strength among the 4 participants. P2 and P11 also mentioned that if they had a much weaker level of strength, their preference might have tilted towards the tactile interface. P11 even explained that he was very interested in the domotic interaction opportunities that the tablet offers. He plans to move in a house equipped with a connected environment and having such interaction device would be very helpful for him.

Finally, the 4 participants felt that despite having achieved a lesser performance with the tactile interface at times, if given more training time, their performance in the tests would be better. Here we can make a case for the tactile interface compared to the joystick: on the plus side, it demanded less force to steer. On the minus side, it demanded more mental load than the joystick. However, we can argue that the mental demand can get lower with time as the user gets more accustomed to it through training. Contrary to the joystick which despite the lighter mental load still needs a certain physical load to be handled that will not get lighter with training. However, we need to conduct further tests over a longer period of time to evaluate this hypothesis.

5   Conclusion

In this paper, we presented a tactile steering interface on smartphones/tablets for wheelchair users suffering from neuromuscular diseases. We have presented a preliminary study that looks at the attitude and acceptance of a group of users to detect tendencies that can guide future clinical trials of this novel interface. The participants were able to quickly grasp the use of this interface. Then, a subgroup participated in a test comparing the performance of the tactile interface to that of their own joysticks. Although the small number of participants does not allow us to make solid generalizations, the tests still suggest that the tactile interface can be an efficient steering device for some people with neuromuscular diseases and its performance is close to that of the joystick. The tests also suggest that users with less physical strength may be inclined to a steering solution if it lowers the physical load even if it has a slightly lower performance than another physically demanding one. Finally, the tests let us hypothesize that given enough training time, users may be able to attain a higher level of performance with the tactile interface. This opens up to further investigation of the tactile steering for power wheelchairs. So, the next step is to accompany users with their tactile interface training, in real life situations, and observe their performance curve.

Acknowledgements. We would like to thank the occupational therapists and the staff of *Le Brasset* functional rehabilitation center for their help, without which this study would not have been possible. Their constructive feedback helped us to carry out our tests. We also extend our warmest thanks to the residents of the center who took part in the study. Despite their motor challenges, they showed great enthusiasm during the tests. They were very cooperative and gave us constructive feedback. Finally, we would like to thank the team of the Technical Aid and Innovation Center of the French Association against Myopathies (AFM TELETHON) for their valuable help in setting up this test campaign with people suffering from neuromuscular diseases.

This work was funded by IDEX Paris-Saclay, ANR-11-IDEX-0003-02.

References

1. Kirkby, R.L.: Wheelchair Skill Assessment and Training. CRC Press, Taylor and Francis Group, Boca Rota (2016)
2. MTES (Ministère de la Transition Ecologique et Solidaire), Ministère la Cohésion des Territoires. Réglementation Accessibilité Batiment (2017). http://www.accessibilite-batiment.fr/. Accessed Mar 2017
3. Guedira, Y., Jordan, L., Favey, C., Farcy, R., Bellik, Y.: Tactile interface for electric wheelchair. In: Proceedings of the 18th International ACM SIGACCESS Conference on Computers and Accessibility (ASSETS 2016). LNCS, Reno, USA, pp. 313–314 (2016). http://www.springer.com/lncs. Accessed 21 Nov 2016
4. Troise, D., Yoneyama, S., Resende, M.B., Reed, U., Xavier, G.F., Hasue, H.: The influence of visual and tactile perception on hand control in children with Duchenne muscular dystrophy. Dev. Med. Child Neurol. 56(9), 882–887 (2014). https://doi.org/10.1111/dmcn.12469. Edited by Bernard Dan. Mac Keith Press. London, UK
5. Guedira, Y., Dessailly, E., Farcy, R., Bellik, Y.: Evaluation cinématique d'une interface tactile pour le pilotage d'un fauteuil roulant électrique: une étude pilote. In: 29ème Conférence Francophone sur l'Interaction Homme Machine (IHM), Poitiers, France (2017). 9p
6. Haslett, J., et al.: Gene expression comparison of biopsies from duchenne muscular dystrophy (DMD) and normal skeletal muscle. Proc. Natl. Acad. Sci. U.S.A. 99(23), 15000–15005 (2002)
7. Jasvinder, C.: Stepwise approach to myopathy in systemic disease. Front. Neurol. 2, 49 (2011)
8. Brooke, J.: SUS: a 'quick and dirty' usability scale. In: Jordan, P.W., Thomas, B., Weerdmeester, B.A., McClelland, A.L. (eds.) Usability Evaluation in Industry. Taylor and Francis, London (1996)
9. Guerrier, Y., Naveteur, J., Kolski, C., Poirier, F.: Communication System for Persons with Cerebral Palsy. In: Miesenberger, K., Fels, D., Archambault, D., Peňáz, P., Zagler, W. (eds.) ICCHP 2014. LNCS, vol. 8547, pp. 419–426. Springer, Cham (2014). https://doi.org/10.1007/978-3-319-08596-8_64
10. Guedira, Y., Bimbard, F., Françoise, J., Farcy, R., Bellik, Y.: Tactile Interface to Steer Power Wheelchairs: A Preliminary Evaluation with Wheelchair Users. In: Miesenberger, K., Kouroupetroglou, G. (eds.) ICCHP 2018. LNCS, vol. 10896, pp. 424–431. Springer, Cham (2018). https://doi.org/10.1007/978-3-319-94277-3_66
11. Adams, M.J., et al.: Finger pad friction and its role in grip and touch. J. R. Soc. Interface 10 (80), 20120467 (2013). https://doi.org/10.1098/rsif.2012.0467
12. Cejudo, P., et al.: Exercise training in mitochondrial myopathy: a randomized controlled trial. Muscle Nerve 32(3), 342–350 (2005). https://doi.org/10.1002/mus.20368
13. Pavlakis, S.G., Phillips, P.C., DiMauro, S., De Vivo, D.C., Rowland, L.P.: Mitochondrial myopathy, encephalopathy, lactic acidosis, and strokelike episodes: a distinctive clinical syndrome. Ann. Neurol. 16(4), 481–488 (1984). https://doi.org/10.1002/ana.410160409

14. Mackay, W.E.: Responding to cognitive overload: co-adaptation between users and technology. Intellectica. **30**(1), 177–193 (2000)
15. Sunrise Medical. MicroPilot: Zero-throw proportional control. Web (2019). https://www.sunrisemedical.com/power-wheelchairs/electronics/mini-proportional-joysticks/switch-it-mi cropilot. Accessed 17 June 2019
16. Pellegrini, N., et al.: Optimization of power wheelchair control for patients with severe Duchenne muscular dystrophy. Neuromuscul. Disord. **14**(5), 297–300 (2004). https://doi.org/10.1016/j.nmd.2004.02.005
17. Sunrise Medical. Switch-it Smart technologies: Alternative Drive Controls. Web (2018). https://www.sunrisemedical.com/getattachment/956dd894-d6ec-4769–9f23-20593e0ad524/Switch-It-Alternative-Drive-Controls-Catalog.aspx. Accessed 17 June 2019
18. Statista.com. Number of smartphone users worldwide from 2014 to 2020 (in billions). https://www.statista.com/statistics/330695/number-of-smartphone-users-worldwide/. Accessed 6 July 2018
19. Quantum Rehab Q-Logic 3 Advanced Drive Control System. https://www.quantumrehab.com/quantum-electronics/q-logic-3-advanced-drive-control-system.asp. Accessed 28 Jan 2020
20. Bodypoint, U-shaped Joystick handle. https://www.bodypoint.com/ECommerce/product/pcpt04/u-shaped-joystick-handles. Accessed 28 Jan 2020
21. Rushton, P.W., Kirby, R.L., Routhier, F., Smith, C.: Measurement properties of the wheelchair skills test questionnaire for powered wheelchair users. Disabil. Rehabil. Assist. Technol. **11**(5), 400–440 (2014). https://doi.org/10.3109/17483107.2014.984778
22. Hart, S.G., Staveland, L.E.: Development of NASA-TLX (task load index): results of empirical and theoretical research. Adv. Psychol. **52**(1988), 139–183 (1988). https://doi.org/10.1016/S0166-4115(08)62386-9
23. Wastlund, E., Sponseller, K., Petterson, O.: What you see is where you go: testing a gaze-driven power wheelchair for individuals with severe multiple disabilities. In: Proceedings of the 2010 Symposium on Eye-Tracking Research & Applications. ETRA 2010, Austin, Texas, USA, pp. 133–136. ACM, New York (2010)
24. Tello, M.G., et al.: Performance Improvements for Navigation of a Robotic Wheelchair based on SSEP-BCI. In: Simpósio Brasileiro de Automação Inteligente, Natal, Brazil, pp. 201–205 (2015)
25. Mog, L.K.: Study on Electronic wheelchair controller with a smartphone's speaker-independent recognition engine. In: Proceedings of the World Congress on Engineering and Computer Science 2015. WCECS 2015, San Francisco, USA, vol. Springer, Germany (2015)

A Methodological Approach to Determine the Benefits of External HMI During Interactions Between Cyclists and Automated Vehicles: A Bicycle Simulator Study

Christina Kaß[1]([✉]), Stefanie Schoch[1], Frederik Naujoks[2] [iD],
Sebastian Hergeth[2] [iD], Andreas Keinath[2], and Alexandra Neukum[1]

[1] Wuerzburg Institute for Traffic Sciences GmbH,
97209 Veitshöchheim, Germany
kass@wivw.de
[2] BMW Group, 80788 Munich, Germany

Abstract. To ensure safe interactions between automated vehicles and non-automated road users in mixed traffic environments, recent studies have focused on external human-machine interfaces (eHMI) as a communication interface of automated vehicles. Most studies focused on the research question *which* kind of eHMI can support this interaction. However, the fundamental question *if* an eHMI is useful to support interactions with automated vehicles has been largely neglected. The present study provides a methodological approach to examine potential benefits of eHMIs in supporting other road users during interactions with automated vehicles. In a bicycle simulator study, 20 participants encountered different interaction scenarios with an automated vehicle that either had the maneuver intention to brake or to continue driving. During dynamically evolving situations, we measured their behavior during interactions with and without eHMI. Additionally, the comprehensibility of the eHMI was measured with a special occlusion method. The results revealed that the eHMI led to more effective and efficient behavior of the cyclists when the automated vehicle braked. However, the eHMI provoked safety-critical behavior during three interactions when the vehicle continued driving. The set-up, experimental design, and behavioral and comprehension measurements can be evaluated as useful method to evaluate the benefits of any given eHMI.

Keywords: External HMI · Automated driving · Methodological approach · Bicycle simulator

1 Introduction

With the introduction of automated vehicles, drivers or rather passengers will be allowed to engage in non-driving related tasks while driving. Related thereto, they will often be unavailable for communication with other road users. Thus, there is a growing body of research on the design of external human-machine interfaces (eHMI) that explicitly convey a certain communication content of the automated vehicle. Prior work

© Springer Nature Switzerland AG 2020
H. Krömker (Ed.): HCII 2020, LNCS 12213, pp. 211–227, 2020.
https://doi.org/10.1007/978-3-030-50537-0_16

has mainly focused on the interaction between automated vehicles and vulnerable road users (VRUs), especially pedestrians [e.g. 1, 2]. Based on the assumption that manual drivers mainly communicate with VRUs using gestures and eye contact, there is a broad consensus among researchers and practitioners that automated vehicles must replace the informal communication of manual drivers with external human-machine interfaces (eHMI) [3, 4].

In recent years, many studies have been published that investigated the effect of various eHMI variants on subjective evaluations and behavior of interaction partners. However, the current state of research provides inconsistent results regarding the effectiveness of eHMI [2, 3, 5]. Most studies have focused on the research question which certain kind of eHMI can support the interaction with VRUs. This question includes considerations concerning the required content of communication (e.g., maneuver intention, detection feedback, automation status) [6] and the concrete interface design (modality, text or symbols, position) [e.g. 6–12]. For example, two studies found that communicating the automation status with light-emitting diodes (LED) strips resulted in a more positive emotional experience [13] and increased perceived safety [14] than without eHMI. In other studies, the information about the automation status did not have an impact on perceived safety [3], perceived stress [15] and reported behavior [16]. Moreover, different studies have shown that eHMI variants that communicate the intention of the automated vehicle led to more positive subjective ratings, such as perceived safety, compared to interactions without eHMI [1, 2, 13, 14]. In contrast, [3] did not find an impact of intention communication via eHMI on pedestrians' perceived safety and trust. In another study, the pedestrians' workload was even higher with eHMI that communicated the vehicle's intentions than without eHMI [17]. Furthermore, the effectiveness of eHMIs that communicate the vehicle's intention differed dependent on the communicated maneuver [2, 18].

Apart from these studies that varied different design aspects of eHMIs, the fundamental question if an eHMI is useful and necessary to support other road users during interactions with automated vehicles has been largely neglected so far. Several studies have shown that pedestrians mainly base their decision to cross a vehicle's trajectory on observed driving behavior, such as speed and deceleration [19–24]. These findings emphasize the need to carefully examine if eHMIs have an improving effect on behavioral decisions of interaction partners. Past research on driver support systems have shown that any driver assistance may lead to non-intended behavioral changes that can eventually decrease or even counteract the intended safety benefit [e.g. 25–27].

Another research gap pertains to the research environments that were used in previous research on eHMIs. Commonly used methods to examine eHMIs, such as virtual reality pedestrian simulators, the Wizard of Oz technique, and photo or videos studies do usually not create dynamic interactions between automated vehicles and interaction partners. With these methodologies, participants indicate their intention to cross the vehicle's trajectory either by reporting their behavior [e.g. 28], pressing a button [e.g. 5], or taking a step forward [e.g. 15]. We argue that conclusions about the usefulness of an eHMI should fundamentally be based on behavioral measurements during dynamically evolving situations rather than on simplified behavioral measurements and subjective assessments. The reasoning behind this is that such an approach puts heavier time constraints on the cognitive and motoric processes that need to take place during the interaction between automated and non-automated traffic participants and, thus, would be more representative of critical interaction situations under realistic traffic conditions. To obtain a complete picture, these behavioral measurements should be complemented by an assessment of the interaction partner's comprehension of the eHMI.

The aim of the present study was to provide a first step towards a methodological approach to examine the benefits of eHMIs in supporting the behavior of interaction partners during interactions with automated vehicles. Regarding the evaluation of HMIs of automated vehicles, including eHMIs, standardized experimental setups regarding the employed test procedures [29] and test cases [30] need to be established. The current paper aims at stimulating this discussion among the technical and scientific community. To overcome the outlined methodological limitations of previous research, the present study was conducted in a bicycle simulator (see Fig. 1). The advantage of this set-up is that it enables the measurement of cyclists' driving behavior during dynamic interactions with automated vehicles.

Fig. 1. Bicycle simulator at WIVW GmbH with 360° surround visualization.

An eHMI should always ensure safe behavioral decisions of the interaction partner. A basic criterion is, thus, the *usability* of the eHMI. Adapting the standard definition of usability [31], it can be considered as useful if it supports the interaction partner in choosing a more effective and efficient behavior during interactions with automated vehicles than without eHMI. In the present study, the eHMI either signaled that the automated vehicle intends to brake to a standstill or to continue driving. When the automated vehicle intends to brake, we assumed that the cyclists would benefit from the eHMI if their minimal velocity is higher with than without eHMI. This finding would mean that the eHMI supports them to anticipate the vehicle's behavior prior to real time. Based on this knowledge, they decide to continue driving. With eHMI, they would be able to make a more effective behavioral decision than without eHMI. When the automated vehicle intends to continue driving, cyclists should reduce their speed significantly to prevent a collision. Therefore, the most important measurement during this maneuver intention of the automated vehicle is the resulting minimal distance between the interaction partners. In the present study set-up, a minimal distance of less than one meter represents a safety-critical situation. This means that the cyclist made the incorrect and highly critical behavioral decision to continue driving at a too high speed.

To prove that the eHMI can support the cyclists' efficiency, it must decrease the time required for the interaction compared to an interaction without eHMI. Independent from the maneuver intention of the automated vehicle, a shorter time to cross signifies that cyclists hesitate less and wait less because the eHMI assist them to anticipate the vehicle's subsequent maneuver.

The described behavioral measurements of the interaction partner in response to the eHMI could be potentially confounded by a guess probability (as the automated vehicle either braked or continued driving) and the possibility to simultaneously observe the vehicle's driving behavior. As a consequence, it cannot be ensured that cyclists' behavioral decisions are exclusively based on a correct comprehension of the eHMI. Therefore, the present research extends behavioral measurements by a method to assess the comprehensibility of the eHMI during dynamic interactions without being influenced by the observed vehicle behavior.

2 Method

2.1 Participants

A total of 20 participants (10 female, 10 male) took part in the study. Participants were recruited from the Würzburg Institute for Traffic Sciences GmbH (WIVW GmbH) driver test panel. Participation was financially rewarded. Subjects were between 21 and 63 years old ($M = 39.6$, $SD = 14.4$). As a prerequisite for participation, they needed to ride a bike at least once a month. Most participants (55%) stated to cycle on a daily basis, 25% at least once a week, and 20% at least once a month. Each of them had received 90 min of simulator training before participating the study in order to get used to the bicycle simulator and to decrease the drop out-rate due to simulator sickness and physical exertion.

2.2 Bicycle Simulator

The study was conducted in the bicycle simulator of WIVW GmbH that uses the driving simulation software SILAB (see Fig. 1). The simulator provides a 360° field of view realized by 12 screens. The mock-up is a real, roadworthy bike that features front and rear brakes, a 11-speed gear shift, and a bicycle computer that displays the current velocity. The bicycle-specific vehicle dynamics integrate road parameters (e.g. road type, slope) and cyclist input data. A realistic cycling experience is achieved by the experience of physical strain via a hardware actuator and the passive movement system that reacts to body movements. To provide auditory input (e.g., surrounding sounds, engine noise), participants wore wireless earphones while driving. Participants had the option to activate a pedelec support with three levels of support via a handlebar button. The experimenter was located on the operator desk behind the simulator and communicated with the participant via intercom.

2.3   External HMI and Automated Driving System

As the present study did not aim to examine the effects of a certain eHMI, the display concept was adapted from eHMI variants which have already been tested and published in prior work. In accordance with concepts published by [1, 13, 32–34], the used eHMI communicated the automated vehicle's intention visually with LED strips. Moreover, the eHMI continuously communicated the automation status [15]. Based on a rec-ommendation of the Society of Automotive Engineers [35], the LED signals had the color blue-green (cyan).

The eHMI signaled the intended driving maneuver of the automated vehicle with four LED strips located vertically above both headlights (see Fig. 2). Each LED light strip consisted of 6–9 single lights. The LED strips were visible from the front and from the side (see Fig. 2). When the vehicle intended to brake, the light strips moved three times from top to bottom before it actually initiated to brake. When the vehicle intended to continue driving a constant speed, the light strips were permanently dis-played without movement to signal a constant speed status. Moreover, the eHMI continuously communicated the automation status. The text "Self-driving vehicle" was displayed at the front above the radiator grill and on both sides below the doors (see Fig. 2).

The automated driving system was a Level 3 system [36]. More specifically, the operational design domain of the system was defined as urban area with a speed range between 0 and 30 km/h. The automated vehicle was always a black BMW model i8 (see Fig. 2).

Fig. 2. Automated vehicle with eHMI (left) and without eHMI (right). (Color figure online)

2.4   Interaction Scenarios

The driving environment looked like a test track without additional surrounding traffic. To achieve ambiguous situations, there were no right-of-way rules on this simulated test track. Cyclists encountered three different interaction scenarios. In the merging scenario (see right part of Fig. 3), the cyclist and the automated vehicle drove in the same direction on parallel roads both with a speed of 17 km/h. At a certain point, the two roads merged and one interaction partner had to brake to avoid a collision. In the intersection scenario, the vehicle approached the cyclist orthogonally from the left side (see left part of Fig. 3). In the third scenario, the vehicle approached the cyclist frontally at the opposite side of a narrow area (see center part of Fig. 3). In the latter two scenarios, the initial speed of the automated vehicle was 30 km/h.

During all interaction scenarios, the automated vehicle became first visible as it drove out from behind sight obstructions such as walls (intersection, merging) and curves (narrow area). The start of the automated vehicle (the moment when it became visible) was activated by the cyclist crossing a predetermined position. When the cyclist crossed this position, he or she would have needed the same time (in seconds) to reach the virtual crossing point (see Fig. 3) as the automated vehicle. This calculation was based on the assumption that the cyclist would continue driving at 17 km/h and the automated vehicle either at 17 km/h (merging) or at 30 km/h (intersection, narrow area).

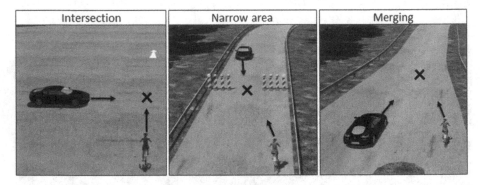

Fig. 3. Interaction scenarios. The arrows indicate the trajectories of automated vehicle and cyclist and the cross indicates their virtual crossing point.

2.5   Experimental Design

The experiment used a repeated measures design with eHMI condition, maneuver intention of the automated vehicle, and interaction scenario as independent factors. In one test block, cyclists interacted with an automated vehicle with eHMI and in another test block without eHMI (eHMI condition). The test block without eHMI served as baseline to be able to compare participants' behavior in response to the eHMI with their behavior during interactions without eHMI. The automated vehicle either had the maneuver intention to brake to a standstill or to continue driving. Participants encountered each maneuver intention during the three different interaction scenarios outlined in Sect. 2.4.

Dependent variables to measure behavioral decisions were the cyclist's minimal velocity, the minimal distance to the automated vehicle, and the time required to cross the trajectory of the automated vehicle during each interaction event. The minimal velocity was assessed in the time window between the moment the vehicle became first visible to the cyclist and the cyclist crossed the virtual crossing point (see Fig. 3 and Sect. 2.4). The minimal distance to the automated vehicle was used as a measure of criticality. Due to the fact that the vehicle was programmed to brake for the cyclist if he or she crosses its trajectory, no collisions were possible in the study set-up. However, a minimal distance of less than one meter signifies that the cyclist was about to cross the vehicle's trajectory even if the vehicle actually intended to continue driving. When the automated vehicle continued driving, the minimal distance between the interaction partners represents the cyclist's distance to the virtual crossing point at the moment that the automated vehicle crossed this crossing point (see Fig. 3). It should be noted that cyclists did not necessarily have to brake to a standstill to handle the situation in an uncritical way. They could also stop to pedal and wait to see what would happen. The time to cross is the time (in seconds) between the moment the vehicle became visible to the cyclist and crossing the virtual crossing point (see Fig. 3).

To assess the comprehensibility of the eHMI, participants encountered all test events during an additional test block with eHMI in which the simulation screen was blanked during each interaction scenario at predefined points in time. This occlusion method was adapted from [37] and intends to achieve an open outcome of the events with the different eHMI signals. The screen was blanked when the eHMI already signaled the maneuver intention (braking, driving) but before the automated vehicle already started to execute the respective maneuver. While the screen was blanked, the cyclist was automatically braked down to a standstill. After 5 s, the screen showed the last scene again while the automated vehicle was removed. The experimenter asked the open-ended question: "What would the automated vehicle have done next?" and categorized the participant's answer as either correct or incorrect. Answers were categorized as correct if participants were able to correctly verbalize the maneuver intention to brake (e.g. "the vehicle brakes", "… stops", "… decelerates") or to continue driving (e.g. "the vehicle continues", "… drives on", "… does not stop"). Answers were categorized as incorrect if participants mentioned a different maneuver intention (e.g. braking instead of driving), answered "I don't know", or gave a different answer that did not fit the question (e.g. "The driver did not intervene"). With the described method, the cyclists' comprehension was exclusively based on the eHMI signal and not influenced by the currently executed driving behavior of the vehicle. This occlusion test block contained all 6 test events and was always encountered after the test block with eHMI in which cyclists experienced the complete interaction scenarios. In order to be defined as comprehensible, we have determined that an eHMI signal must be correctly understood by at least 85% of the participants.

To control for transition effects, the sequence of test blocks with eHMI (behavior measurement and occlusion test block) and without eHMI was counterbalanced. Subjects were quasi randomly distributed the two sequences with a cell size of $n = 10$. The test blocks with and without eHMI contained of all possible factor combinations of maneuver intention and interaction scenario ($2 \times 3 = 6$ events). We permutated the sequence of these events within and between the test blocks.

2.6   Procedure

Upon arrival, participants signed a consent form that informed them about the method of data anonymization, their right to decline to participate, and to withdraw from the study at any point. Participants were informed about the procedure, that the study examines interactions between automated vehicles and cyclists, about the target speed of 17 km/h, the possibility to use the pedelec support, and that there are no right-of-way rules during the experiment. They were told that automated driving systems perform the entire dynamic driving task and that the driver can perform other tasks than driving. The instruction did not mention the eHMI. To achieve a constant cycling speed of 17 km/h, the current speed was displayed on the bicycle computer and participants received an auditory feedback on the earphones if they drove slower than 16 km/h (low frequency beep) or faster than 18 km/h (high frequency beep).

The study started with a five-minute familiarization drive without any interactions with the automated vehicle where participants could practice to drive at 17 km/h supported by the auditory feedback. After the familiarization drive, participants went through a learning phase in which they already encountered all interaction scenarios with automated vehicles with eHMI signals. In a short interview after this learning phase, participants were asked if they noticed anything during the interactions and if they have seen the signals of the automated vehicle. Without mentioning the meaning and intention of the signals, the experimenter explained that the aim of the study is to test signals that automated vehicles can use to communicate with other road users. It was important that all participants were aware of the signals for the upcoming test blocks. After the learning phase, participants encountered the test blocks with eHMI (block with behavioral measurements and occlusion test block) and the test block without eHMI in a counterbalanced sequence. After each test block, participants had the opportunity to take a break for physical recovery. Overall, the experiment took about 1.5 h per participant.

3   Results

To analyze the comprehensibility of the eHMI (measured during the occlusion test block), the absolute and relative frequencies of correct answers grouped by maneuver intention and interaction scenario were determined (see Table 1). When the automated vehicle braked, 65 to 90% of participants were able to understand the eHMI signal and gave the correct answer. When the vehicle continued to drive, 85 to 90% of the answers were categorized as correct.

Table 1. Absolute and relative frequencies of correct and incorrect answers to comprehension question grouped by maneuver intention and interaction scenario.

Maneuver intention	Interaction scenario	Frequency of correct answers		Frequency of incorrect answers	
		Absolute	Relative	Absolute	Relative
Automated vehicle brakes	Intersection	13	0.65	7	0.35
	Narrow area	18	0.9	2	0.1
	Merging	14	0.7	6	0.3
Automated vehicle continues driving	Intersection	18	0.9	2	0.1
	Narrow area	17	0.85	3	0.15
	Merging	17	0.85	3	0.15

Analyses of behavioral data were performed using the software "R", version 3.6.1 [38]. The significance level was set at $\alpha = 0.05$.

To test the effect of the eHMI on cyclists' minimal velocity when the automated vehicle braked, a 3x2 repeated-measures ANOVA with eHMI condition and interaction scenario as independent factors, and minimal velocity as dependent variable was carried out. The minimal velocity was significantly higher with eHMI ($M = 9.75$, $SD = 6.12$) than without eHMI ($M = 6.85$, $SD = 5.65$), $F (1, 19) = 8.50$, $p < .01$, $\eta^2 = 0.08$. Additionally, there was a significant main effect of interaction scenario, $F (2, 38) = 27.97$, $p < .001$, $\eta^2 = 0.25$. In the merging scenario, cyclists reduced their velocity to a lower extent than in the other scenarios. Figure 4 shows boxplots of minimal velocities grouped by eHMI condition and interaction scenario.

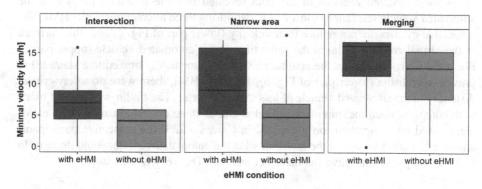

Fig. 4. Cyclists' minimal velocities (when the automated vehicle braked) grouped by eHMI condition and interaction scenario.

When the automated vehicle continued driving, a 2×3 repeated-measures ANOVA was conducted to determine the effect of eHMI condition and interaction scenario on the minimal distance to the automated vehicle. The analysis revealed that the minimal distance was equally high with eHMI ($M = 29.43$, $SD = 19.91$) and without eHMI ($M = 27.54$, $SD = 14.54$), $F (1, 20) = 0.45$, $p = .51$. Due to the different environments of the interaction scenarios, the ANOVA revealed a significant main effect of interaction scenario, $F (2, 38) = 43.73$, $p < .001$, $\eta^2 = 0.43$. Figure 5 presents boxplots that show the minimal distances grouped by eHMI condition and interaction scenarios.

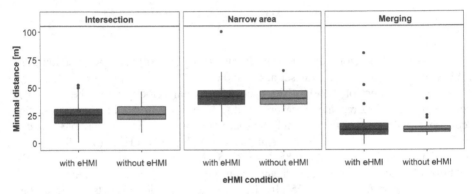

Fig. 5. Minimal distance between cyclist and automated vehicle (that continued driving) grouped by eHMI condition and interaction scenario.

A more detailed analysis of the data revealed that the minimal distance to the automated vehicle has fallen below one meter during three interactions with eHMI. The respective cyclists did not reduce their velocity (lower part of Fig. 6) and, thus, arrived at the virtual crossing point at the same time as the automated vehicle (upper part of Fig. 6). As a consequence, the automated vehicle immediately braked to a standstill to prevent a collision (lower part of Fig. 6). Without eHMI, there were no safety-critical distances to the automated vehicle of less than one meter. The findings of the test block with comprehension measurements revealed that all three critical situations with eHMI were based on comprehension problems. In Case 1 and 3, the respective participants answered "I don't know" to the question what the automated vehicle would do next. In Case 2, the participant gave the incorrect answer "The vehicle will stop".

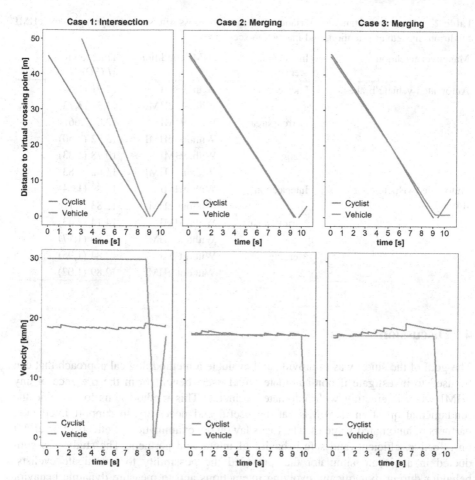

Fig. 6. Distance to virtual crossing point (upper part) and velocity of automated vehicle and cyclist (lower part) during three test cases in which the minimal distance between interaction partners fell below one meter.

To determine the effect of eHMI condition and interaction scenario on the cyclists' time to cross, two 2x3 repeated-measures ANOVAs were carried out separately for both maneuver intentions of the automated vehicle. The respective means and standard deviations are displayed in Table 2. When the vehicle braked, cyclists required significantly less time to cross the vehicle's trajectory with than without eHMI, $F(1, 20) = 7.90$, $p < .05$, $\eta^2 = 0.06$. Moreover, the time to cross differed significantly between the interaction scenarios, $F(2, 38) = 10.04$, $p < .001$, $\eta^2 = 0.07$. When the vehicle continued driving, the time to cross did not differ with and without eHMI, $F(1, 20) = 0.09$, $p = .76$. There was a significant main effect of interaction scenario, $F(2, 38) = 6.58$, $p < .01$, $\eta^2 = 0.09$.

Table 2. Means and standard deviations of time to cross (in seconds) grouped by eHMI condition, maneuver intention, and interaction scenario.

Maneuver intention	Interaction scenario	eHMI condition	Time to cross (s) M (SD)
Automated vehicle brakes	Intersection	With eHMI	11.91 (1.54)
		Without eHMI	13.38 (3.73)
	Narrow area	With eHMI	12.0 (2.66)
		Without eHMI	13.72 (3.50)
	Merging	With eHMI	10.78 (1.83)
		Without eHMI	11.62 (1.83)
Automated vehicle continues driving	Intersection	With eHMI	12.89 (1.84)
		Without eHMI	12.83 (1.11)
	Narrow area	With eHMI	14.14 (2.43)
		Without eHMI	13.80 (1.17)
	Merging	With eHMI	12.84 (1.95)
		Without eHMI	12.89 (1.07)

4   Discussion

The goal of the study was to provide and evaluate a methodological approach that can be used to investigate if non-automated road users benefit from the presence of any eHMI when interacting with automated vehicles. This method aims to examine the fundamental question if eHMIs can be useful and necessary to support interaction partners of automated vehicles. The focus lay on the examination of effects of an eHMI on effective, efficient, and safe behavior of interaction partners. The study was conducted in a bicycle simulator that provided the possibility to investigate cyclists' behavior during dynamically evolving interactions and to measure dynamic behavior patterns. To be able to compare behavioral decisions in response to an eHMI to behavior during interactions without eHMI, participants encountered the same interaction scenarios with automated vehicles with and without eHMI. In order to ensure that effective and efficient behavior of the cyclists is caused by the supporting effect of the eHMI and that incorrect behavioral decisions are based on usability problems of the eHMI (and not, for example, misuse), we have additionally assessed the comprehensibility of the eHMI. For this purpose, the study used an occlusion method that made it possible to measure the cyclists' comprehension of the eHMI without being influenced by the observed driving behavior of the automated vehicle.

In the present study, 20 participants encountered three different interaction scenarios (intersection, narrow area, merging) with an automated vehicle that either had the maneuver intention to brake or to continue driving. When the automated vehicle intended to brake, the results revealed that cyclists' minimal velocity was higher with eHMI than without. Moreover, cyclists required a shorter time to cross the vehicle's trajectory during its braking maneuver with than without eHMI. As a conclusion, the eHMI apparently helped cyclists to anticipate the vehicle's unobserved future braking behavior prior to real time and, thus, they decided to continue driving. This intention prediction led to a more effective and efficient behavior than during interactions without eHMI. As the eHMI helped cyclists to reduce their current velocity to a lower extent, the course of the interaction became more fluent and faster. During interactions without eHMI, cyclists had to wait until they could observe the braking behavior of the automated vehicle. Therefore, they had to reduce their speed to a higher extent while taking a wait-and-see attitude. As a consequence, they needed more time to cross the vehicle's trajectory. The findings of the comprehension measurements supported the assumption that the majority of cyclists understood the meaning of the eHMI that signaled the vehicle's intention to brake. The interaction scenario with the best comprehensibility rate (narrow area, 85% correct answers) showed the greatest difference in minimal velocity and time to cross between interactions with and without eHMI. However, the braking intention of the automated vehicle during the intersection and merging scenario did not fulfill the predefined comprehensibility criterion of at least 85% correct answers. Overall, the comprehension measurements seem to correspond to the behavioral data.

The most important requirement of an eHMI should be to ensure safe behavioral decisions of the interaction partner. When the automated vehicle intends to brake, interactions would never lead to critical situations. In contrast, when the automated vehicle intends to continue driving, cyclists must reduce their velocity to avoid safety-critical situations. During interactions without eHMI, all cyclists reduced their velocity sufficiently. Thus, there were no critical distances between the interaction partners of less than one meter. During three interactions with eHMI, however, the respective cyclists did not brake and, thus, provoked minimal distances of less than one meter. Driving data have shown that all interactions would have resulted in a collision if the automated vehicle had not braked in response to the cyclist (see lower part of Fig. 6). According to the controllability guidelines defined in the RESPONSE Code of Practice [39], the system must be evaluated as non-controllable as soon as even one in 20 participants provokes a safety-critical event. The comprehension measurements confirmed that the identified critical incidents were caused by problems to understand the meaning of the eHMI signal. Alternatively, it would have been conceivable that cyclists wanted to try out whether the automated vehicle would stop for them. With the comprehension measurements, such alternative explanations for their safety-critical behavior could be rejected. Such a result strongly questions the benefit of using eHMIs. It is more desirable that interaction partners behave in a risk-averse manner because they are uncertain about the intention of the automated vehicle than if they behave in a safety-critical manner due to misunderstandings of the eHMI signals. When the automated vehicle continued driving, further analyses have shown that cyclists needed the same amount of time to cross the vehicle's trajectory with and without eHMI. It must be considered that the eHMI could not lead to great efficiency gains in these scenarios as cyclists always had to wait until the automated vehicle crossed the crossing point before they themselves could continue driving. Therefore, we recommend to no longer use the time to cross as a measure of efficiency when the automated vehicle continues driving.

Furthermore, the analyses revealed significant effects of the different interaction scenarios on minimal velocity, minimum distance, and time to cross. The result concerning the minimal velocity emphasizes the need to examine eHMIs during different use cases. For example, even without being able to anticipate the vehicle's maneuver intention, cyclists could handle the merging scenario without reducing the velocity to a high extent. As the interaction scenarios were of different lengths, the effect of scenario on time to cross did not have a substantive significance. Significant interaction effects of eHMI condition and interaction scenario would have indicated that the benefits of eHMI signals differ dependent on the interaction scenario. However, the present study did not reveal such a finding.

The occlusion method can be evaluated as a useful tool to assess the comprehensibility of eHMIs during dynamic interactions. This method can be also applied to pedestrian simulation or classic driving simulations with manual drivers who interact with automated vehicles. There were no shortcomings with regard to simulator sickness. We deliberately decided to ask for the next maneuver of the automated vehicle instead of explicitly asking for the meaning of the eHMI. We expected that participants might have problems to verbalize the concrete meaning of the signal.

Similar to the procedure used in the present study, we recommend a learning phase at the beginning of an eHMI study before assessing effects of the system on behavior and comprehension. Participants must have the opportunity to learn the association between the eHMI signal and the subsequently executed driving maneuver of the automated vehicle.

Future research can further develop the methodology to investigate the benefits of an eHMI. The proposed behavioral parameters could be supplemented by further measurements. Moreover, the method could be complemented by self-reported measurements, such as usefulness, trust, or satisfaction. In further work, it could be examined whether the information of interacting with an automated vehicle influences the behavior of interaction partners compared to the assumption to interact with manually driven vehicles.

We conclude that the experimental design in combination with the bicycle simulator provide a suitable method to determine if interaction partners benefit from the presence of any eHMI prototype during interactions with automated vehicles. An eHMI should increase the safety of interaction partners. As soon as it results in safety-critical behavior, its effectiveness must be rejected. Moreover, the results demonstrate that the method can reveal positive effects of eHMI on cyclists' behavior when the vehicle signaled a braking maneuver. Before focusing on concrete interface design concepts, practitioners should carefully examine the effectiveness of new technological developments.

Acknowledgments. We thank Stefanie Ebert, Thomas Stemmler, and Florian Fischer for their technical support during study preparation and conduction.

References

1. Böckle, M.-P., Brenden, A. P., Klingegård, M., Habibovic, A., Bout, M.: SAV2P: exploring the impact of an interface for shared automated vehicles on pedestrians' experience. In: Proceedings of the 9th International Conference on Automotive User Interfaces and Interactive Vehicular Applications Adjunct, pp. 136–140. Oldenburg, Germany (2017)
2. de Clercq, K., Dietrich, A., Núñez Velasco, J.P., de Winter, J., Happee, R.: External human-machine interfaces on automated vehicles: effects on pedestrian crossing decisions. Hum. Factors 61(8), 1353–1370 (2019)
3. Hensch, A.-C., Neumann, I., Beggiato, M., Halama, J., Krems, Josef F.: How should automated vehicles communicate? – Effects of a light-based communication approach in a Wizard-of-Oz Study. In: Stanton, N. (ed.) AHFE 2019. AISC, vol. 964, pp. 79–91. Springer, Cham (2020). https://doi.org/10.1007/978-3-030-20503-4_8
4. Merat, N., Louw, T., Madigan, R., Wilbrink, M., Schieben, A.: What externally presented information do VRUs require when interacting with fully automated road transport systems in shared space? Accid. Anal. Prev. 118, 244–252 (2018)
5. Song, Y.E., Lehsing, C., Fuest, T., Bengler, K.: External HMIs and their effect on the interaction between pedestrians and automated vehicles. In: Karwowski, W., Ahram, T. (eds.) IHSI 2018. AISC, vol. 722, pp. 13–18. Springer, Cham (2018). https://doi.org/10.1007/978-3-319-73888-8_3
6. Schieben, A., Wilbrink, M., Kettwich, C., Madigan, R., Louw, T., Merat, N.: Designing the interaction of automated vehicles with other traffic participants: design considerations based on human needs and expectations. Cogn. Technol. Work 21(1), 69–85 (2019). https://doi.org/10.1007/s10111-018-0521-z
7. Deb, S., Strawderman, L.J., Carruth, D.W.: Investigating pedestrian suggestions for external features on fully autonomous vehicles: a virtual reality experiment. Transp. Res. Part F Traffic Psychol. Behav. 59, 135–149 (2018)
8. Mahadevan, K., Somanath, S., Sharlin, E.: Communicating awareness and intent in autonomous vehicle-pedestrian interaction. In: Proceedings of the 2018 CHI Conference on Human Factors in Computing Systems, Montréal, Canada, pp. 1–12 (2018)
9. Eisma, Y., van Bergen, S., ter Brake, S., Hensen, M., Tempelaar, W., de Winter, J.: External human-machine interfaces: the effect of display location on crossing intentions and eye movements. Information 11(1), 13 (2020)
10. Fridman, L., Mehler, B., Xia, L., Yang, Y., Facusse, L.Y., Reimer, B.: To walk or not to walk: crowdsourced assessment of external vehicle-to-pedestrian displays. arXiv preprint arXiv:1707.02698 (2017)
11. Kooijman, L., Happee, R., de Winter, J.: How do eHMIs affect pedestrians' crossing behavior? A study using a head-mounted display combined with a motion suit. Information 10(12), 386 (2019)
12. Otherson, I., Conti-Kufner, A.S., Dietrich, A., Maruhn, P., Bengler, K.: Designing for automated vehicle and pedestrian communication: perspectives on eHMIs from older and younger persons. In: de Waard, D., et al. (eds.) Proceedings of the Human Factors and Ergonomics Society Europe Chapter 2018 Annual Conference, Berlin, Germany, pp. 135–148 (2018)
13. Lagstrom, T., Malmsten Lundgren, V.: AVIP-Autonomous vehicles interaction with pedestrians. Master's thesis, Chalmers University of Technology (2015)
14. Habibovic, A., et al.: Communicating intent of automated vehicles to pedestrians. Front. Psychol. 9, 1–17 (2018)

15. Rodríguez Palmeiro, A.: Interaction between pedestrians and Wizard of Oz automated vehicles. Master's thesis, TU Delft, Netherlands (2017)
16. Hagenzieker, M.P., et al.: Interactions between cyclists and automated vehicles: results of a photo experiment. J. Transp. Saf. Secur. **12**(1), 1–22 (2019)
17. Gruenefeld, U., Weiß, S., Löcken, A., Virgilio, I., Kun, A.L., Boll, S.: VRoad: gesture-based interaction between pedestrians and automated vehicles in virtual reality. In: Proceedings of the 11th International Conference on Automotive User Interfaces and Interactive Vehicular Applications Adjunct, Utrecht, Netherlands, pp. 399–404 (2019)
18. Weber, F., Chadowitz, R., Schmidt, K., Messerschmidt, J., Fuest, T.: Crossing the Street Across the Globe: A Study on the Effects of eHMI on Pedestrians in the US, Germany and China. In: Krömker, H. (ed.) HCII 2019. LNCS, vol. 11596, pp. 515–530. Springer, Cham (2019). https://doi.org/10.1007/978-3-030-22666-4_37
19. Ackermann, C., Beggiato, M., Bluhm, L.-F., Löw, A., Krems, J.F.: Deceleration parameters and their applicability as informal communication signal between pedestrians and automated vehicles. Transp. Res. Part F Traffic Psychol. Behav. **62**, 757–768 (2019)
20. Beggiato, M., Witzlack, C., Krems, J.F.: Gap acceptance and time-to-arrival estimates as basis for informal communication between pedestrians and vehicles. In: Proceedings of the 9th International Conference on Automotive User Interfaces and Interactive Vehicular Applications Adjunct, Oldenburg, Germany, pp. 50–57 (2017)
21. Beggiato, M., Witzlack, C., Springer, S., Krems, J.: The right moment for braking as informal communication signal between automated vehicles and pedestrians in crossing situations. In: Stanton, N.A. (ed.) AHFE 2017. AISC, vol. 597, pp. 1072–1081. Springer, Cham (2018). https://doi.org/10.1007/978-3-319-60441-1_101
22. Dey, D., Terken, J.: Pedestrian interaction with vehicles: roles of explicit and implicit communication. In: Proceedings of the 9th International Conference on Automotive User Interfaces and Interactive Vehicular Applications Adjunct, Oldenburg, Germany, pp. 109–113 (2017)
23. Fuest, T., Michalowski, L., Träris, L., Bellem, H., Bengler, K.: Using the driving behavior of an automated vehicle to communicate intentions-a Wizard of Oz Study. In: Proceedings of the 21st International Conference on Intelligent Transportation Systems, pp. 3596–3601. IEEE (2018)
24. Schneemann, F., Gohl, I.: Analyzing driver-pedestrian interaction at crosswalks: a contribution to autonomous driving in urban environments. In: 2016 IEEE intelligent vehicles symposium (IV), pp. 38–43. IEEE (2016)
25. Muhrer, E., Reinprecht, K., Vollrath, M.: Driving with a partially autonomous forward collision warning system how do drivers react? Hum. Factors **54**(5), 698–708 (2012)
26. Naujoks, F., Totzke, I.: Behavioral adaptation caused by predictive warning systems – the case of congestion tail warnings. Transp. Res. Part F **26**, 49–61 (2014)
27. Rudin-Brown, C.M., Parker, H.A.: Behavioural adaptation to adaptive cruise control (ACC): implications for preventive strategies. Transp. Res. Part F Traffic Psychol. Behav. **7**(2), 59–76 (2004)
28. Li, Y., Dikmen, M., Hussein, T.G., Wang, Y., Burns, C.: To cross or not to cross: Urgency-based external warning displays on autonomous vehicles to improve pedestrian crossing safety. In: Proceedings of the 10th International Conference on Automotive User Interfaces and Interactive Vehicular Applications Adjunct, Toronto, Canada, pp. 188–197 (2018)
29. Naujoks, F., Hergeth, S., Wiedemann, K., Schömig, N., Forster, Y., Keinath, A.: Test procedure for evaluating the human–machine interface of vehicles with automated driving systems. Traffic Inj. Prev. **20**(Suppl 1), 146–151 (2019)

30. Naujoks, F., Hergeth, S., Wiedemann, K., Schömig, N., Keinath, A.: Use cases for assessing, testing, and validating the human machine interface of automated driving systems. Proc. Hum. Factors Ergon. Soc. Annu. Meet. **62**(1), 1873–1877 (2018)
31. International Organization for Standardization: ergonomics of human-system interaction - Part 11: Usability: Definitions and concepts (ISO 9241-11), Geneva, Switzerland (2018)
32. Faas, S.M., Baumann, M.: Yielding light signal evaluation for self-driving vehicle and pedestrian interaction. In: Ahram, T., Karwowski, W., Pickl, S., Taiar, R. (eds.) IHSED 2019. AISC, vol. 1026, pp. 189–194. Springer, Cham (2020). https://doi.org/10.1007/978-3-030-27928-8_29
33. Volvo Cars. https://www.media.volvocars.com/global/en-gb/media/pressreleases/237019/volvo-360c-concept-calls-for-universal-safety-standard-for-autonomous-car-communication1. Accessed 28 Jan 2020
34. Ford Media Center. https://media.ford.com/content/fordmedia/fna/us/en/news/2017/09/13/ford-virginia-tech-autonomous-vehicle-human-testing.html. Accessed 28 Jan 2020
35. Tiesler-Wittig, H.: Light Signaling and Lighting Requirements for ADS Vehicles. GTB Document No. CE-5523 (2018)
36. SAE International: Taxonomy and definitions for terms related to on-road motor vehicle automated driving systems (No. J3016). (2018). https://saemobilus.sae.org/content/j3016_201806
37. Kaß, C., Schmidt, G.J., Kunde, W.: Towards an assistance strategy that reduces unnecessary collision alarms: an examination of the driver's perceived need for assistance. J. Exp. Psychol. Appl. **25**(2), 291–302 (2018)
38. R Core Team: R: A language and environment for statistical computing. R Foundation for Statistical Computing, Vienna, Austria (2019). https://www.R-project.org/
39. Response Consortium: Code of Practice for the Design and Evaluation of ADAS; A PReVENT Project: Response 3 (2006)

Mobility-as-a-Service: Tentative on Users, Use and Effects

I. C. MariAnne Karlsson[✉]

Chalmers University of Technology, Gothenburg, Sweden
mak@chalmers.se

Abstract. Mobility-as-a-Service has been argued to lead to more sustainable mobility, but dissemination has hitherto been slow. Private and public actors have raised concerns as to the actual 'market' for MaaS as well as the desired effects. Based on an analysis of an excerpt of available literature, the paper attempts to provide tentative answers to the following questions: Who are the (potential) users of MaaS? And Does MaaS lead to any changes in users' travel behaviour? Prospective studies propose that some user categories (e.g. urban, digitally mature) are more positive than others. The same studies indicate that the services should not be offered as packages but customised to the individual's or household's particular needs for transport and their present travel patterns. Evaluations of pilots reveal a slightly broader user profile. Changes in travel behaviours are reported but also imply that MaaS must offer a higher level of multimodal integration in order for the service to result in noticeable changes in users' travel behaviours.

Keywords: Mobility-as-a-Service · Users · Usage

1 Introduction

1.1 Background

The need for transportation is predicted to continue to rise, resulting in an even further increase in emissions, noise, and congestion. Different more or less successful schemes have been implemented in order to support a shift from less to more sustainable travel including for example economic and legal measures (e.g. congestion charging), awareness campaigns, ICT-based information services (e.g. travel planners, real-time information), development of public transport (PT) vehicles etc., as well as investments into physical infrastructure (e.g. cycle paths). Along with societal trends such as digitalisation, servicification, and the sharing economy, Mobility-as-a-Service (or MaaS) has been argued as part of the solution to reduce the use of private cars and instead increase the use of more sustainable alternatives, such as for example PT and bicycle or car sharing services.

Fundamentally any transport service is a mobility service, i.e. for example public transport and taxi. However, the 'new' concept of 'Mobility-as-a-service' or MaaS includes some additional elements, described as, for example *"... mobility distribution model in which a customer's major transportation needs are met over one interface*

H. Krömker (Ed.): HCII 2020, LNCS 12213, pp. 228–237, 2020.
https://doi.org/10.1007/978-3-030-50537-0_17

and are offered by a service provider ... //... The central element of Mobility-as-a-Service requires a mobility platform that offers mobility services across modes." (Hietanen 2014) or as *"... a digital interface to source and manage the provision of a transport related service(s) which meets the mobility requirements of a customer."* (Catapult 2016)

However, the implementation and dissemination of MaaS have until now been slow. Analyses have identified a number of barriers including legislation and regulations (e.g. König et al. 2016) as well as a lack of appropriate business models (e.g. Catapult 2016; König et al. 2016). However, another barrier among private as well as public actors concerns an uncertainty regarding the actual 'market' for MaaS (e.g. Kamargianni et al. 2015; Karlsson et al. 2020) as well as the actual effects of MaaS in term of for example a reduction of private car use (e.g. Karlsson et al. 2020).

1.2 Purpose and Method

The purpose of the paper is to present results from a review of an excerpt of publicly available MaaS literature. Included in the review was literature that describes empirical studies in which the service was referred to as an example of MaaS. Grey literature was not included. The following questions guided the review: Who are the (potential) users of MaaS?; What MaaS offers (of any) appear to be most attractive to which users? and Have MaaS been found to lead to any changes in users' travel behaviour?

2 Findings

A substantial part of the still limited empirical literature on MaaS is based on prospective studies with the intention to capture (i) travellers' idea of MaaS and (ii) their assumed willingness-to-pay for the service.

2.1 Prospective Studies

A substantial part of the (still limited) empirical literature on MaaS is based on studies with the intention to capture either travellers' idea of MaaS and/or their assumed willingness-to-pay for the service.

In 2014, ITSEC in Finland conducted a study to investigate people's attitude to MaaS - although no questions were asked about MaaS specifically (Sochor and Sarasini 2017). Instead, respondents were asked about their attitude to different scenarios. Some of these scenarios were perceived as positive, such as one "ticket" for all types of transport, mobility on-demand instead of regular PT and car sharing to save money. Other alternatives were perceived more negatively, such as replacing the private car with taxis and that all trips would be by PT. Differences in attitudes were found between groups. Women were generally more positive to the different alternatives than men, younger people were more positive than older people, frequent public transport users and non-car owners as well as those who used the car more rarely more positive than those who used the car often and those living in cities were more positive than rural residents. A majority of the 1305 respondents were car owners (78%) and the majority also used the car frequently (66%).

Another study is a survey performed by Intermetra in Sweden (Intermetra 2018). The share of respondents (n = 1528 in total) who found the idea of MaaS to be an attractive alternative was 42%. Approximately 50% did not consider MaaS an alternative for commuter trips – but a possibility for other trips related to daily activities. Most positive were younger adults, women, people living in cities and those with a certain level of "digital maturity".

A common approach has been Stated Preference (SP) studies. An SP-based survey was for example distributed in the Helsinki area, Finland (Ratilainen 2017). The survey presented different MaaS "packages" including subscriptions of different combinations of public transport, bicycle sharing, car sharing, and taxi at different prices etc. Approximately half of the respondents (n = 252 in total) were considered positive to MaaS. The highest interest was found among respondents with PT tickets, younger respondents and those with lower incomes. The respondents were willing to pay for unlimited access to PT in combination with access to bicycle sharing, whereas taxi was of limited interest as was car sharing. In additional information collected by means of e-mails, one of the questions raised was the reason for pre-packaging of the service offer and several of the (limited number) of respondents would rather choose their own package as the suggested ones did not really fit their transport needs.

In another study, personal interviews were held with the 252 respondents in Sydney, Australia (Hensher et al. 2017; Ho et al. 2017). Also in this case different MaaS scenarios were presented but the alternatives were adapted to the respondent's actual possibilities (i.e. no driving license, no alternative included car sharing). One of the conclusions was that the group most prone to subscribe to a MaaS service was the non-frequent car users whereas the least interested were the frequent car users and those who already used PT, bicycle, etc. For these respondents, car sharing and discounts on taxi trips added to the value of the service offer.

The MaaSLab in London, UK, has in several studies addressed people's attitudes to MaaS in general and different MaaS offers. In one of the studies targeting citizens in London (Kamargianni et al. 2017), 70% of the respondents claimed that they would consider subscribing to a MaaS service provided that it offered certain discounts. However, approximately half reported a concern that the subscription would not cover their travel needs and 40% meant that they would feel 'locked in'. However, as many as 50% stated that they would try new modes of transport if provided by the MaaS. In this case the respondents had clear preferences for offers which included PT.

In summary, based on these studies MaaS appear to attract some user groups more than others (Table 1). What type of MaaS was perceived as the most attractive differed between different categories of users.

Table 1. Summary

Users more positive to MaaS	Users more negative to MaaS
Women	Men
Younger	Older
Non-frequent car users	Public transport users
	Frequent car users
Urban households	Rural households
Low income households	Higher income households
Individuals with a higher level of digital maturity	Individuals with a lower level of digital maturity

2.2 MaaS Pilots

Documentation from systematic and more thorough evaluations of MaaS pilots are (as yet) even more scarce than are the literature describing prospective studies.

MaaS type 1. An example of an early MaaS is Kutsuplus in Helsinki, Finland (2012-2015). A number of minibuses (with a capacity of nine passengers) drove a predefined route from one 'stop' to another. Users used a website to specify a trip and received an offer including time for departure, arrival time and cost. The service was based on advance payment. In a survey distributed after the cancellation of the service, a majority of the users were between 31 and 65 years old, of whom more than half owned a car or had access to a car (Weckström et al. 2018). A majority used the service to a limited degree. The service was used primarily for socio-recreational trips, less for commuting between for example home and work. A majority of users were less frequent car users. Motives for using the service were for example a lack of good PT connections and that the price was lower than for a regular taxi journey. Motives for non-use were that the service was not available at certain times (night) and/or the distance to the "stop" (Weckström et al. 2018).

Another service is Kyyti which today offers on-demand ridesharing for different organisations in different parts of the world. When introduced in 2017, the service was offered to travellers in Helsinki. Customers could choose between travelling alone according to a fixed timetable and at a higher price or to share trip with others at a lower price but then also with a more flexible timetable (Taskinen et al. 2017). Early investigations of who the users were, three different types were found: frequent users (8%), semi-frequent users (46%) and non-frequent users (46%). More than half owned a car and were frequent car users. The share of mode-mixers was larger than for the Finnish population at large (34% compared to 13%). The service was primarily used for trips during evenings and weekends and for transport to/from for example airports, railway stations, etc. in most cases replacing taxi trips rather than other modes of transport. The service did not appear to have affected everyday travel. However, the most frequent users (8%) used the service also for commuting.

In summary:

- Many of the users were car owners and car users although the services were mainly used by less frequent car users;

- The services were primarily used during evenings and weekends and for socio-recreational trips or for trips to/from airport, railway stations, etc.;
- The services were by many users perceived as affordable alternatives to ordinary taxi and more flexible than public transport;
- The services did not appear to significantly have affected everyday travel/commuting.

MaaS type 2. Two examples of multimodal mobility services and some form of evaluation has been documented and made available are SMILE in Vienna, Austria and UbiGo in Gothenburg, Sweden.

The SMILE pilot (2014–2015) aimed to test a prototype for information on and booking and payment of multimodal trips; PT, taxi as well as car and bicycle sharing according to a pay-as-you-go principle. The number of registered users was 1200, but when a survey was sent out to find out more about users and use this was answered by only approximately 25%. The respondents were mainly men (79%), between 20 and 40 years old, residents of Vienna and well educated with relatively high income (smile-einfachmobil.at; Karlsson et al. 2016b). The majority (60%) of these owned a car. Thus, the respondents differed from the average traveller in Vienna and its surroundings and possibly, but not necessarily, from the average SMILE users. The service was used daily by 6% of the respondents and by another 30% several times per week, mostly for private purposes (64%) and for leisure trips (59%). According to the respondents, access to the SMILE service had resulted in that approximately 48% had changed their mobility behaviour, 55% that they more often combined different modes of transport, one out of four (26%) had increased their use of public transport and 21% had reduced their use of the private car. One in ten stated that they often used the bicycle sharing service.

Although the results of the UbiGo pilot carried out in Gothenburg in 2013–2014 have been presented in many contexts, it is still one of the few pilots of an integrated service where there is fairly rich information on users, motives, and possible changes in travel patterns (e.g. Karlsson et al. 2016a; Sochor et al. 2016; Strömberg et al. 2018;). For example, data was collected before, during as well as after the pilot by means of surveys, travel diaries and personal interviews. Households subscribed monthly to a customized subscription, which included trips by public transport, bicycle and car sharing, rental cars and taxi. The trips were "cancelled" via an "app" where users could also check household travel balance, etc. UbiGo households were to greater extent families with children compared to the average Gothenburg citizen, but to a lesser extent single households, students and pensioners. The minimum cost of subscribing to UbiGo was probably a decisive factor for the latter groups, but the UbiGo households otherwise considered the service to be an economically advantageous alternative. Forty-two percent were downtown residents (compared to 23% when considering the entire city). At the same time, it was precisely these centre dwellers with good access to public transport and car sharing that were the actual target group for the pilot. Forty-eight percent of households had one or more cars, which on the other hand compared relatively well with Gothenburg as a whole. An important motive for becoming a UbiGo user was curiosity but over time this changed. Instead, it was the benefits of the service in terms of simplicity, increased accessibility, flexibility and economy that made them want to remain users. The service was perceived as an alternative by households who considered investing in a car and especially those who would otherwise have invested in a "second car". At the end of the pilot a majority (64%) reported changes in their travel habits and 43% also reported changes in the choice of means of transport. The reported use of private cars decreased while the use of other modes of transport (including active modes: walking and bicycle) increased. Overall, the participants also became less positive towards private car use and more positive towards other means of transport. Furthermore, the participants became more satisfied with their transport solution after becoming UbiGo users, even though some more planning was required.

In summary:

- Many of the users were (also) car owners;
- Users were men and women, single people and families, well-educated with higher incomes and primarily living in urban areas;
- Both services led to increased (reported) use of public transport and other public transport, somewhat more in UbiGo compared to SMILE;
- Both services led to reduced (reported) use of the private car, a little less in the case of SMILE compared to UbiGo;
- The services were used for different trips including everyday travelling /commuting;
- In the case of UbiGo, changes in attitudes and increased satisfaction with available transport options were noted - even though the actual supply of transport options was not really changed.

3 Discussion and Implications

3.1 Users of MaaS

If one compares actual user profiles with the profiles generated in the prospective studies, there are both similarities and differences In the prospective studies (Sect. 2.1) women were more positive than men to the idea of MaaS, younger people more positive than older, urban households more positive than rural, and non-car owners more positive than car owners. In the pilots (Sect. 2.2), the profiles were more diverse. Users included men and women, car owners and non-car owners, families with children as well as without children. Among the users were also PT users. However, the main part of pilot participants were not really young people or older citizens. In addition, there were no rural households – but at the same time the described services were not designed for or targeted this group.

Based on the findings one can neither define 'the MaaS user', nor is it possible to conclude whether those who are attracted by or become users of a type A service differ from those users of a service type B. The studies suggest, however, that in order to be attractive to users MaaS must be customised to the individual's or household's and their particular needs for transport.

Nevertheless, two interesting groups emerge from the available data. One is mode-mixers, i.e. those travellers who already used different modes of transport (including public transport). It could be argued that a MaaS which offers multimodal integration facilitates and reinforces an already established behaviour rather than requires radical behavioural changes. Another group includes users who experience a need for access to a car, or to a second car, but who do not necessarily need to own the car. In these cases, a MaaS service that offers access to various modes of transport, including car access, can be an alternative to becoming a "car owner" and thus the lock-in effects and perceived problems that this entails (cf. Strömberg et al. 2018).

3.2   Changes in Travel Behaviour

Does MaaS lead to changes in users' travel behaviour? Again, in order to understand the effects on travel behaviour, it is important to relate the results of different studies to what type of MaaS that was offered. The studies summarised in Sect. 2.1 and Sect. 2.2. describe for example unimodal and multimodal services, different principles of payment (in advance, pay-as-you-go, subscription) and services in which information services are integrated and services where it is not.

In order to describe different services, a structure consisting of 5 'levels', or rather typologies, was proposed by Sochor et al. (2017) (Fig. 1). Level 0 refers to services which offer no integration; Level 2 refers to information services, providing integrated information in terms of for example multimodal travel planners etc.; Level 2 services focus on single trips but offer a one-stop shop where users can find, book and pay through the same user interface. Level 3 offers an alternative to car ownership, focusing on a user's complete mobility needs. Level 4 represents a level where for example public authorities influence the impacts of the service by setting conditions for the operators so that they will create incentives for desired behaviours (ibid.)

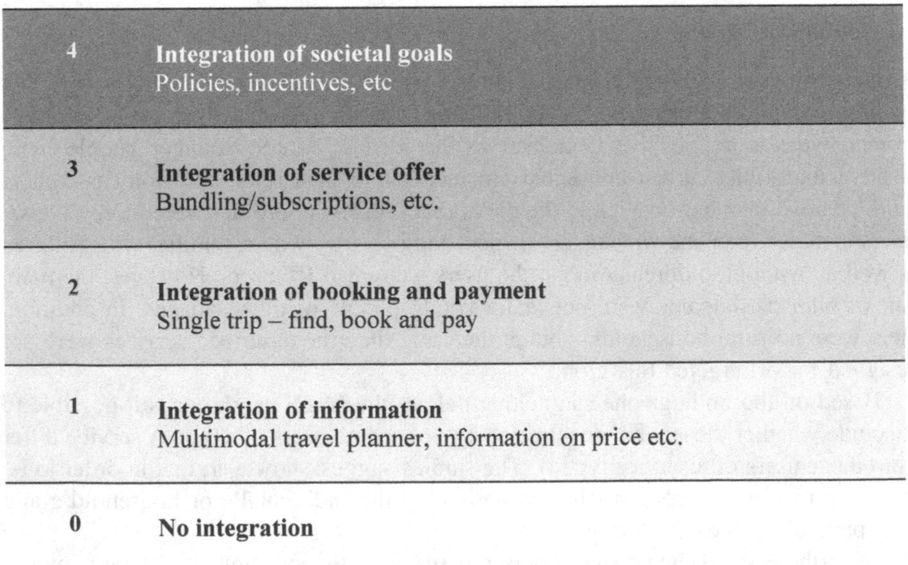

4	**Integration of societal goals** Policies, incentives, etc
3	**Integration of service offer** Bundling/subscriptions, etc.
2	**Integration of booking and payment** Single trip – find, book and pay
1	**Integration of information** Multimodal travel planner, information on price etc.
0	**No integration**

Fig. 1. A proposed typology of MaaS. Source: Sochor et al. 2017

Another typology – or taxonomy - has been proposed by Lyons et al. (2019). This taxonomy concerns "operational, informational and transactional integration that is suggested to reflect a hierarchy of user needs." (ibid.) and consists of five levels. Level 0 refers here (again) to No integration (no operational, informational or transactional integration across modes) whereas Level 5 describes Full integration under all conditions (fully operational, informational and transactional integration across modes for all journeys).

However, even though both typologies attempt to facilitate descriptions of different MaaS and comparisons of their respective effects on, for example, users' travel behaviour, it is not evident how to classify different examples of MaaS. For example, according to the typology proposed by Sochor et al. (2017) the Kutsuplus service could be categorised as Level 0 (no integration of different modes of transport) but also as Level 2 (integration of booking and payment) whereas the UbiGo pilot could be categorised as a level 3 service as it offered a subscription to an integrated multimodal service but at the same time it did not include any integration of travel information (cf. Level 1). Hence, it is difficult to assess whether the impacts on travel behaviour of a service on a "lower integration level" differ from that of a service of a "higher" integration level.

Nevertheless, the Kutsuplus and Kyyti services appear to have changed users' choice of means of transport, but at the same time the changes seem limited to some situations and types of trips. One could even argue that one mode of transport has merely been replaced by another. The SMILE and UbiGo services appear to have had a greater impact on everyday travel as a whole. One interpretation is that a MaaS must include more than one modality in order for travellers to have access to the alternatives that are perceived best suited for the specific situation and for different types of travel and further that MaaS must offer a higher level of integration in order for the service to result in noticeable changes in users' travel behaviours.

However, one can also contemplate that some types of MaaS may create important mental as well as actual "steps" between the two endpoints "100% private car use" - "100% use of other modes of transport". It can for example be argued that Kyyti (as described in the literature referred) may have appeared as another form of taxi, a type of service which has similarities to something well known – but which is at the same time something else. Hence, the 'perceived risks' are likely to be small compared to other and more radical alternatives. Nevertheless, by daring to try this service (and given that it works satisfactorily), the individual's perceived action space can change (cf. Strömberg 2015) and in the long run lead them to dare try also other service options.

Considerably more empirical data is needed to draw but tentative conclusions regarding what type of MaaS attract which type of users and what the effects of different types of MaaS may have people's travel behaviour. This will require more pilots and pilots in which systematic and thorough evaluations are made.

References

Catapult: Mobility as a Service. Exploring the opportunities for Mobility as a Service in the UK. Catapult Transport Systems, London (2016). https://ts.catapult.org.uk/wp-content/uploads/2016/07/Mobility-as-a-Service_Exploring-the-Opportunity-for-MaaS-in-the-UK-Web.pdf

Hensher, D., Mulley, C., Ho, C., Wong, Y.: What are the prospects of switching out of conventional transport services to mobility as a service (MaaS) packages? In: Proceedings on ICoMaaS, 1st International Conference on Mobility as a Service, Tampere, 28–29 November 2017

Hietanen, S.: Mobility as a Service - the new transport model? Eurotransport 12(2) (2014). http://www.fiaregion1.com/download/events/its_supp_et214.pdf

Ho, C., Hensher, D., Mulley, C., Wong, Y.: Prospects for switching out of conventional transport services to mobility as a service subscription - a stated choice study. In: Proceedings on Thredbo15 - International Conference Series on Competition and Ownership in Land Passenger Transport, Stockholm, 13–17 August 2017

Intermetra: Kombinerad Mobilitet. Kundperspektiv. (Combined mobility. A customer perspective) (2018)

Kamargianni, M., Matyas, M., Li, W.: Londoners' attitudes towards car-ownership and Mobility-as-a-Service: impact assessment and opportunities that lie ahead. MaaSLab-UCL, Energy Institute Report. Prepared for Transport of London (2017). http://www.maaslab.org

Kamargianni, M., Matyas, M., Li, W., Schäfer, A.: Feasibility Study for "Mobility as a Service" Concept in London. UCL Energy institute, London (2015)

Karlsson, M., Sochor, J., Strömberg, H.: Developing the 'service' in Mobility as a Service: experiences from a field trial of an innovative travel brokerage. Transp. Res. Procedia 14, 3265–3273 (2016a)

Karlsson, M., Sochor, J., Aapaoja, A., Eckhardt, J., König, D.: Deliverable 4 – Impact Assessment. Report. Deliverable to the MAASiFiE project funded by CEDR (2016b)

Karlsson, M., Mukhtar-Landgren, D., Koglin, T., Kronsell, A., Sarasini, S., Sochor, J.: Development and implementation of Mobility-as-a-Service A qualitative study of barriers and enabling factors. Transp. Res. Part A: P and Practice 1831, 283–295 (2020)

König, D., Sochor J., Eckhardt, J.: State-of-the-art survey on stakeholders' expectations for Mobility-as-a-Service (MaaS) – Highlights from Europe. Paper presented at the 11th ITS European Congress, Glasgow, Scotland, 6–9 June 2016

Lyons, G., Hammond, P., Mackay, K.: The importance of a user perspective in the evolution of MaaS. Transp. Res. Part A: Policy Pract. 121, 22–36 (2019)

Ratilainen, H.: Mobility-as-a-Service. Exploring consumer preferences for MaaS subscription packages using a Stated Choice experiment. MSc thesis. Delft University of Technology (2017). https://julkaisut.liikennevirasto.fi/pdf8/lr_2017_maas_diplomityo_web.pdf

Sochor, J., Sarasini, S.: User' motives to adopt Mobility as a Service. Presentation at the 1st International Conference on Mobility as a Service (ICoMaaS), Tampere, 28–29 November 2017

Sochor, J., Karlsson, M., Strömberg, H.: Trying out Mobility as a Service: experiences from a field trial and implications for understanding demand. Transp. Res. Rec. 4(2542), 57–64 (2016)

Sochor, J., Arby, H., Karlsson, I.C.M., Sarasini, S.: A topological approach to Mobility as a Service: a proposed tool for understanding requirements and effects and aiding policy integration. Res. Transp. Bus. Manag. 27, 3–14 (2017)

Strömberg, H.: Creating space for action – Supporting behaviour change by making sustainable transport opportunities available in the world and in the mind. Dissertation. Chalmers University of Technology, Gothenburg (2015)

Strömberg, H., Karlsson, M., Sochor, J.: Inviting travelers to the smorgasbord of sustainable urban transport: evidence from a MaaS field trial. Transportation **45**, 1655–1670 (2018)

SMILE: smile-einfachmobil.at

Taskinen, J., Karvonen, R., Salonen, A.O.: Why do people switch to a modern on-demand ride service based on sharing? Background and motivation of Kyyti Rideshare passengers in Finland. In: Proceedings of ICoMaaS, 1st International Conference on Mobility as a Service, Tampere, 28–29 November 2017, pp. 52–57 (2017)

Weckström, C., Mladenovic, M., Waqar, U., Nelson, J.D., Givoni, M., Bussman, S.: User perspectives on emerging mobility services: ex post analysis of Kutsuplus pilot. Res. Transp. Bus. Manag. **12**, 84–97 (2018)

A Passenger Context Model for Adaptive Passenger Information in Public Transport

Christine Keller(✉), Waldemar Titov, and Thomas Schlegel

Institute of Ubiquitous Mobility Systems, Karlsruhe University of Applied Sciences,
Moltkestraße 30, 76133 Karlsruhe, Germany
{Christine.Keller,iums}@hs-karlsruhe.de
http://iums.eu

Abstract. Passengers in public transport expect passenger information to be exact, timely and appropriate to their situation. Therefore, future passenger information systems should adapt to the passenger's context as precisely as possible. In this paper, we present a context model and describe our architecture for an adaptive, multi-device passenger information system. We will also present adaptation scenarios that show the application of our context model.

Keywords: Context model · Adaptation · Public transport

1 Introduction

A good customer experience and trust in public transport are essential in the development of future public transport systems [4,8]. Passengers need simple and instant access to all information regarding their itinerary. Furthermore, some passengers require additional support, for example to find free seats on a train, as reported by Oliveira et al. [18]. Depending on their experience using public transport and their travel purpose, passengers need different information regarding their trip. In addition to passenger's properties, the public transport situation also influences the information needs of users. If, for example, a vehicle will take a detour after the next stop because of an accident, passengers in this vehicle will need to know if and how they should adapt their trip. In current passenger information systems, information is usually not filtered and passengers need to find out by themselves, which information is relevant to them.

At the same time, current technologies are being adopted more and more frequently for future passenger information systems. Wearable devices are finding their way into public transport systems, for example for measuring the user's walking speed and considering it in routing decisions, in the system of Chow et al. [6]. Public displays are utilized for passenger information at stations or in public

H. Krömker (Ed.): HCII 2020, LNCS 12213, pp. 238–248, 2020.
https://doi.org/10.1007/978-3-030-50537-0_18

transport vehicles [16,17]. Bluetooth, Near Field Communication, GPS and Wifi create possibilities for locating passengers and vehicles, as well as possibilities for mobile fare systems [1,11].

These systems allow passenger information to be much more personalized and specific than before. However, passenger information systems need a basis on which to make decisions about which information to present to the user and how to present it. For this purpose, we have modeled a passenger context model for adaptive passenger information systems. We have also developed a context management architecture and have outlined several adaptation scenarios to illustrate the application of our context model.

2   Adaptive Passenger Information Systems

Public transport is benefitting from digitalization in several aspects, ranging from infrastructure, public transport vehicles to passenger information systems [8]. Public transport data is made available digitally, for example sensor data on vehicle occupancy or real-time information on delays or disruptions. Several data sources are integrated for public transport information and some data is provided as linked data, as well [2,13]. These developments in public transport are key requirements to improve passenger information systems. Several approaches towards passenger information systems that adapt to the passenger's situation as well as the public transport situation have been developed recently and we will describe some of them in the following paragraphs.

De Amorim et al. developed and evaluated a mobile ticketing application, that implements a check-in/be-out scheme [1]. Their application utilizes location context information provided by the user by checking in a public transport vehicle, using Near Field Communication (NFC). GPS and bluetooth beacons are used to locate the user and to automatically detect the user's alight station. Fares are calculated and charged automatically each month. Using this application, the passenger does not have to understand the ticketing information of the public transport service to choose the correct and cheapest ticket any more.

The approach of Chow et al. uses walking speed as a context factor that is measured by a wearable device [6]. A smartphone application provides personalized public transport information. The user's location and walking speed, as well as real-time public transport data are used to recommend public transport routes. Depending on the user's walking speed, the application computes, which buses the user should be able to catch and which route alternatives are therefore relevant to them. Additionally, the wearable devices can give haptic feedback, if the user should increase their walking speed to catch a certain bus.

An Internet-of-Things approach is presented by Handte et al. [11]. The authors implemented a system that uses Wifi in buses, a mobile application and a crowd-sensing platform. The system uses the user's location, the vehicle's location, public transport real-time information and data on how many passengers are on which bus, to improve passenger information during a trip. The mobile application is able to identify if a user is entering the correct vehicle and when they should alight. It also takes data on crowd-levels in buses into account, to suggest less crowded vehicles to the user.

The mobile application that Chowdhury and Giacaman present uses the passenger's location, real-time public transport information and the purpose of the trip the passengers indicate while using the app, to support trip planning [7]. The application continuously informs the user about their trip. The user is able to re-plan or add additional stops during the trip.

Keller et al. developed a mobile application specifically for tourists, that is able to compute tourism-related information based on the user's context [14]. The application considers the user's location, their interests and local weather in order to propose points of interest to the user. It also proposes a route to each point of interest, using public transport. Additionally, the user can state a time interval and the application computes an appropriate sightseeing route via public transport that takes the user's context into account.

Besides location information, the context factors these approaches use are as heterogeneous as the implemented adaptations of the applications. The approaches also focus on different tasks in public transport, ranging from catching a bus to trip planning. We therefore chose a systematic approach towards adaptive passenger information systems to analyze passenger's information need and relevant context factors throughout several stages of public transport usage.

3   Context and Context-Awareness in the Public Transport Domain

The usage of context data to adapt applications and to better support its users is a topic of ubiquitous computing and discussed, for example, in Mark Weiser's work on ubiquitous computing, by Schilit et al., as well as by Dey and Abowd [9,20,24]. Dey's and Abowd's definition of context is widely used, but also very open. They include all information that characterizes the situation of an entity as context. In order to use this context definition, it has to be specified further for each application domain and specific application.

Early work on context-aware applications very often only used the user's location as relevant context factors, such as the active badge system by Want et al. or the context-aware tourist guide application GUIDE by Cheverst et al. [5,23]. For systems that support people's mobility, like public transport, location is, of course, crucial. However, as Schmidt et al. argued in 1999, location is not the only context factor that can be utilized to achieve a ubiquitous computing experience and to improve usability of ubiquitous systems [22].

Schlegel and Keller discussed context for public systems, such as public transport and built upon context definitions by Schilit et al., Schmidt et al. and Dey and Abowd to define and describe context dimensions for public systems [10,20–22]. The dimensions of time, location, user activity, physical environment and user identity were extended by, for example, a socio-technical context that captures specific context factors relevant in public systems, such as, for example, sociological context or operational context.

An analysis of context factors in public transport from a usability point of view is presented by Hörold et al. [12]. They identified a number of context factors, from environmental factors, such as light or noise, to location and social factors, such as the presence of others. The authors use these factors to evaluate stop points and to visualize the user's need for information on several categories of bus stops, as a basis for the design of passenger information systems. They do not focus on context-aware applications.

Boudaa et al. also describe a context model for the mobility domain, specifically for the application in context-aware recommender systems [3]. They identify *user*, *mobile device*, *activity* and *environment* as relevant context dimensions. The authors use a high-level semantic modeling approach and model an OWL ontology, using SWRL rules for reasoning.

3.1 A Passenger Context Model for Adaptive Passenger Information

In our research project on contextual passenger information, we modeled context in an OWL ontology, in order to implement adaptations based on gathered context information. To develop our context model, we chose a two-fold method. In a top-down approach, we used the context dimensions of Schlegel and Keller as a basis and extended them by additional user context, as depicted in Fig. 1 [21]. We further detailed several of the context dimensions into context categories.

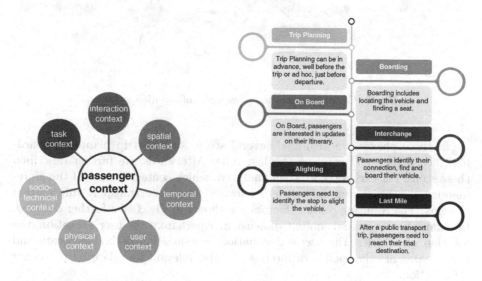

Fig. 1. Context dimensions after Schlegel et al. [21].

Fig. 2. Phases in a travel chain in public transport.

At the same time, we implemented a bottom up approach and detailed context factors along the travel chain of public transport passengers, a method, that was also used by Hörold et al., for example [12]. Another example for this approach is by Oliveira et al., who developed a customer journey map for public transport usage and conducted passenger interviews to identify, at which stages of a public transport journey, technology could support passengers [18]. The results provided us with indications as to which context factors are relevant in which situations. For our work, we divided the travel chain of public transport usage into three phases: *pre-trip*, *on trip* and *post trip*, as shown in Fig. 2.

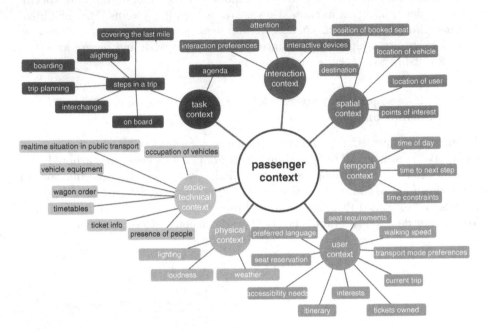

Fig. 3. Context factors for passenger information systems.

In these phases, we identified several steps, such as trip planning, boarding, interchange and covering the last mile. Afterwards, we further described these steps. For each step, we then analyzed, which context factors of the aforementioned context dimensions were relevant for public transport information systems. The resulting context factors are shown in Fig. 3. We further detailed, for example, the spatial context dimension. Apart from the user's location, the location of vehicles, the user's destination, the location of booked seats and the location of other points of interest are also relevant for adapting passenger information.

3.2 An Approach Towards Adaptive Passenger Information Systems

Context information not only needs to be modeled, but must also be recorded and stored. Perera et al. described a context lifecycle for context-aware systems in the IoT domain [19]. We adapted this lifecycle for our project, as presented in

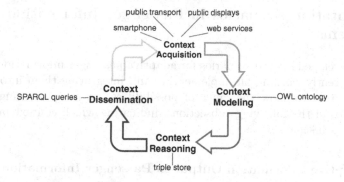

Fig. 4. An tailored context lifecycle for adaptive passenger information, after [19].

Fig. 4. The first step is the context acquisition step. We implemented a mobile application for passenger information, that is used as a context source and provides information about the user's itinerary, seat, language or transport mode preferences and the user's location, among others.

Additionally, we use public transport information systems as a context source. We are able to access servers of the local public transport provider and can query schedule information as well as real-time information and traffic reports concerning public transport. We also query other web services for weather information and data on points of interest as well as data about bicycle and car sharing. As we use public displays, specifically semi-transparent displays instead of regular vehicle windows as a way to inform passengers, we can use them as context source, as well. In our prototype, we deployed bluetooth beacons next to the displays in order to recognize the presence of users close to the display.

For the context modeling step, we modeled an OWL ontology that reflects the relevant context factors for our prototypical adaptation scenarios. For context sources that do not provide semantic data, we implemented semantic adpater components, that transfer the data into our semantic representation.

As for now, we use only limited context reasoning. We store our data in a triple store and utilize its reasoning capabilities for simple reasoning steps. We are planning to extend the reasoning capabilities of our approach in future work, to include additional high level context information.

The last step in one context lifecycle is the dissemination step. In order to adapt to current context, the context data needs to be available to adaptation components. We use SPARQL queries to query context data and to decide upon adaptations. Depending on the type of adaptations, different components are responsible to execute them. Several adaptations are implemented on our data platform and concern the processing or enrichment of user queries. Some adaptations are also implemented at user-interface level in our smartphone application or our display application.

4 Adaptation Scenarios in Passenger Information Systems

We have developed several scenarios for adaptive passenger information systems and are currently working on implementing and integrating them into our prototype. In order to show the range of possibilities, we are presenting some of the scenarios in the following subsections, indicating which context factors are utilized in each scenario.

4.1 Adaptive Information Output in Passenger Information

A prototype that is in part already implemented, is our adaptive information output scenario [15]. We have implemented an adaptation component that uses our context data to decide, which output device and output modality should be used to inform the passenger. The algorithm uses a location filter to match which output devices are available at the user's location (e.g. a display in a public transport vehicle, their smartphone or their smartwatch). Ability and preferences filters are used to match the user's perceptive abilities and preferences to the available output modalities and after that, a general filter tries to determine, for example, if the user's smartphone is available or if the user has put it away. It also checks for headphones. This approach is implemented data-wise, the implementation on devices is still work in progress.

4.2 Adaptive Boarding Assistant

An adaptive boarding assistant supports passengers in finding the right vehicle and helps them navigating to their coach and seat, as shown in Fig. 5. In a big station, it is often hard for passengers to identify the platform their vehicle departs from and to navigate there, especially when platforms change. The assistant provides indoor navigation in stations and navigates passengers to the right platform, using realtime information on the vehicle's departure. If the passenger has booked a seat in advance, the assistant guides them to the right spot on the platform, where they can easily board at the right door. Inside the vehicle, the assistant helps finding the seat and identifies other vehicle equipment such as toilets, wifi, dining car or spaces for extra luggage. If passengers did not book a seat, the adaptive boarding assistant can guide them to free seats. Seats and other equipment, like power sockets, tables and toilets can be adapted to the passenger's seat requirements and accessibility needs.

4.3 Adaptive Public Display at a Bus Stop

Apart from displays in vehicles, we are also researching public displays that can be deployed at stop points and stations. The usage of interactive public displays at stations can improve passenger information and decrease maintenance in contrast to static, paper-based passenger information. Mayas et al., for example,

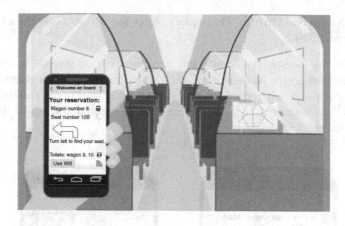

Fig. 5. An adaptive boarding assistant.

have evaluated public displays as passenger information systems in a field trial
[17]. However, these were not context-aware. We are conceptualizing a public
display for a bus stop or other stop points that is reacting to the presence of
users. An example for such a public display at a bus stop is shown in Fig. 6. If
nobody is waiting at the stop, the public display shows general information on
the next buses on the left, and some advertisements on the right that are targeted
at passers-by. The display reacts to the presence of a person standing in front
of it by showing additional realtime information on delayed buses, announcements on disruptions or, for example, realtime posistions of buses instead of the
advertisements.

Fig. 6. Mockups of an adaptive public display based on user presence.

4.4   An Adaptive Public Transport App for Proactive Passenger Information

As apps like the Google Assistant[1] are implemented for general traffic-related information, a proactive passenger information application is learning from regular trips of the user, their regular places, such as their workplace and information about their appointments to provide them with personalizes, up-to date information about public transport.

Fig. 7. Mockups of a proactive passenger information app.

Figure 7 shows mockups for such an application. It confirms the location it identified as the user's workplace, in order to provide proactive information about the public transport situation. If there are delays or disruptions, the application is able to inform the user even prior to their trip, not only for planned trips, but also for trips the user is using regularly.

5   Outlook

In this paper, we have presented our context model for adaptive passenger information systems. We described our approach to context modeling in public transport and the resulting context factors in several context dimensions. We also presented our context lifecycle and adaptation approach, which is based on an OWL ontology and a triple store as context data storage. The components of

---

[1] https://www.google.com/search/about/.

our context lifecycle are already implemented as prototypes. Subsequently, we introduced several adaptation scenarios that can be implemented using our context model and adaptation approach. The adaptation scenarios are at different stages of implementation within our prototype. We are currently planning the evaluation of our implemented adaptations through user studies and are looking forward to report our results.

Acknowledgements. This work was conducted within the scope of the research project "SmartMMI - model- and context-based mobility information on smart public displays and mobile devices in public transport" and was funded by the German Federal Ministry of Transport and Digital Infrastructure as part of the mFund initiative (Funding ID: 19F2042A).

References

1. de Amorim, D.M., Dias, T.G., Ferreira, M.C.: Usability evaluation of a public transport mobile ticketing solution. In: Ahram, T., Karwowski, W., Taiar, R. (eds.) IHSED 2018. AISC, vol. 876, pp. 345–351. Springer, Cham (2019). https://doi.org/10.1007/978-3-030-02053-8_53
2. Beutel, M.C., et al.: Heterogeneous travel information exchange. In: Mandler, B., et al. (eds.) IoT360 2015. LNICST, vol. 170, pp. 181–187. Springer, Cham (2016). https://doi.org/10.1007/978-3-319-47075-7_23
3. Boudaa, B., Hammoudi, S., Benslimane, S.M.: Towards an extensible context model for mobile user in smart cities. In: Amine, A., Mouhoub, M., Ait Mohamed, O., Djebbar, B. (eds.) CIIA 2018. IAICT, vol. 522, pp. 498–508. Springer, Cham (2018). https://doi.org/10.1007/978-3-319-89743-1_43
4. Camacho, T.D., Foth, M., Rakotonirainy, A.: Pervasive technology and public transport: opportunities beyond telematics. IEEE Pervasive Comput. **12**(1), 18–25 (2013). https://doi.org/10.1109/MPRV.2012.61
5. Cheverst, K., Davies, N., Mitchell, K., Friday, A.: Experiences of developing and deploying a context-aware tourist guide: the guide project. In: Proceedings of the 6th Annual International Conference on Mobile Computing and Networking (MobiCom 2000), pp. 20–31. ACM, New York (2000). https://doi.org/10.1145/345910.345916. http://doi.acm.org/10.1145/345910.345916
6. Chow, V.T.F., et al.: Utilizing real-time travel information, mobile applications and wearable devices for smart public transportation. In: 2016 7th International Conference on Cloud Computing and Big Data (CCBD), pp. 138–144, November 2016. https://doi.org/10.1109/CCBD.2016.036
7. Chowdhury, S., Giacaman, N.: En-route planning of multi-destination public-transport trips using smartphones. J. Public Transp. **18**(4), 31–45 (2015)
8. Davidsson, P., Hajinasab, B., Holmgren, J., Jevinger, Å., Persson, J.A.: The fourth wave of digitalization and public transport: opportunities and challenges. Sustainability **8**(12) (2016). https://doi.org/10.3390/su8121248. http://www.mdpi.com/2071-1050/8/12/1248
9. Dey, A.K., Abowd, G.D.: Towards a better understanding of context and context-awareness. Technical report (1999)
10. Dey, A.K., Abowd, G.D., Salber, D.: A conceptual framework and a toolkit for supporting the rapid prototyping of context-aware applications. Hum.-Comput. Interact. **16**(2-4), 97–166 (2001). https://doi.org/10.1207/S15327051HCI16234_02. http://www.tandfonline.com/doi/abs/10.1207/S15327051HCI16234 02

11. Handte, M., Foell, S., Wagner, S., Kortuem, G., Marrón, P.J.: An internet-of-things enabled connected navigation system for urban bus riders. IEEE Internet Things J. **3**(5), 735–744 (2016). https://doi.org/10.1109/JIOT.2016.2554146
12. Hörold, S., Mayas, C., Krömker, H.: Analyzing varying environmental contexts in public transport. In: Kurosu, M. (ed.) HCI 2013. LNCS, vol. 8004, pp. 85–94. Springer, Heidelberg (2013). https://doi.org/10.1007/978-3-642-39232-0_10
13. Keller, C., Brunk, S., Schlegel, T.: Introducing the public transport domain to the web of data. In: Benatallah, B., Bestavros, A., Manolopoulos, Y., Vakali, A., Zhang, Y. (eds.) WISE 2014. LNCS, vol. 8787, pp. 521–530. Springer, Cham (2014). https://doi.org/10.1007/978-3-319-11746-1_38
14. Keller, C., Pöhland, R., Brunk, S., Schlegel, T.: An adaptive semantic mobile application for individual touristic exploration. In: Kurosu, M. (ed.) HCI 2014. LNCS, vol. 8512, pp. 434–443. Springer, Cham (2014). https://doi.org/10.1007/978-3-319-07227-2_41
15. Keller, C., Schlegel, T.: How to get in touch with the passenger: context-aware choices of output modality in smart public transport. In: Adjunct Proceedings of the 2019 ACM International Joint Conference on Pervasive and Ubiquitous Computing and the 2019 International Symposium on Wearable Computers, Ubi-Comp/ISWC 2019 Adjunct. ACM, New York (2019, to appear). https://doi.org/10.1145/3341162.3349321. http://doi.acm.org/10.1145/3341162.3349321
16. Keller, C., Titov, W., Sawilla, S., Schlegel, T.: Evaluation of a smart public display in public transport. In: Mensch und Computer 2019 - Workshopband. Gesellschaft für Informatik e.V., Bonn (2019)
17. Mayas, C., Steinert, T., Krömker, H.: Interactive public displays for paperless mobility stations. In: Kurosu, M. (ed.) HCI 2018. LNCS, vol. 10902, pp. 542–551. Springer, Cham (2018). https://doi.org/10.1007/978-3-319-91244-8_42
18. Oliveira, L., Bradley, C., Birrell, S., Davies, A., Tinworth, N., Cain, R.: Understanding passengers' experiences of train journeys to inform the design of technological innovations. In: Re: Research - the 2017 International Association of Societies of Design Research (IASDR) Conference, Cincinnati, Ohio, USA, pp. 838–853 (2017)
19. Perera, C., Zaslavsky, A., Christen, P., Georgakopoulos, D.: Context aware computing for the internet of things: a survey. IEEE Commun. Surv. Tutor. **16**(1), 414–454 (2014). https://doi.org/10.1109/SURV.2013.042313.00197
20. Schilit, B., Adams, N., Want, R.: Context-aware computing applications. In: Workshop on Mobile Computing Systems and Applications, pp. 85–90, December 1994. https://doi.org/10.1109/MCSA.1994.512740
21. Schlegel, T., Keller, C.: Model-based ubiquitous interaction concepts and contexts in public systems. In: Proceedings of the 14th International Conference on Human-Computer Interaction (2011)
22. Schmidt, A., Beigl, M., Gellersen, H.W.: There is more to context than location. Comput. Graph. **23**(6), 893–901 (1999). https://doi.org/10.1016/S0097-8493(99)00120-X. http://www.sciencedirect.com/science/article/pii/S0097849399 00120X
23. Want, R., Hopper, A., Falcao, V., Gibbons, J.: The active badge location system. ACM Trans. Inf. Syst. **10**(1), 91–102 (1992). https://doi.org/10.1145/128756.128759. http://portal.acm.org/citation.cfm?doid=128756.128759
24. Weiser, M.: The computer for the 21st century. Sci. Am. **265**, 94–104 (1991)

An Evaluation Environment for User Studies in the Public Transport Domain

Christine Keller[✉], Waldemar Titov, Mathias Trefzger, Jakub Kuspiel,
Naemi Gerst, and Thomas Schlegel

Institute of Ubiquitous Mobility Systems, Karlsruhe University of Applied Sciences,
Moltkestr. 30, 76133 Karlsruhe, Germany
{Christine.Keller,iums}@hs-karlsruhe.de
http://iums.eu

Abstract. User studies to evaluate public transport systems are often hard to set up. While field tests provide important insight into real-world usability of public transport systems, they are also complex and expensive. Especially in early development stages of public transport related systems, field tests are not appropriate. However, usability of public transport systems is often depending on "real-life" context factors that are hard to reproduce in lab-based user studies. We have developed a mockup of a tram or train compartment that can be flexibly used to create a public transport experience in user studies. In this paper we will describe our experiences and recurring challenges with user studies in public transport, the design and set-up of our mockup, as well as give an insight into its applications in studies we conducted and lessons we learned.

Keywords: User studies · Public transport · Evaluation · Passenger information system · Smart public transport

1 Introduction

Public transport is an important part of many people's mobility. Usable passenger information systems and public transport systems in general are a core interest of public transport providers. Camacho et al. for example, argue for passenger-centric innovation and advocate continuous evaluation of passenger-related systems for future public transport [6]. Systems that are relevant for passengers in public transport are, among others, fare systems, seat booking systems and passenger information systems that inform about schedules, but also about incidents, delays and other real-time information. User studies can give valuable insight into usability and user experience of public transport systems. Habermann et al. describe, for example, a usability study evaluating a mobile application for public transport [12]. De Amorim et al. performed a field test in Porto, Portugal, to evaluate a mobile ticketing application [3]. Another field study is, for example, described by Mayas et al., evaluating public displays for passenger information at stations [23].

© Springer Nature Switzerland AG 2020
H. Krömker (Ed.): HCII 2020, LNCS 12213, pp. 249–266, 2020.
https://doi.org/10.1007/978-3-030-50537-0_19

New technologies and devices, such as smartphones, the Internet of Things, wearables or public displays are adopted for public transport systems. Those technologies transform user interfaces, which requires careful development and evaluation of the system's usability and user experience. Some examples in the following paragraph show the breadth of possible systems.

Chowdhury and Giacaman report on a smartphone application that uses public transport real-time and schedule information to support users in planning trips with possibly multiple stops [8]. The application specifically was designed to enable ad hoc stops during a route. The application implements en-route information to support users during the trip.

Handte et al. present an application called Urban Bus navigator that is designed and implemented as an Internet of Things application and supports passengers on bus journeys, including short walking distances [13]. Their smartphone application supports ad hoc trip planning from the user's current location and helps users to find the correct bus and bus stop.

Chow et al. developed a mobile application for planning trips in Hong Kong public transport, that utilizes wearable devices [7]. Wearable devices are used to measure the walking speed of their user and the application takes that speed into account when planning a trip and to remind the user to adapt their walking speed in order to catch the next bus.

Garcia et al. present a route travel assistant, which provides real time information about the user's route based on Android technologies [11]. The system uses the bus infrastructure (for example the bus positioning system and vehicle bluetooth network) and relevant elements of the public transport network (for example stations, equipped with bluetooth and Wifi) for information exchange. The Android-based smartphone application uses push notifications for passenger information en route.

We argue that in public transport systems, usability and user experience is highly affected by the user's situation and therefore, by their context. The examples above show, that public transport systems are becoming ubiquitous, which also means that the usage context is gaining importance while evaluating and assessing these systems. The user's experience of planning a trip using a smartphone application probably changes whether they use it at home or in a lab or on the road, under time pressure to reach the next bus. To consider this context in user studies is complex. Field studies can provide the necessary context but are hard to control and very costly. We therefore propose an evaluation environment that replicates a compartment of a public transport vehicle and can be used to create a public transport experience in user studies. We will describe the challenges we address with our mockup, its development and some of its possibilities, including some exemplary studies we performed using it, in the following sections.

2 Usability Evaluation in Public Transport

Expert-based evaluation methods, such as heuristic or model-based evaluations as well as cognitive walkthroughs are one possibility to measure the usability of public transport systems. In this paper, we focus on user-based evaluation methods. We believe that for public transport systems, a mix of expert-based and user-based evaluation methods is preferable. In public transport, there are various different user groups with very different backgrounds, which calls for evaluation methods that can incorporate this diversity, for example user studies. Evaluation methods that include users can be case studies, user observations, lab experiments, field tests, surveys and others, specified for mobile and ubiquitous applications, for example, by Reis et al. or Kjeldskov and Paay [18,28]. In the following subsection, we will summarize some evaluations of mobile or ubiquitous public transport systems from related work.

2.1 Examples of Evaluations of Public Transport Systems

De Amorim et al. report on a usability study testing a mobile application for public transport ticketing that applied a check-in/be out scheme, using near field communication, Bluetooth and GPS [3]. The study was implemented as a field test in the city of Porto, using a full working prototype of the developed application. In this study, eight users tested the app in real public transport, performing a given set of tasks. The users started at a station they chose and their trip required at least one interchange. The study included a pre-test questionnaire and during performing the tasks, the users were asked to think aloud. All tasks were performed using the mobile application. After the tasks were finished, the authors conducted a structured interview, focusing on the app's usability.

A system for passenger information at public transport stations was evaluated by Mayas et al. [23]. Interactive public displays were installed at a public transport station to replace paper-based information on schedules, tickets or network plans. Mayas et al. evaluated these displays in the field, over the course of one year. In their field evaluation, the authors examined extent of use, context of use and degree of acceptance. They used several methods, combining the analysis of logfiles with an online questionnaire and observations of users with post-observation interviews. In the logfiles, the authors were able to analyze over 250.000 user interactions with the displays. The online questionnaire yielded 162 completed questionnaires and during the observation, 55 usages were observed and 25 of these people were interviewed afterwards.

Habermann et al. describe the usability evaluation of a smartphone application that was developed as part of a project concerned with multimodal transport [12]. A platform supporting mobility with several modes of transport was also developed in this project. The core functions of the applications were routing and booking of trips with public transport or other transport modes, such as car sharing. The authors report on the evaluation of an initial prototype for the smartphone application. 32 participants were asked to perform a set of tasks using the prototype in a lab-based study. The authors evaluated the perceived ease of navigation and perceived interface quality in a questionnaire that was filled out by participants after performing the tasks. Additionally, the authors asked the participants about their mobility habits, their experience with mobility applications and their technical expertise.

The smartphone application that Chowdhury and Giacaman report on, mentioned above, was also evaluated in the field [8]. The application enables the user to plan trips with one or more destinations and supports the user during their trip. The authors perfomed a study with 21 engineering students that were asked to travel four routes in total, of which the participants should travel one set of two routes using the given application and one set of two routes without the application, using information sources they normally would use during such a task. The study was unsupervised and the participants were given forms to record at which time they started their trip, how long they waited and how long it took them to board a vehicle, as well as their time of arrival and how long it took them to plan their trips. The participants were also asked to fill out a questionnaire focused on the application's usability, after completing their tasks.

2.2 Challenges of Usability Evaluation in Public Transport

Lyytinen and Yoo characterized computing systems based on their level of embeddedness and their level of mobility [22]. Most passenger-oriented public transport systems have a high level of mobility and some also have a high level of embeddedness, qualifying them as mobile or ubiquitous systems. Both embeddedness and mobility result in a highly dynamic context of use that impacts the system's usability. Challenges for usability evaluation of ubiquitous and mobile systems therefore also apply to public transport systems [5,26].

Bezerra et al. identified several challenges for conducting user studies for ubiquitous systems that extend to public transport systems [5]. Based on literature reviews and their own experience with evaluation, they identified the evaluation environment as the main challenge. Ubiquitous systems are highly dependent on "real-life" context factors, therefore a traditional laboratory test was found not suitable by many researchers [9,17]. Generally, evaluation of interactive and especially ubiquitous or mobile systems should be performed in authentic settings, in the context of authentic use, as Abowd and Mynatt described [1].

Additional challenges stem from special characteristics of ubiquitous systems. Bezerra et al. and Santos et al. identified the following five characteristics [5,29]:

- Context-Awareness (the ability of the application to collect context information from the real world and use it to adapt information and services)
- Transparency (the user should be minimally distracted, so the system needs to act proactively)
- Focus (the user should concentrate mainly on their activity and not on the system)
- Calmness (the system should not overload the user)
- Availability (the ability to provide access to the system anywhere at any time)

A context-aware application adapts information and services based on the context of the environment, so it is important to include context in the evaluation environment. It is also important to involve all relevant adaptations in the test cases, so every important function is evaluated [5]. In traditional laboratory environments, the particpants can fully focus on the usability tasks and the evaluated system, but in real environments, such focus is not probable [17]. Most public-transport systems are context-aware in some way, since public transport services and information adapts to traffic situations, delays and disruptions very often. Some systems are also designed as context-aware in a narrower sense, implementing different adaptations based on the user's context [14]. Context factors are often very relevant in the execution of public transport systems and therefore should be considered in a usability evaluation. Other of the above-mentioned characteristics apply to public transport systems, too. Users of public transport focus on other tasks than their mobility and a public transport system should therefore not need their full attention, but its availability anywhere and at any time is essential.

Apart from the initial planning task, all tasks of the users in public transport involve the user's mobility. The user's situations are determined by their location and often also by their surroundings, for example while finding a seat on a tram, waiting at a bus stop in the rain or orienting oneself at a huge train station, catching a connecting train. These are factors that also affect the evaluation of such a system. The question therefore is, how can these characteristics be addressed in a usability evaluation of public transport systems?

Field trials have a very authentic context and can therefore give good insights into a public transport system's usability. In a field trial, all relevant context factors are available and the user is exposed to the environment the system is supposed to be used in. The user can focus on their task getting from one location to another. However, field trials require a fully functional prototype, such as de Amorim et al., Mayas et al. or Chowdhury and Giacaman evaluated [3,8,23]. Such a prototype must use real public transport information and therefore use, for example, interfaces to realtime data servers of local public transport agencies. In most cases, this requires a lot of effort. Early design variants can hardly be evaluated in field trials, for example.

Context factors also can not be completely controlled in field trials. Weather or, for example, ambient noise are hardly controllable in the field. Public transport itself can not be controlled or altered for the sake of one field trial, at least in most settings. Delays or disruptions may be unwanted in a trial or, if such situations should be evaluated, delays or disruptions can not be caused for the sake of a field trial. However, taking control over certain context factors and study parameters can be desirable. This has often been an argument for laboratory-based usability studies and against field trials even for public systems, such as public displays, for example [2].

There are case studies that evaluate ubiquitous applications with creating an authentic environment. They built an evaluation environment based on the real environment, so they developed a fully equipped house to test smart home technologies [16].

We decided to follow this example. Since our research project focuses on passenger information in public transport vehicles, we decided to build a vehicle for our lab.

3   A Public Transport Vehicle in the Lab

We developed and constructed a mockup of a tram or train, that simulates a realistic public transport context for study participants while enabling us to control context factors and data.

The tram compartment comprises a wood casing painted in the original colors of the local public transport network and features a window and handrails that frame its interior. Original tram seats are installed on metal rails that enable us to change the distance between them, as seen in an advanced version of our mockup in Fig. 1. The arangements of the seats can easily be changed from facing in the same direction to facing each other. We also can replace them with folding seats in several orientations. This flexibility allows us to reconstruct several vehicle configurations most commonly used in public transport, including long-distance trains. The described mockup and its current setup is a replica of a railway compartment of local trams, including, as mentioned before, the coloring, floor covering and original seats.

Fig. 1. Tram compartment with adjustable seat configuration.

Since we are developing passenger information user interfaces on semi-transparent displays for tram windows, we wanted to use the mockup to evaluate these interfaces. Unfortunately, the actual semi-transparent displays are hard to come by at the moment and we are focusing on installing one of these in an actual tram during our project. For our mockup, we therefore had to prototype a subsitute. In the first stage, the demontrator was equipped with a semitransparent plastic plane. A short-throw projector was used to display passenger information or any other image on to the semitransparent plane. We used prototypes of our passenger information visualization combined with a video that shows the sight out of the window during an actual tram ride in our local public transport network. This video is shown in the background of the prototype and creates the illusion of sitting next to a window of a moving tram. Since most of our study participants are familiar with local public transport, we also synchronized the display of next stops and departures at these stops with the stops of the tram ride shown in the video, so that no discrepancies disrupt our studies. The display of next stops changes at the exact moment, when the tram leaves a stop. The projected tram ride video in combination with the matching sound of the same tram ride and the matching departure monitor displayed on the plane we were able to relocate the participants into the context of using real public transport. This way, study participants we able to perceive a almost realistic tram ride and were able to interact with the window display showing passenger information on top of the reality right in front of them.

The rubber outline of the window that holds the plastic plane enables us to be very flexible by removing and to replace it. In order to add multi-touch interaction, a mountable multi-touch frame was used.

In a second stage of our mockup development, we used a pair of two multi-touch displays of 44 in. each, that were mounted on a display rack in front of the window. On top of the two interactive displays, we mounted a third non-interactive display as shown in Fig. 2. This non-interactive display shows information on the size of a 1/3 cutted 32 in. display, which matches the construction of our prototype that will be implemented in a real tram in real public transport.

While both interactive displays can be used to display dynamic and interactive content, like a geographical map, a schematic route map or the passenger's choosen route, the third display represents a kind of passenger information display that is already used in several public transport vehicles, often also located on top of the side windows. In most configurations, it shows static information, sometimes complemented by realtime information. Our standard visualization on this display shows the four next stops of the vehicle with the corresponding departing lines. The interactive displays can either be removed, if a current version of a tram or train should be simulated, or used as mockups of a window, for example using the above-mentioned video of a tram ride, or they can be used as the interactive public displays we are developing in our project, displaying additional passenger information of any kind [15].

We further added two bluetooth beacons on both upper edges of the window frame, that can be used to detect passengers with smartphones and to locate if

Fig. 2. Tram compartment with two interactive and one non-interactive displays.

they are seated near this window and on which side. They are currently used to trigger passenger information on the interactive displays as soon as a passenger is seated next to the window.

In this close-to-real situaitonal public transport setup we conducted several user studies. Mostly positive feedback of participants that participated in our studies using our mockup shows its benefit as an evaluation environment. We are convinced that it can improve lab studies in public transport settings by providing context of use, while being more cost efficient and controllable than field tests.

4   Study Examples

In this section, we want to highlight some of the usability studies for public transport systems we conducted in our train compartment mockup. We describe each study and the lessons we learned.

4.1   Drawing Attention to an Interactive Window - An Eye-Tracking Study

A previous evaluation of our design showed us that our existing menu design for our interactive tram window is incomprehensible to some people [15]. A button with an index finger at the lower middle edge of the screen of the interactive window was implemented as a starting point. Tapping this icon opened a menu with further passenger information options.

Utilizing eye tracking technology, we wanted to verify these findings and to test and compare additional variants that were developed as alternatives. The goal of this study was to find a design that is quickly grasped and that attracts attention. We used the tram compartment mockup for this study, in order to give the participants the correct orientation towards the interactive windows and to incorporate the context of a public transport vehicle in their line of sight.

For this study, we invited participants of two age groups, in order to compare differences between the age groups of digital natives and digital immigrants. Digital natives are people born after 1980 who grew up with today's technologies, such as the internet, while digital immigrants are people born before 1980 [27]. Each group comprised twelve people.

In the study, four menu variants were compared including our initial version, as described above and three alternatives. For each type of menu design, we had three images showing different content, for example the geographic map.

The different variants were displayed in random order on the interactive window as non-interactive images. Each variant was displayed for 30 s. The participants were equipped with head mounted Tobii Glasses 2.0 and were asked to sit in the vehicle compartment and look at the display next to them. We installed video cameras and recorded the study on video.

After the study, we analyzed the eye tracking and video data. All fixations of all participants on one of our menu variants are shown in Fig. 3 as a heatmap. Each menu and content variant was evaluated in the same way. It is noticeable that the fixations of the digital natives are much more erratic and faster than those of the digital immigrants.

The evaluation in Table 1 shows that the time to first fixation is with 3.36 or 3.42 s by far the longest for variant 1. All other variants were already viewed after 0.89 to 1.87 s. A look at the total duration of all visits shows that variant 1 was with 2.31 respectively 3.28 s the variant that was looked at shortest. This indicates that the button is not conspicuous enough or that the function of the button is not visualized by the icon in a sufficiently understandable way. One variant involving an animated, moving hand attracted the most attention, being looked at for 12.81 s and 17.66 s in the user groups. With the exception of variant four, digital immigrants focused all variants longer.

We could reproduce the results of our earlier studies, that our first design prototype does not draw enough attention. In contrast, the animated version could attract a lot of attention. Especially when looking at the large surface of

Table 1. Time to first fixation and total visit duration for each design variant

	Time to first fixation (s)		Total visit duration (s)	
Design	Digital natives	Digital immigrants	Digital natives	Digital immigrants
1. Existing design	3,36	3,42	2,31	3,28
2. Moving hand icon	1,36	1,04	12,81	17,66
3. Icon with text	0,89	1,87	8,66	9,33
4. Preview	1,11	1,41	7,71	7,09

Fig. 3. A heatmap, displaying the fixations of all participants on one of the menu variants.

the interactive window, movements seem to be useful to arouse the interest of the train drivers. The use of text is also a useful option - Kukka et al. have shown that text animates more interaction than single icons [19]. We are currently working on improvements to our designs that incorporate these findings.

We are convinced that eye tracking is a good way to analyze the distribution of attention and gaze patterns and is very useful in the evaluation of designs for passenger information in public transport. Our laboratory setup ensures a high density of gaze points, in contrast to the application of eye tracking in field trials. On sunny days during field trials, the sunlight can interfere with the recording of the gaze points and lead to poor eye tracking data.

4.2   Usability Evaluation of an Interactive Window Prototype

The aim of this study was to identify potential usability problems of older persons (digital immigrants) abd digital natives in public transport passenger information systems. In this study, we used a clickable prototype for passenger information on interactive windows and a three-step evaluation. We implemented a usability test and used a standardized usability questionnaire and an additional individual interview afterwards.

The study involved 12 digital natives ($\varnothing 22{,}42 \pm 2{,}71$ years) and 12 digital immigrants ($\varnothing 64{,}17 \pm 10{,}74$ years). We were specifically interested in differences between the groups of participants. None of the test persons knew the interactive window prototype in advance.

Since a large proportion of tram passengers are older than 40 years, it is of particular interest to develop a design that is usable to this age group. Since people born before 1980 did not grow up with the current technology of, for example, digital displays, they have different mental models and behaviour which can negatively influence the usability of digital systems [27].

Many approaches deal with the usage motivation or fears of older groups of people. While researchers such as Czaja et al. concluded that older people have no interest in using new technologies, researchers in more recent studies often drew contrary conclusions [10]. Kurniawan or Page cite fear of failure as the main reason for non-use of digital technologies [20, 25]. Murata and Iwase name cognitive and physical impairments as the cause of these fears [24]. Typical usability problems of older people are also small font sizes and small icons and a lack of colour contrast. In addition, many older people have difficulty understanding the standardized icons of smartphones [21]. Foreign words and technical terms can also be an obstacle, especially in public transport [4].

In the course of the study, the test persons were given five typical tasks for the use of the passenger information on our interactive window, which were becoming increasingly complex. The tasks were result-based and could be completed with a gesture or a statement. At the beginning of the test, the respondents were asked not to speak during the tasks, since the tasks were timed. After the usability test, the participants were asked to fill out a standardized SUS questionnaire. Finally, the test persons were interviewed individually, with regard to any difficulties that arose during the test.

It can be summarized that the digital immigrants rated the protype worse than the digital natives. As shown in Fig. 4, digital immigrants required more time to complete the tasks than digital natives. The efficiency of digital natives is about twice as high as that of digital immigrants, regardless of their tasks. Although the digital immigrants handled the assignments much less effectively and efficiently than the digital natives, there are individual digital immigrants who achieved better results than the younger comparison group. This reflects that competence with interactive media can be learned throughout life.

In the interview afterwards, the digital immigrants reported that they felt insecure when using the system and found it difficult to solve the tasks. Nevertheless, the willingness to use is comparable to that of digital immigrants.

The evaluation of the individual questions provided important information about usability problems of our prototype and about possible solutions. In particular, the labelling of symbols seems to be unreliable for digital immigrants, whereas this is not a problem for digital natives. Furthermore, the survey showed that digital immigrants need information about internal system processes, to feel more secure while using the system. Despite the comparatively poor results during the usability test, the majority of digital immigrants shared the opinion that people can quickly learn the functionalities of the prototype. In order to make

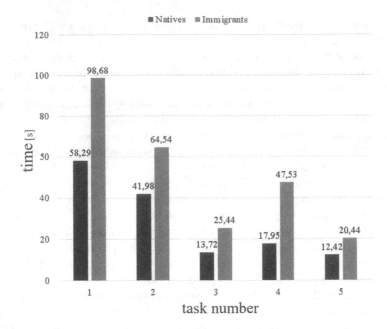

Fig. 4. Required time to solve the five variable tasks by the different user groups.

it easier for older people to get started, the implementation of a tutorial would be a useful aid.

4.3 User Study on Passenger Information in Disruption Situations

In another study we conducted in our mockup tram compartment, we examined passenger information systems specifically for disruptions in public transport. We developed a clickable prototype of an Android application that manages the itinerary of a user and is able to receive and display notifications. We developed notification schemes for disruption situations that are supposed to inform a passenger of their options, if they can't continue their planned trip due to detours or other events. Using a web-based interface, we could trigger predefined notifications that were delivered to the Android application. The notification's modality could be configured via the web interface. The application would use only the visual display of a notification, additionally use a ringtone or vibration to alert the user. In our study, we were interested in which modality the users preferred for urgent messages concerning their trip.

We additionally implemented a prototyical visualization of disruption information for the interactive windows in our tram compartment. The displays showed information on the tram line, next stops and a route map in normal mode, distributed on the available three displays. From the outside, we could then trigger the display of disruption information on each screen. On the small screen on top, the information on next stops could be changed, according to an imaginary detour of the vehicle. On the big screens, detail information on the detour and information about the changed routes on a route map, as well as detail information about connecting trains was shown. We wanted to determine, which device and modality would be preferred in different settings.

The study was conducted with 12 participants and they were split in three groups with four participants each. All participants had the same tasks in two different scenarios - to take the "tram" and to exit at the right stop, in order to reach the destination they were told to reach. One group of participants was given a smartphone with the Android application. For this group of participants, the displays only showed the regular information and no update information on the disruption. In many public transport vehicles, there either are no information displays, or they are not showing any realtime updates or information on disruptions which is simulated for this group. The second group was given the smartphone, too but when the disruption occurred in the scenario, the disruption information was not only shown via smartphone notifications and app, but also on the displays in the tram. The modalities of smartphone notifications varied for each participant that used a smartphone. A third group of participants only received disruption information on the displays.

The participants were asked to fill out a questionnaire on demographic data and their public transport usage at the beginning of their session. After they performed the tasks in the two scenarios, they were given a usability questionnaire and were asked about the information they received, the devices and modalities that were used. The study was filmed with a video camera, in order to detect and understand possible usability problems.

Interestingly, many of the participants could not correctly identify the modality of the notifications they received on the smartphone after they finished the tasks, a finding we will try to confirm in future work. Through observation and answers in the questionnaire, we could determine that most participants who used the smartphone dismissed the notifications very quickly. They also did not read information texts to the full extent, a fact that will influence our design of future applications for passenger information. Participants clearly favoured the smartphone over the display information. The study also revealed some usability issues of our design for disruption information that will be revised and fixed in our future work.

4.4   Pre-testing Evaluation Methods for a Field Test Concept

In our current research project on passenger information in public transport that includes the development of a working prototype of a semi-transparent, interactive display for public transport vehicles, we were able to test several design prototypes in our mockup tram compartment and gained valuable insight on the usability of several aspects of our prototypes. Additionally, towards the end of our project, a field test is planned. Two prototypical semi-transparent displays will be implemented in a tram of the local public transport provider. We are currently planning the usability evaluation of our system in the field and in order to do that, we conducted a study to test various evaluation methods in the mockup tram department. We wanted to determine the best suitable evaluation methods for subsequent studies in real public transport environments. Based on the results of this study, we developed our evaluation concept for the upcoming field test.

For this study, we used the mockup configuration that closely matches the configuration of a local tram. We arranged the seats as four, two facing the direction of travel and two opposite to the direction of travel, which will be the seat arrangement present in the vehicle that will be used in our field test. The semi-transparent displays will be installed next to such a seat configuration, in the same dimensions as our interactive displays are installed in the tram mockup.

As can be seen in Fig. 2, information for all passengers is displayed in the upper area, on the small screen. This screen displays the line number, destination, upcoming stops, transfer options, the arrival time at the following stations, the current time and weather, including a forecast. The two lower screens that will be semi-transparent in the field test are multi-touch enabled and, due to the seat configuration, can be operated by several people.

A network map and a geographical map can be accessed through the menu item at the bottom of the display. The geographical map is automatically centered on the location of the tram the user is currently sitting in. The maps can be explored interactively and display points of interest, other stations and additional detail information. On the opposite side of the tram department a 55 in. screen simulated the opposite window of the vehicle as shown in Fig. 5. During the study, a video shot along the entire route of the simulated tram line is shown on this display. The sound of opening and closing doors as well as the information displaying the next stops is synchronized with the video for consistent information.

Prior to the study itself, study participants were asked to fill out an online survey determining their technological affinity and public transport usage. Based on the survey results the participants were grouped into three typical public transport usage groups (commuters, occasional users and tourists), based on public transport personas. Three specific scenarios were designed for each group.

While participants classified as commuter were asked to complete tasks being situated in a simulated delay scenario, participants classified as tourists were faced with tasks associated with more touristic destinations and time irrelevant tasks.

A combination of pressed keys were used as a trigger for specific information displays on each display. This triggering approach allowed us to display different messages independent from each other and independent from time. The participants in different age groups between under 18 years and over 60 years, needed a little more time to familiarize themselves with the scenario and the interactive window, which is why the timing of additional information display was not synchronized for all participants. Even though this might lead to some discrepancies between the content of the notification and the timing of the video running in the background, the potential synchronicity error was classified as negligible. This approach of triggering events during a study by hand seems also more applicable

Fig. 5. Tram department with simulated opposite window.

for field studies, since the number of variables affecting time only increases when swapping the simulated environment for an actual moving vehicle.

The first step of the interaction with the interactive window, meaning the mere detection of a possible interaction came up as a potential usability issue throughout the study. Though the participants knew roughly what the study was about, some claimed that some sort of call to attention might be helpful to communicate the possibility to interact with the window. Due to the nature of the simulated tram and the overall concept of the studies, which have to be explained at least to a certain degree to the participants beforehand, it is difficult to measure the recognition of operability and the motivation to try it out. This should be easier to investigate further through field studies that will observe casual tram passengers, who sit beside the interactive window and may or may not use it unbeknownst to any study taking place, which is why we are planning general observations in our field test.

The results of this study informed our field test concept, but also showed the limitations and possibilities of our tram mockup. We are planning more usability studies on public transport systems in our mockup and are also planning to further develop possible applications of the mockup.

5 Outlook

In this paper, we presented a tram mockup as an evaluation environment for usability studies in public transport. We have developed and constructed a tram compartment mockup that mimics a real public transport vehicle. We have argued, that it presents an opportunity to conduct controlled usability studies in a lab, but also creates a more authentic context of use than a simple lab-based study. In contrast to a field test, several context factors can be controlled, such as public transport information (line, next stops, delays or disruptions), social situation (one person in a four-seat compartment versus four persons on four seats) or ambient conditions, such as loudness. In this tram mockup, it is possible to study early prototypes, while still maintaining a public transport atmosphere, as well as more mature prototypes.

We are planning several user studies in our tram mockup in the future and are additionally planning to further enhance the mockup itself. We will add original signage that will be provided by the local public transport provider and will further add to the realistic impression of our mockup. We have also discussed, to extend the mockp by speakers to mimic acoustic passenger information in the vehicle. Further plans involve installing additional displays, for example on a separating wall next to the handrails.

In our studies, we have determined that the current solution of an additional screen on the window above the two interactive screens ist not very usable, because, depending on the seating of a user, the screen can't be seen very well. We are therefore working on an alternative solution to display information for all passengers independently from the interactive displays. We are currently experimenting with a projection with a short-throw beamer for this task.

We are convinced that our tram mockup can improve our findings on the usability of passenger information systems for public transport in future studies. We are especially looking forward to comparing our results on the usability of our interactive, semi-transparent windows from our lab-based studies using the tram mockup to the results of our upcomping field test.

Acknowledgments. This work was carried out as part of the research project "Smart-MMI - model- and context-based mobility information on smart public displays and mobile devices in public transport" and was funded by the German Federal Ministry of Transport and Digital Infrastructure as part of the mFund initiative (Funding ID: 19F2042A). We would like to thank Johannes Bauer, Fabian Boschert-Hennrich, Kai-Lukas Schwägerl, Michael Wagner and Felix Werner for their contributions.

References

1. Abowd, G.D., Mynatt, E.D.: Charting past, present, and future research in ubiquitous computing. ACM Trans. Comput.-Hum. Interact. **7**(1), 29–58 (2000)
2. Alt, F., Schneegaundefined, S., Schmidt, A., Müller, J., Memarovic, N.: How to evaluate public displays. In: Proceedings of the 2012 International Symposium on Pervasive Displays (PerDis 2012). Association for Computing Machinery, New York (2012). https://doi.org/10.1145/2307798.2307815

3. de Amorim, D.M., Dias, T.G., Ferreira, M.C.: Usability evaluation of a public transport mobile ticketing solution. In: Ahram, T., Karwowski, W., Taiar, R. (eds.) IHSED 2018. AISC, vol. 876, pp. 345–351. Springer, Cham (2019). https://doi.org/10.1007/978-3-030-02053-8_53

4. Beul-Leusmann, S., Habermann, A.L., Ziefle, M., Jakobs, E.M.: Unterwegs im öv. mobile fahrgastinformationssysteme in der usability evaluation. In: Prinz, W., Borchers, J., Jarke, M. (eds.) Mensch und Computer 2016 - Tagungsband. Gesellschaft für Informatik e.V., Aachen (2016). https://doi.org/10.18420/muc2016-mci-0052

5. Bezerra, C., et al.: Challenges for usability testing in ubiquitous systems. In: Proceedings of the 26th Conference on l'Interaction Homme-Machine (IHM 2014), pp. 183–188. Association for Computing Machinery, New York (2014). https://doi.org/10.1145/2670444.2670468

6. Camacho, T., Foth, M., Rakotonirainy, A., Rittenbruch, M., Bunker, J.: The role of passenger-centric innovation in the future of public transport. Public Transp. 8(3), 453–475 (2016). https://doi.org/10.1007/s12469-016-0148-5

7. Chow, V.T.F., et al.: Utilizing real-time travel information, mobile applications and wearable devices for smart public transportation. In: 2016 7th International Conference on Cloud Computing and Big Data (CCBD), pp. 138–144, November 2016. https://doi.org/10.1109/CCBD.2016.036

8. Chowdhury, S., Giacaman, N.: En-route planning of multi-destination public-transport trips using smartphones. J. Public Transp. 18(4), 31–45 (2015)

9. Consolvo, S., Arnstein, L., Franza, B.R.: User study techniques in the design and evaluation of a ubicomp environment. In: Borriello, G., Holmquist, L.E. (eds.) UbiComp 2002. LNCS, vol. 2498, pp. 73–90. Springer, Heidelberg (2002). https://doi.org/10.1007/3-540-45809-3_6

10. Czaja, S., et al.: Factors predicting the use of technology: findings from the Center for Research and Education on Aging and Technology Enhancement (CREATE). Psychol. Aging 21(2), 333–352 (2006)

11. García, C.R., Candela, S., Ginory, J., Quesada-Arencibia, A., Alayón, F.: On route travel assistant for public transport based on Android technology. In: 2012 Sixth International Conference on Innovative Mobile and Internet Services in Ubiquitous Computing, pp. 840–845, July 2012. https://doi.org/10.1109/IMIS.2012.103

12. Habermann, A.L., Kasugai, K., Ziefle, M.: Mobile app for public transport: a usability and user experience perspective. In: Mandler, B., et al. (eds.) IoT360 2015. LNICST, vol. 170, pp. 168–174. Springer, Cham (2016). https://doi.org/10.1007/978-3-319-47075-7_21

13. Handte, M., Foell, S., Wagner, S., Kortuem, G., Marrón, P.J.: An Internet-of-Things enabled connected navigation system for urban bus riders. IEEE Internet Things J. 3(5), 735–744 (2016). https://doi.org/10.1109/JIOT.2016.2554146

14. Keller, C., Struwe, S., Titov, W., Schlegel, T.: Understanding the usefulness and acceptance of adaptivity in smart public transport. In: Krömker, H. (ed.) HCII 2019. LNCS, vol. 11596, pp. 307–326. Springer, Cham (2019). https://doi.org/10.1007/978-3-030-22666-4_23

15. Keller, C., Titov, W., Sawilla, S., Schlegel, T.: Evaluation of a smart public display in public transport. In: Mensch und Computer 2019 - Workshopband. Gesellschaft für Informatik e.V., Bonn (2019)

16. Kientz, J.A., Patel, S.N., Jones, B., Price, E., Mynatt, E.D., Abowd, G.D.: The Georgia Tech aware home. In: CHI'08 Extended Abstracts on Human Factors in Computing Systems (CHI EA 2008), pp. 3675–3680. Association for Computing Machinery, New York (2008). https://doi.org/10.1145/1358628.1358911

17. Kim, S.H., Kim, S.W., Park, H.: Usability challenges in ubicomp environment. In: Proceedings of the International Ergonomics Association (2003)
18. Kjeldskov, J., Paay, J.: A longitudinal review of mobile HCI research methods. In: Proceedings of the 14th International Conference on Human-Computer Interaction with Mobile Devices and Services (MobileHCI 2012), pp. 69–78. Association for Computing Machinery, New York (2012). https://doi.org/10.1145/2371574.2371586
19. Kukka, H., Oja, H., Kostakos, V., Gonçalves, J., Ojala, T.: What makes you click: exploring visual signals to entice interaction on public displays. In: Proceedings of the SIGCHI Conference on Human Factors in Computing Systems (CHI 2013), pp. 1699–1708. Association for Computing Machinery, New York (2013). https://doi.org/10.1145/2470654.2466225
20. Kurniawan, S.: Older people and mobile phones: a multi-method investigation. Int. J. Hum.-Comput. Stud. **66**(12), 889–901 (2008). https://doi.org/10.1016/j.ijhcs.2008.03.002. http://www.sciencedirect.com/science/article/pii/S1071581908000281. Mobile human-computer interaction
21. Leung, R., McGrenere, J., Graf, P.: Age-related differences in the initial usability of mobile device icons. Behav. Inf. Technol. **30**(5), 629–642 (2011). https://doi.org/10.1080/01449290903171308
22. Lyytinen, K., Yoo, Y.: Issues and challenges in ubiquitous computing. Commun. ACM **45**(12), 62–65 (2002). https://doi.org/10.1145/585597.585616
23. Mayas, C., Steinert, T., Krömker, H.: Interactive public displays for paperless mobility stations. In: Kurosu, M. (ed.) HCI 2018. LNCS, vol. 10902, pp. 542–551. Springer, Cham (2018). https://doi.org/10.1007/978-3-319-91244-8_42
24. Murata, A., Iwase, H.: Usability of touch-panel interfaces for older adults. Hum. Factors **47**(4), 767–776 (2005). https://doi.org/10.1518/001872005775570952. pMID: 16553065
25. Page, T.: Touchscreen mobile devices and older adults: a usability study. Int. J. Hum. Factors Ergon. (IJHFE) **3**(1), 65–85 (2014)
26. Poppe, R., Rienks, R., van Dijk, B.: Evaluating the future of HCI: challenges for the evaluation of emerging applications. In: Huang, T.S., Nijholt, A., Pantic, M., Pentland, A. (eds.) Artifical Intelligence for Human Computing. LNCS (LNAI), vol. 4451, pp. 234–250. Springer, Heidelberg (2007). https://doi.org/10.1007/978-3-540-72348-6_12
27. Prensky, M.: Digital Natives, Digital Immigrants Part 1, vol. 9, pp. 1–6. MCB University Press, Bingley (2001)
28. Reis, R.A.C., de Fontão, A.L., Gomes, L.L., Dias-Neto, A.C.: Usability evaluation approaches for (ubiquitous) mobile applications: a systematic mapping study. In: 9th International Conference on Mobile Ubiquitous Computing, Systems, Services and Technologies (2015)
29. Santos, R.M., de Oliveira, K.M., Andrade, R.M.C., Santos, I.S., Lima, E.R.: A quality model for human-computer interaction evaluation in ubiquitous systems. In: Collazos, C., Liborio, A., Rusu, C. (eds.) CLIHC 2013. LNCS, vol. 8278, pp. 63–70. Springer, Cham (2013). https://doi.org/10.1007/978-3-319-03068-5_13

Design Guidelines for the Simulation
of the Usage Context "Station"
in VR Environment

Regina Koreng[✉]

Technische Universität Ilmenau, Ilmenau, Germany
regina.koreng@tu-ilmenau.de

Abstract. Virtual environments are best suited for developing new information concepts for public transport. The paper deals with this possibility and examines immersion and presence for the two end devices CAVE and HMD. As an example a virtual system with dynamic and interactive elements is examined. The paper shows the relevant guidelines that a virtual environment has to fulfill in order to perform a meaningful test. For a realistic virtual environment, the travel chain and the individual elements in stations were examined in advance. Thus, a virtual model was created which contained real elements such as sound, avatars, information points, trains and ticket machines. The test persons were thus able to fully immerse themselves in the station.

Keywords: Design guidelines · Mobility Services · Virtual Reality

1 Introduction

Information is provided in public transport through the use of various media and systems. The train station represents an important hub in the travel chain. It's not only regarded as a departure and arrival point, but also serves as a point of orientation for changing trains. The passenger should be provided quickly with all important information on various digital media, such as public displays, dynamic train destination displays and available on the smartphone too. For this the Mobility Services concepts are regularly revised and prepared in a user-centered manner, which so far can only be evaluated in complex field test. In order not to paralyze the entire operating process with integration and testing of new concepts, they can also be tested in VR (virtual reality).

For the user's point of view, there are many factors that influence and promote the immersion and presence of passengers in VR environment, especially in the "station" usage context. These are summarized in a catalog with design guidelines for the "station" usage context. In addition, it is analyzed whether the design guidelines are equally valid for the CAVE (cave automatic virtual environment) and HMD (head mounted display) end devices. This results in the following research question for this paper:

Which design guidelines have to be fulfilled by VR environment for the evaluation of mobility services in the usage context "station"?

© Springer Nature Switzerland AG 2020
H. Krömker (Ed.): HCII 2020, LNCS 12213, pp. 267–281, 2020.
https://doi.org/10.1007/978-3-030-50537-0_20

2 Mobility Services

The paper deals with the test environment "station". The travel chain should have its focus on public transport and passenger information. [1] Not only print media have to be considered, but also the various digital distribution channels available to the passenger during the journey, Especially, with the digital information path, it is important to remember that the information is available on the move and is therefore always up-to-date for the passenger. [1] For the user, the different stages of the journey are coupled with different information needs. [2] During the planning and before the start of the journey, the user can be at any location. The want to receive information on travel times, prices and route. Afterwards, they go to the station, which they reach by mean of signposts and guidance information. **Before starting the journey**, the user decides which vehicle they want to use to cover their route. This requires information that they need to plan their journey, such as a clear route, times and price. This information should be easily accessible and confirmed or correctly indicated, as incidents during the journey are an obstacle. This basic information is relevant at the beginning of the journey, so that detailed planning can provide specific content. For a high level of immersion and presence, the station should receive this information so that the user can obtain information at the station. [1, 2] If the traveler is then **on his journey**, they need rather continuous information about the course of the journey. The user has to be able to determine their current position during their journey. Further announcements can be made directly at station, for example by means of train destination displays, display board and loudspeaker announcements. The same announcements can also be displayed directly in the means of transport. The dynamic display enables the user to react to possible delays or disturbances. In this way they can search for alternative routes. As shown in Sub-Sect. 3.1, the virtual station is both visually and audibly relevant for the user. The VR environment thus adapts to the real user experience and lets the traveler immerse themselves in the station. During the journey or possible transfer situation, the passenger is in a means of transport or at a corresponding station. In order to find their way around the respective station, the user need vehicle or stop information to help them find their way. [1, 2] At the **end of his journey**, the passenger needs information for the destination station. Signposting systems and maps of the surrounding area are particularly important here [1, 2].

The station plays a central role during the journey with public transport. All real elements for information retrieval at the station should be implemented and extended by real additional objects. In VR, the traveler should feel as if they were in a real train station. The paper deals with the virtual implementation of this mobility service and the design guidelines for a successful evaluation.

First, the specifics of VR test environment should be worked out and then these should be mapped in an application.

3 VR as Test Environment

3.1 VR

"Virtual Reality refers to immersive, interaction, multi-sensory, viewer-centered, three-dimensional computer generated environments and the combination of technologies required to build these environments." [3].

For the design of virtual test environment "station" the same criteria apply as for general applications in VR. For the traveler, the virtual usage context should be immersive and interactive. The paper deals specifically with the station, such as waiting areas, route and travel information as well as.

Information on means of transport. Basically, the sense of the human being are to be addressed, which are relevant in the context of the execution of mobility service tasks. The main goal is to give the user the feeling that they interact in a real environment. [4, 5] An important part of the research is to analyze the specific factors that ensure immersion and presence in the mobility environment.

Immersion refers to the immersion in the VR environment. In the process, the reality of the recipient is overlapped. According to Rittmann [6], the user is immersed on a medium in which the real world is combined with the virtual world and merges with it. Immersion describes the necessary technical requirements to enable a complete fusion. Other sources use the metaphor of swimming in water. The feeling of being completely surrounded by a matter, that is not air, can be compared to the merging with the VR. [5–7] In the literature on VR research it is specified even more precisely. Thon [7] differs between perceptive and psychological immersion. Perceptive immersion means that reality is covered by the sensory perception. Of the created environment, so that a new reality is given to the recipient. In psychological immersion, attention is shifted from the real world to the media content of the virtual world. In contrast, Rittmann [6] differs the term according to the type of presentation. Immersion can result from the creation of a space that represents VR in the media. Secondly, immersion can be made possible without predefined images by immersing the user in VR through the acoustic perception of texts. The imagination of the recipient creates individual images. The environment into which the viewer is to be immersed does not have to correspond completely with the real world. What is required, is a credible and logically accessible atmosphere, because people find it easier to immerse themselves in a virtual world if they do not have to constantly question it. Furthermore, the laws of physics have to apply. In addition, an environment with obvious gaps in narration or representation leads to a logical break that limits or even prevents immersion. [6, 7] Slater and Wilbur [8] highlight another immersion concept. They consider the possibilities of technology to integrate the user into the scene in such a way that they forget their real surroundings. These can be divided into different categories:

- inclusive: The ability to fade out reality enough so that the generated reality is exclusively computer generated. [8]
- extensive: Number of senses addressed. As many senses as possible should be addressed simultaneously. [8]
- surrounding: The extent that describes the viewing angle. The imaginable range lies between a narrowly limited field of vision and complete panoramic view. [8]
- vivid: The measure of the liveliness and detail of the model [8].

With these four classes, Slater and Wilbur [8] realize different stages of immersion, so that a classification from low- to high-immersion media is made. Immersion is a subjective perception that is felt differently by each person. [3, 8] To achieve this, the scenic representation "station" is chosen. As an important feature, a logical and for the recipient credible VR is created, which allows a separation from the outside world. [6] In order to create a realistic usage context for the travelers, real-time functionality, interactivity and immersion are of particular importance [5].

Immersion is the requirement that the traveler to feel a high **presence** in the context of use. A shift in spatial perception results, which is achieved by the user's complete involvement in VR. [5, 9] Dörner et al. [3] show that presence happens when a high degree of immersion is achieved. In this case the person acknowledges the VR environment and accepts the corresponding technical requirements. However, it is relevant that they can control the VR themselves and interact with the individual objects and subjects in the scene. [3] Lombard and Ditton [9] make an additional distinction, speaking of a "perceptual illusion of non-mediation". They assume that the interface, the connection between people and the system disappears or transforms into a social unit. Accordingly, presence is indirectly measures by the extent to which the environment is excluded. Consequently, the recipient's imaginative capacity in the specially constructed world and their connection to it can be regarded as presence. [9] In order to make this possible, the user need to be willing to put themselves in the virtual world and allow the imaginary world. The state of presence can only enter when the representation of the scene as a VR world is recognized and accepted. Many stimuli are addressed that allow the user to interpret the scene. In the usage context "station" not only the visual stimuli are addressed, by a realistic remodeling of the environment and the integration of avatar. For the auditory stimuli, background noises and real announcements are integrated. This space is created to consciously direct the intention of the user. Additional interpretations that lie outside those if the designer are prevented. The presence is influenced by the biological, psychological and cultural experience of the person and implies basic user experiences in the projected area. [5, 6, 9, 10] Biocca [10] also describes presence by means of various criteria. He differentiates between physical, virtual and imaginary environment, whereby the environment is regarded as the buzzer of these three levels. Accordingly, Slater and Wilbur [8] see presence as a psychological state. In this state, the virtual environment should be closer to the recipient than the real world around them. This placing in the artificial environment implies that the VR corresponds to the experiences of the user. The presence is an interplay of subjective and objective representations, which should not contradict the personal experiences of the user. This means that the operation and actions in the virtual world correspond to those in the real world. The degree of immersion of the recipient in the virtual environment is decisive for the subjective representation [8].

The analysis of the subjective presence is carried out by evaluating questionnaires in which the recipient expresses their feelings in the virtual situation through yes/no questions. [9] In the objective representation, the realistic actions of the recipient are examined. The objective presence thus describes the probability of successfully solving the set task. This may overlap with immersion. For the orientation of the user it's important that there is a realistic description of VR, but this increases the probability of immersion. Thus, the scene is no longer viewed only from the outside, but also from the inside. In principle, a distinction is made between the application- and task-specific context. It's mainly measured on the basis of observations [3, 8–10].

Table 1 summarizes the different statements on immersion and presence of the above mentioned authors. For the usage context "station" it becomes clear that immersion is caused by the user being able to put themselves in the environment. For this purpose, the room should be designed realistically and contain elements that exist. In the case of a station, for example these are navigation instructions, information displays, ticket machines or pedestrian flows. This should also be reflected in the objects, textures and operation in the VR environment. Further possibilities for this is the integration of avatars. Visual stimuli are enhanced by other factors during presence. A station is characterized by its specific sounds, which should be reflected in the virtual world. Important factors for immersion are the creation of space, no gaps in the narration in the travel chain at the station, the psychological immersion through the objects relevant in the travel chain as well as the typical sounds of a station. On this basis, the presence comes about through the recognition of the VR environment, the test person's imagination performance leads to the fact that they are closer to the virtual than to the real world.

Table 1. Comparison of literature analysis of immersion and presence

Authors	Immersion	Presence
Rittmann [6]	• Immersion in the virtual environment • Technical requirements, for a comprehensive implementation • Space generation with physical laws	• Influenced by the biological, psychological and cultural experience of the person
Rademacher et al. [5]	• Merging with matter like "swimming in water"	• Shift in spatial perception • Recognition of the VR environment and the technical requirements • Influenced by the biological, psychological and cultural experience of the person
Thon [7]	• Merging with matter like "swimming in water" • Distinguishes between perceptive and psychological immersion • Logical break: an environment with obvious gaps in the narration	
Slater et al. [8]	• To integrate the user into the scene so that they forget their real environment • Subjective perception, which is felt differently by each person	• Psychological state: recipients should be closer to the virtual than the real world • Interplay of subjective and objective representation
Regenbrecht [9]		• Shift in spatial perception • Recognition of the VR environment and the technical requirements • Perceptual illusion of non-mediation • Imagination performance of the recipient

3.2   CAVE vs. HMD

The technical implementation of VR plays a significant role in the presence. The VR can be realized in different end devices. The paper deals with a CAVE as well as an HMD. The CAVE is a multi-sided projection screen, which can vary in its position. By parallelizing different PC components, projectors and the tracking system can be connected to the VR input devices and the audio rendering. [11] With an HMD, a mobile visualization and interaction system directly on the head. A much smaller display generated an image that is displayed to the user n a larger format. [3] Table 2 shows the comparison of the two devices. Based on this comparison, criteria are extracted that are relevant for a high level of immersion and presence in VR.

Table 2. CAVE vs. HMD

	Pro	Contra
CAVE	• Large FOV/FOR; higher resolution per square-degree of visual angle [12, 13] • Makes use of peripheral vision [13] • User's entire visual system is stimulated [12] • Promote a very natural physical and visual environment [12] • Real and virtual objects easily mixed in VR application [13] • Multiple user can be in a CAVE at the same time [12]	• Users still must wear glasses [12] • Typically limited to one active tracked viewer [13] • Requires large amount of physical space; interactive devices generally awkward and spatially inaccurate [12, 13] • Loses immersion whenever they look up or down [14] • Physical/virtual object occlusion problem [13] • Can be expensive [12, 13]
HMD	• 360-degree FOR; higher resolution [12, 13] • Immersion: be better, since VR is visible all around the subject [14] • Portable; smaller footprint [12, 13] • No physical/virtual object occlusion problems [13] • Unlimited tracked number of users in stereo (one display per user) [13] • Much more affordable entry into VR, inexpensive [12, 13]	• Medium FOV compared to surround-screen display; limited peripheral vision [13, 14] • Not afford interaction with "real" objects aside from control devices [12] • Natural interaction within the virtual space [12] • Unable to see other users or the real world directly [13] • Have to wear the device; weight and cables cause ergonomic issues [13]

Table 3 shows that the two technologies have no obvious advantages or disadvantages for the implementation of virtual station test environment. In the following, a comparative usability test is therefore to be carried out, in which the respective presence is measured in relation to the related technologies.

4 Method

For the investigation of design guidelines in VR environment with the usage context "station", the method of Collaborative Usability Inspection (CUI) and Comparative Usability Testing (CUT) is applied.

To ensure a high ergonomic quality of VR environment, a heuristic evaluation in form of CUI with travelers, mobility and usability experts was carried out. [15]

Collaborative Usability Inspection of the Virtual Station
During the investigation, users express their thoughts about the system out loud. The various survey groups express themselves independently of each other without commenting on each other. Individual screens of the software, the navigation between them, their relevance for task fulfillment and the fault tolerance of the user interface are examined. During the evaluation, the participants take in different roles to ensure efficiency and effectiveness. [15]

Table 3. Advantages and disadvantages of CUI [15]

Advantages	Disadvantages
• Display of station screen displays; cheaper and less time consuming • Involving travelers and experts with different perspectives and knowledge • Developers can use the knowledge gained during the evaluation for the usability improvement of software, even beyond the current project • (…)	• Disagreements between the travelers and experts can't be processed • No precise indication or measures of the increase in effectiveness compared to related methods • Numerous travelers and experts are involved, so that their availability and motivation must be ensured • (…)

Comparative Usability Testing of the Virtual Station
The comparative study is carried out during the evaluation of the systems. A comparison is made between the individual display variants and between the two systems CAVE and HMD. [15] The respondents evaluate and compare the different systems in terms of the criteria of immersion and presence, their strengths and weaknesses (Table 4). [15]

Table 4. Advantages and disadvantages of CUT [15]

Advantages	Disadvantages
• Evaluation of station variants by features test persons before independent implementation • Creating added value by identifying new features desired by test persons in a station • Strengths of features in the virtual station model are adopted and doubtful features are not implemented • (…)	• Extensive station scenarios require more time and resources than testing a single user interface • If existing station concepts are adopted by other, this can lead to "bad design replication across many different websites" [15, p. 252] • Comparability of results may be impaired in the station model if identical tasks are not compared • (…)

The black box method is used for the evaluation. [9] The cognitive processes of the user's imaginative performance from the black box by which immersion and presence can be measured. The feeling of being present at a certain place can be presented as a purely psychological measure. The immersion factors can be measured objectively. According to Regenbrecht [9], there are three measurements that allow data collection for psychological variables: observation of expressive behavior, recording of physiological indicators and recording of subjective experience. Two of the methods are used for the investigation with the virtual station.

- observation of expressive behavior
  During the interaction in VR, the respondent is observed by the expert. Their movements, actions, facial expressions and gestures are analyzed.
- recording of the subjective experience
  In written form, subjective values can be recorded before or after a usability test with a questionnaire. Orally this can be done by an interview after the test.

5   Conception

5.1   Criteria of Evaluation

Immersion and presence are the key concepts of this paper. As a basis for the mobility services to be created and for the subsequent usability test, a summary of the criteria of immersion and presence into categories is helpful. For both characteristics, four higher categories emerged, which are to be retained in the evaluation (immersion => Table 5; presence => Table 6)

Table 5. Summary of the criteria of immersion in four categories

Category	General guidelines	Station specific guidelines
Sensory impressions	• User is located in the middle of the simulated world • Multimodal stimulation of human perception (auditory and haptic) • Kinesthetic, objects work and function realistically in VR • Spatial situation (obstacles) • Dynamic changes	• A comprehensive station model supports the natural FOV of the users • Virtual station model should be realistically reproduced • Natural motion sequences, for example in virtual ticket machines
Basic requirements	• Large FOV • Perspective and stereoscopic vision • Motion parallax • Basic human abilities (perception of form, objects and depth) • Shadow projection • Symptoms of fatigue	• Human models strengthen the feeling of reality in the station • Virtual integration of station noises and audio announcements
Navigation/interaction	• Natural, intuitive interaction • Use of objects that are used in reality • Computer-generated; computer system that recognizes the actions of the user	
Hide environment	• Direction of view and position detected by input device • Realistic, attractive visual design • Hide real environment • Relevant, plausible contents • Output devices should completely surround the user	

The first group includes all sensory impressions. The immersion should address as many sensations (visual, auditory, haptic) of the users in the station as possible. To give the user the feeling of being part of VR, all object and elements in the station should look and function realistically. Furthermore, every person has an inner sense of balance, which is decisive for the balance in the virtual environment (kinesthetic). As a second category, the basic requirements include elements that enable immersion. The station model should be created in such a way that perspective vision, perception of form, objects and depth are implemented as well as motion parallax. In order for the station model to be spatially classified, attention should be paid to occlusions and shadow projections. The third category navigation/interaction includes aspects of the technical view of VR. In the virtual world, immersion is characterized by the fact that all elements are computer-generated. The interaction with the station model should be natural and intuitive. In order to achieve this, the objects used, such as the ticket machine, should be based on reality and be available to the user in the real world. The last group, hide environment, includes the criteria that should be fulfilled to enable the user to completely hide their real environment and achieve a high level of immersion. The output devices should completely surround the user and the input devices should be recognized by their position. The interaction of the respective input device with the station scene shall consider the current FOV of the person into account and adapt to it. Likewise, the content in the station model of the VR should be relevant and plausible for the test person.

These listed criteria intensify the immersive feeling in the user, but can be individually different. For the implementation of the station model, they were taken into account and asked of the respondent during the usability test.

Table 6. Summary of the criteria of presence in four categories

Category	General guidelines	Station specific guidelines
Being in VR	• Subjective feeling of being in VR are in a place, the body is part of a defined space • To be able to move freely in VR; self-controlled movement • Manipulate persons and objects • Interaction with objects and subjects • Impression of the room and the number of users	• The user wants to get involved in the virtual station model • The user accepts that all objects of the station are created virtually • Influenced by the biological, psychological and cultural experience of the users at the station • The user must be able to manipulate and interact with the station objects in VR • The user can move freely and independently in the station model
Imagination	• Behavior as in the real world; result of a cognitive process • Willingness to provide an imaginative service; perceiving media mediation • Real and virtual stimuli match • Motion simulation	
Basic Requirements	• Immersion as a prerequisite • Unrestricted view • High technical quality of input and output devices • No real ambient noise	
User experience	• Experiences of the user • Sum of the real, virtual and imaginary environment • Projectable properties predominate (recorded by sensors)	

The first group includes the factors that create the feeling of being in VR. It describes the subjective feeling of being psychologically in a station while the body is moving in a defined examination space. In addition, the user should be able to move freely in the VR as well as interact with the ticket machines and information. The interaction is self-determined, the user can move alone in the station scene.

Table 7. Listing of the information concept and the variable components

Concept: System with dynamic/interactive elements

Elements	
static	none
dynamic	<ul><li>clock (real speed)</li><li>hall signs (change is time-depending)</li><li>train destination displays (change is time-depending)</li><li>guide strips (change is time-depending)</li><li>persons (change is time-depending)</li><li>sound (background noises, announcement)</li><li>trains (change is time-depending)</li></ul>
interactive	<ul><li>information terminals (individually operated, travel information)</li><li>ticket machine (individually operated, travel information, train ticket)</li></ul>

An important factor is the imagination of the user. The willingness of the person to engage with the virtual station model is a relevant basis for increasing the feeling of presence. In addition, there is the willingness to perceive and ignore the mediation. In the third category, basic prerequisites are listed that are not directly related to the VR. Immersion is thus a basic for the feeling of presence. However, it can also lead to the user feeling part of the station environment and not looking at it from the outside. As with immersion, a high technical quality of the input and output devices as well as a large FOV is a factor that increase the sense of presence. At the same time, external influences such as real ambient noise are a real draft should be prevented. In the last group the user experience is combined. These criteria play a significant role, since the feeling of presence can be understood as the sum of the real, virtual and imaginary environment. The personal experience of the users in a station and thus the transfer situation is thus decisive for the mental evaluation of VR.

As a unit, the various factors distinguish presence from immersion. However, the subject's own experience is more strongly integrated and are taken into account accordingly in the evaluation.

5.2 Display Concept – Dynamic/Interactive System

Different alternative workflow variants and scenarios for the interaction of the virtual usage context "station" are developed. These range from an almost static to a highly dynamic representation. As an example, Table 7 shows the concept of a dynamic and interactive system.

The mobility information concepts vary in terms of their functionality. The system with dynamic/interactive objects represents a completely active model. Static elements are not used throughout the station. For the acquisition of information, the hall board, train destination displays and guide strips are available to the passenger as time-dependent components. A public display and a ticket machine are provided as further options. The user can independently and specifically train information about a train connection through interaction with the vending machine. The ticket machine also enables the passenger to purchase a train ticket. This concept uses other people, different sound variations and departing trains to create a realistic environment.

6   Evaluation

Test Design
The paper examines guidelines for the implementation of a VR environment in the context "station". For a high level of immersion and presence, the factors from the above sections must be observed and implemented. As can be seen in Sect. 3, there are different ways to collect data. For the query of perception, the test persons are asked four questions about the mobility service in the virtual usage context "station".

1. How realistic was your impression of the virtual station?
2. How appealing did you find the people who were integrated into the station?
3. How intuitive did you find the interactions you were able to carry out in the virtual station?
4. How did you feel during your tour of the virtual station?

The virtual usage context "station" creates a test environment that can be used flexibly in early phases for the evaluation of mobility services. The paper shows the development of design guidelines for the context "station" for the end devices CAVE and HMD. These design guidelines ensure a high degree of immersion and presence during the test.

Test Persons
To achieve a high level of immersion and presence, it's important to virtually integrate user experiences and expectations. With this in mind, we were looking for test persons who are experiences and users of the mobility service. Both everyday users and exceptional users were interviewed. There are no restrictions with regard to experiences in the field of VR, both test persons with and without experiences participate.

Test Execution
Various test series are being carried out to investigate the two terminals. The test persons are evaluated independently of the tests. After a short introduction to the aim of the test, the respondent is given the opportunity to interact with the terminal device in a practice situation. Subsequently, the test is carried out according to the methods in Sect. 4 and the results are noted down both by means of a questionnaire and by observation. The test persons have no time limit and can look around the station and analyses train connections.

7 Results

7.1 CAVE

During the study in the Cave 15 persons were interviewed. Enclosed is the evaluation for the system with dynamic/interactive objects.

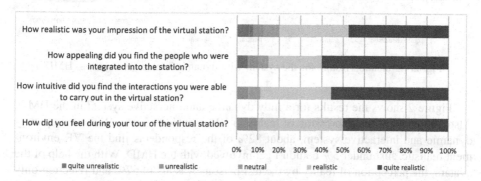

Fig. 1. Evaluation results for the system with dynamic/interactive objects – CAVE

Figure 1 shows the results for a fully dynamic and interactive system in CAVE. First impression of reality in VR is evaluated. In the case of a completely dynamic and interactive system, about 80% of the respondents find VR environment realistic. Only 12% couldn't get involved in VR in the CAVE. With the help of the avatars, the perception of reality among the test persons was 85% and rejection only 10%. Furthermore, in Fig. 1 the possibility of interaction is evaluated. At this point, the test persons also rated the VR environment as realistic with ca. 90%. And only with 5% as unrealistic.

The results from the evaluation with the CAVE shows that the test persons were able to put themselves in the environment well to very well due to the realistic representation, the ambient noises as well as the avatars and interaction possibilities. The possibility to immerse oneself in the scene is a sign of high immersion among the test persons. A high level of presence can also be evaluated, as the test persons consciously dealt with the interaction possibilities.

7.2 HMD

During the study with the HMD 44 persons were interviewed. For comparability with the CAVE, the system with dynamic/interactive objects is also evaluated.

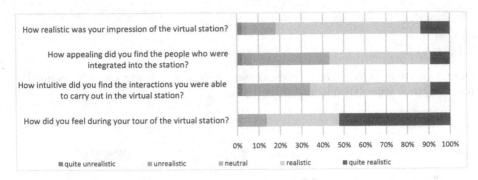

Fig. 2. Evaluation results for the system with dynamic/interactive objects – HMD

Figure 2 shows the results for a fully dynamic and interactive system in the HMD. Also, at this point the impression of reality in VR is evaluated first. With a completely dynamic and interactive system, about 82% of the respondents find the VR environment realistic. Just under 5% couldn't get involved with the HMD. With the help of the avatars, the perception of reality by the test persons was about 57% and rejection only 7%. The large number of neutral voices reduce the positive evaluation here. The rejection is significantly lower than with the CAVE. Finally, in Fig. 2 the possibility of interaction is evaluated. At this point, the test persons also rated VR as realistic with circa 66%. Only with 5% as unrealistic.

The results from the evaluation with the HMD shows that the test persons were able to put themselves in the environment well to very well due to the realistic representation, the ambient noises as well as the avatars and interaction possibilities. Even, if some respondents abstained with a neutral evaluation, the number of negative evaluations shows that there is an improvement in contrast to the CAVE.

The possibility to immerse oneself in the scene is a sign of high immersion among the respondents. A high presence can also be evaluated, as the test persons have consciously dealt with the interaction possibilities.

8 Summary

The paper deals with the research question which design guidelines for the evaluation of mobility services in the usage context "station" fulfill. This is done in a VR environment using a CAVE or HMD. A CAVE describes a projection room with at least three screens. With an HMD, however, a display is placed directly in front of the eyes. The usage context of the paper refers to train station. These are the central connection point during the journey. For the integration of new elements, an evaluation in VR is useful, as this avoids work stoppages at the station. For a successful evaluation, VR must have a high degree of immersion and presence. The paper shows different criteria that the virtual mobility services should contain. In the following, both VR are tested with a test scenario in a short evaluation. The evolution is then performed in relation to a complete dynamic and interactive system. In the evaluation CAVE vs. HMD, it becomes clear that if the evaluation criteria are met, the test persons will receive a high level of immersion and presence. Both systems are suitable for the use of evaluations of mobility services. Only the negative votes show an advantage with the HMD. By attaching the end devices to the head, the test persons feel more integrated in the VR.

References

1. Daduna, J.R., Voß, S.: Informationsmanagement im Verkehr, pp. 7–9. Physica-Verlag, Heidelberg (2000)
2. Mattfeld, D.C., Schneidereit, G., Suhl, L.: Dynamische Fahrgastinformation im Öffentlichen Personennahverkehr-Grundstrukturen und kundenorientierte Funktionalitätsanforderungen. In: Informationssysteme in Transport und Verkehr, pp. 47–69. BoD - Books on Demand (2006)
3. Dörner, R., Broll, W., Grimm, P., Jung, B. (eds.): Virtual und augmented reality (VR/AR), vol. 46, pp. 13–14. Springer, Heidelberg (2013). https://doi.org/10.1007/978-3-642-28903-3
4. Preim, B., Dachselt, R.: Interaktive Systeme. Band 2: User Interface Engineering, 3D-Interaktion, Natural User Interfaces, 2nd edn. Springer, Heidelberg (2015). https://doi.org/10.1007/978-3-642-45247-5
5. Rademacher, M.H., Krömker, H., Klimsa, P.: Virtual Reality in der Produktentwicklung. Instrumentarium zur Bewertung der Einsatzmöglichkeiten am Beispiel der Automobilindustrie. SM. Springer, Wiesbaden (2014). https://doi.org/10.1007/978-3-658-07013-7
6. Rittmann, T.: MMORPGs als virtuelle Welten. Immersion und Repräsentation, pp. 47–48. Verlag Werner Hülsbusch, Boizenburg (2008)
7. Thon, J.-N.: Immersion revisited. Varianten von Immersion im Computerspiel des 21. Jahrhunderts. In: Medien - Zeit - Zeichen. Beiträge des 19. Film- und Fernsehwissenschaftlichen Kolloquiums, pp. 125–132, Marburg (2006)
8. Slater, M., Wilbur, S.: A framework for immersive virtual environments (FIVE): speculations on the role of presence in virtual environments. Presence: Virtual Augmented Reality 6, 603–616 (1997)
9. Regenbrecht, H.: Faktoren für Präsenz in virtueller Architektur, pp. 29–31, 73–74. Bauhaus Universität Weimar, Weimar (1999)
10. Pietschmann, D.: Das Erleben virtueller Welten Involvierung, Immersion und Engagement in Computerspielen, pp. 42–43. Verlag Werner Hülsbusch, Boizenburg (2009)
11. Bente, G., Krämer, N.C., Petersen, A.: Internet und Psychologie. Neue Medien in der Psychologie. In: Virtuelle Realität, vol. 5, pp. 2, 10. Hogrefe-Verlag GmbH & Co. KG, Göttingen (2002)
12. Havig, P., McIntire, J., Geiselman, E.: Virtual reality in a cave: limitations and the need for HMDs? Proc. SPIE Int. Soc. Opti. Eng. 3, 1–6 (2011)
13. LaViola Jr., J.J., Kruijff, E., McMahan, R.P., Bowman, D.A., Poupyrev, I.: 3D User Interfaces. Theory and Practice, 2nd edn, pp. 172–174. Addison-Wesley, Munich (2017)
14. Mestre, D.R.: CAVE versus head-mounted displays: ongoing thoughts. In: The Engineering Reality of Virtual Reality 2017, pp. 31–35. Electronic Imaging (2017)
15. Künnemann, L.: Entwicklung eines kontextabhängigen Instrumentariums für Usability-Evaluation, pp. 213–217, 250–254 (2019)

UI Proposal for Shared Autonomous Vehicles: Focusing on Improving User's Trust

Minhee Lee[✉][iD] and Younjoon Lee[iD]

Graduate School of Industrial Arts, Service Design,
Hongik University, Seoul, Korea
minhee@mail.hongik.ac.kr, younjoonlee@hongik.ac.kr

Abstract. The automotive industry is rapidly evolving into automated vehicles by integrating cutting-edge technologies. This paper focuses on shared autonomous vehicles that are currently highly feasible in terms of commercialization and proposes a proper UX design that improves the factors that hinder the formation of trust in the user's shared autonomous vehicles experience. The research method is largely divided into three processes. The first is to review the literature to address the importance of trust-building in the user's autonomous vehicle experience and to derive sub-factors to evaluate it. The second is an empirical study, in which the participants are given an indirect experience of watching shared autonomous vehicles service video. This study has academic implications in that it has found that the formation of a user's trust is an important factor for a user to accept new technology and showed that the factors that form trust in autonomous vehicles must be identified from various angles. When people think of fully autonomous vehicles, they are still hesitant and not completely comfortable. This is a critical point of how important it is to gain user trust in developing and simulating autonomous vehicles. In this respect, this study has academic implications in that it has found that the formation of a user's trust is an important factor for a user to accept new technology and showed that the factors that form trust in autonomous vehicles must be identified from various angles.

Keywords: Automated vehicles · Shared autonomous vehicles · User trust · Design for user trust · User Interface

1 Introduction

1.1 Research Background and Goal

The automotive industry is rapidly evolving into a connected vehicle and an automated vehicle by integrating cutting-edge technologies such as ICT, sensors, and satellite navigation. As focusing on the realization of fully autonomous vehicles or fully shared autonomous vehicles, most of the previous studies have dealt with the development of driving technology and the infrastructure to which it can be applied. Yet the goal of this paper is to identify the cognitive and emotional responses experienced by the user in the process of shared autonomous vehicle experience from the perspective of trust

© Springer Nature Switzerland AG 2020
H. Krömker (Ed.): HCII 2020, LNCS 12213, pp. 282–296, 2020.
https://doi.org/10.1007/978-3-030-50537-0_21

formation and to present the UI design for enhancing the user's trust in the shared autonomous vehicle service.

1.2 Research Question

In order to achieve the research goal, the research questions are set as follows.

RQ 1. What are the factors that affect negatively the user's trust in the ride experience of shared autonomous vehicles?

RQ 2. How should the UI of shared autonomous vehicles be designed to mitigate the negative factors?

2 Theoretical Reviews

2.1 Autonomous Vehicles

Autonomous vehicles are cars that recognize the surrounding environment and determine the route and risk factors even if the driver does not directly control the steering wheel, brakes, or accelerator pedals. The role of the driver is replaced by sensors, semiconductors, and software. Taxonomy and definitions for terms related to driving automation systems for on-road motor vehicles [1] were released by Society of Automotive Engineers (SAE). Although it is not included in the SAE standard, "Autonomous Vehicles (AV)" is also frequently used in various research areas including media reports. Based on this fact, this paper adopts the term "Autonomous Vehicles (AV)" which focuses more on the concept and utilization rather than the technology of the system.

2.2 Trust

Trust has been considered as a major determinant of acceptance of new technology [2, 3]. A common description of trust in Lee and See [3] is that trust plays a key role in shaping the attitude toward trustee; the object of trust, which is new technology.

Sub Factors of Trust. There are many studies that derive trust-building factors regarding various services and technology. Lee and See [3] divided dimensions that describe the basis of trust which have been investigated differently by various researchers. The researchers categorized trust attributes derived from studies dealing with automation and organizational relations according to the three factors that constitute trust presented by Lee and Moray [4]. The key basis of trust is summarized by the three dimensions as follows (Table 1).

The dimensions of trust related to technology were divided into performance, process, and purpose. According to the researchers, the basis of performance was the competence, ability, and expertise of technology in various studies. The basis of the process was predictability, accessibility, understanding, availability, reducing uncertainty and confidentiality. Finally, the basis of purpose was intention and motives, benevolence, loyalty, and faith.

Table 1. Factors describing the basis of trust

Performance	Process	Purpose
Competence	Persistence	Fiduciary responsibility
Ability	Integrity	Loyalty
Functional/specific competence	Consistency	Intention
Interpersonal competence	Openness	Motives
Business sense	Discreetness	Motivation to lie
Judgment	Predictability	Benevolence
Expertise	Accessibility	Concern
Reliability	Availability	Faith
Congeniality	Understanding	Generalized value congruence
Context-specific reliability	Willingness to reduce uncertainty	Leap of faith
Trial and error experience	Confidentiality	Fiduciary responsibility

3   Research Methods

This study consists of two processes; first, the process of deriving research questions through the Empathy map technique and second, the process of finding solutions through the co-creation workshop.

3.1   Empathy Map

Definition of Empathy map. Empathy Map (EM) means drawing the user's level of empathy for an object, product or service [5]. It starts with the premise that if an operator or service provider understands the consumer, even small design changes can have a big impact on the consumer. This method helps to design business models from the consumer's point of view, goes beyond demographics and gives a better understanding of the consumer's environment, behaviors, aspirations, and interests [6].

In the first version of EM, four different areas were addressed when creating an EM of a person; think & feel, hear, see, and say & do. Since then, it was improved including the Pain and Gain areas. As a result, EM consists of six areas as below [5] (Fig. 1).

See. what the user sees in their environment

Say & do. words and actions – the way the users say and act

Think & feel. thoughts and feelings – what happens to the user's mind

Hear. how the environment affects the user

Pain. frustration, pitfalls, and risks experienced by te user

Gain. what the user actually wants and can do to achieve the goal.

Experimental Procedure. In the experiment, participants indirectly experienced Waymo, a shared autonomous vehicle service by Google. At this time, the participants

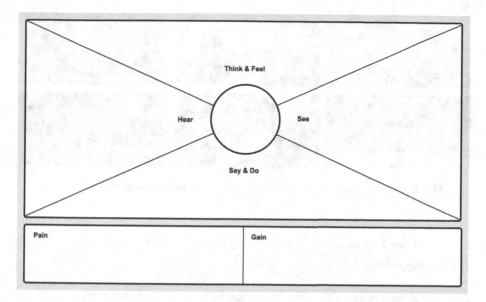

Fig. 1. Template of the EM [6]

encounter the shared autonomous vehicles for the first time, and this process aims to collect various responses of the participants to the shared autonomous vehicles service. The experimental procedure for developing an empathy map is as follows.

First: Watching a Video. Participants watch a video of a prototype service of shared autonomous vehicles in the first-person view. In order to derive the response of the participant related to trust, it was determined that the degree of sophistication and the quality of the prototype that the participant indirectly experienced would influence the outcome. As a result, the researcher adopted a prototype service video of a representative company, which is open to the public and actively develops shared autonomous vehicles services, without implementing a prototype (Fig. 2).

Second: Creating an Empathy Map. After viewing the video, the participant creates the first Empathy Map. This process is divided into five steps for the user to book and board a shared autonomous vehicle, and then create an empathy map according to each step. The five steps of using the shared autonomous vehicles service are as follows (Table 2).

Third: Imaginary Technique. Imaginary techniques are used to induce participants a realistic commitment to shared autonomous vehicles services. Based on the indirect experience through the video provided above, the participants are asked to imagine a rainy situation in the city center of Korea where they currently reside. In order to materialize the situation and collect additional responses from the participants, the researcher lets the participant imagine a busy and complex road condition when using the shared autonomous vehicles service. Participants imagined the situation where the

Fig. 2. Images of shared autonomous vehicle in the video viewed by participants

Table 2. User journey of experiencing the shared autonomous vehicles

Step no.	Step division	Description
1	Calling a vehicle	User requests a car service from the app
2	Waiting	While the user is waiting, he or she can check the vehicle's movement on the map in real time. The shared car service app ensures that the requested vehicle is correct before boarding for safety
3	Boarding (Confirming and getting in the vehicle)	The user recognizes his or her face in the mobile app to confirm that he or she is a passenger in the assigned shared autonomous vehicles
4	Moving	–
5	Getting off	Upon arrival, the user pays with a credit card stored in the app account or with online cash

traffic jam occurred, and only the THINK & FEEL, PAIN & GAIN were written because the specificity of the situation was limited.

Last: Discussion. The researcher asks the participants to share their opinions freely. After gathering enough feedback, the experiment is completed.

Participants Information. There were 8 participants with an average age of 33 (SD = 1.05). In order to gather in-depth information, the mobile device UX designers were selected. Therefore, they know the definition of a technology called autonomous vehicles. All the participants have experienced driving. The preliminary test inquired whether the participants know and trust the company that provides shared autonomous vehicles services in the video presented in the experiment.

When asking participants opinions about the video in this experiment, the research topic of trust was not specifically mentioned. The reason for this is to confirm that trust is an important issue in the acceptance of participants' shared autonomous vehicles services. It is also to measure how much comments mentioning trust are submitted voluntarily (Table 3).

Table 3. Participants information

	Participants information
Age	• The average age of the 8 participants was 33
Job	• Title: Mobile UX designers • Career: the average years of career is 8.2
Prior knowledge on autonomous vehicles	• I know the definition of autonomous vehicles • I know how the stages of autonomous driving are distinguished • I know how far autonomous vehicles are currently being developed • I know an shared autonomous vehicle service is being developed * Recruited only those who answered yes to all of the above questions
Attitude on the company	• I know a company (Google and Alphabet) That offers a prototype of the shared autonomous vehicles service in this experiment • I have trust in the company presented in this experiment

3.2 Co-creation Workshop

In the 2nd step, a workshop is conducted by collecting participants with work experience in the field of autonomous vehicles. The second workshop focuses on collecting the participants' professional views. Based on their views and the industry trends, the goal of the second workshop is to create feasible UI of shared autonomous vehicles service that involves all participants.

The co-creation process means designing with others. Others here may mean experts in other fields or non-design experts [7]. Emerging issues on design practices more focus on 'designing for a purpose' rather than 'designing of products' [8]. The UX design of autonomous services also applies to the views described by Sanders and Stapper. From the point of view of the formation of consumer trust, the UX design of shared autonomous vehicles service needs to solve a problem through co-creation. Co-creation empowers the whole group of people, who are determined to participate in design together, to help designers in their creative activities through discussion.

4 Results

4.1 The Result of the 1st Experiment

Empathy Map Results. From the first experiment with 8 participants of mobile UX designers, an Empathy Map was drawn according to the six factors of EM. The results are presented as follows according to the sequence.

SEE. At the exterior of the vehicle, the participants carefully looked at LiDAR (radar), an element not found in existing vehicles. Inside the vehicle, they mainly saw operation buttons and display devices that can be operated by the user. The operation buttons consist of the functions that the user can use in an emergency situation. The participant thoroughly examined the configuration of the emergency button.

The display device provides different functions for each situation. When driving, the display device provides real-time modeling of navigation (maps, landmarks), surrounding vehicles, and pedestrians. It also shows traffic infrastructure including signs and real-time signal information and digitized blinking signals. This is more complicated and the amount of information to be delivered is much more than the navigation device of the existing vehicles. When not driving, the display provides interaction elements, such as announcements and buttons, for users' use of the service.

SAY & DO. Since the participants boarded alone in the shared autonomous vehicles, the behavior of Say never appeared. However, there was an expectation for voice interaction in later discussions (5 participants). Boarding-moving-getting off, which is directly related to autonomous vehicles, has been discussed as the key User Journey. Among the core tasks, "moving", many participants performed the following tasks.

Checked the display device information (8 participants)

Watched the outside scenery (8 participants)

Observed the internal elements such as emergency button, camera, steering wheel, etc. (8 participants).

It was shown that all - participants focused on driving-related information displayed on the display device. All of the participants answered that they were curious about the modeling of vehicles and pedestrians around them in real-time. Accordingly, the participants compared and confirmed in real-time whether the display information and the information outside the car are accurate (6 participants). Behaviors of the participants that find the emergency button and verify the function of the emergency button can be inferred as a prerequisite to knowing the response in a hazardous situation.

In the case of cameras, 3 participants said they wanted to verify that the camera was working, rather than simply verifying that the camera was in place.

HEAR. The voice announcement was recognized by 7 participants, but there was no further discussion because it was similar to the navigation application of the existing vehicles. As for the sound, many participants said that it was very quiet overall and small sounds were noticeable.

"I would be sensitive to the sound from the car until I get used to the autonomous cars."
(Participant 3)

"Because this car is very quiet, the blinking sounds are loud." (Participant 4)
"I can hear a small mechanical sound inside and outside." (Participant 5)

Interpretation of the vehicle sound differed from participant to participant. While one participant mentioned it as a characteristic of an electric vehicle (Participant 4), others expressed anxiety about having a louder mechanical sound than a normal engine vehicle (2 participants). While the participants did not respond very much to the information that they could infer, they had a tendency to feel anxious about the machine due to the differences from their existing experience. In addition, there was a discussion regarding the blinking sound, and all participants were aware of the blinking sound. Four said the blinking sound felt loud. Participant 5 said it was awkward to hear the blinking sound without lights being turned on. Three participants mentioned that no blinking is needed.

THINK & FEEL. In discussing the thoughts and feelings, many different opinions of the participants were expressed, but mostly negative thoughts or feelings. All participants judged that the driving of the shared autonomous vehicles was immature and incomplete.

The results of the participant's discussion by dividing the Think & Feel into negative feelings, positive feelings, curiosities, concerns about specific situations that did not occur, wishes, and others were as follows.

First, regarding negative feelings, although no abnormalities occurred during driving, all participants considered the driving ability of shared autonomous vehicles to be incomplete. Also, participants imagined various edge cases or specific situations that did not occur and were concerned about how shared autonomous vehicles would handle those kinds of imaginable dangerous situations. Many commented that it was too slow and frustrating (7 participants). In addition, the comments on self-moving steering wheels and flashing were negative. Some commented that the overall atmosphere or user environment was either quiet or dry. Some participants naturally recalled autonomous driving accidents encountered in the media (3 participants). Two participants said they wanted to sit in a seat with a steering wheel, which reveals their anxiety. Some female participants expressed fear of taking an unmanned shared autonomous vehicle alone on a dark night.

"I think it would be scary to ride alone at night."(Participant 3)
"When a car enters a dark and quiet road at night, it's scary to see me alone in the car through the glass."(Participant 5)

Second, regarding positive feelings, most of the participants felt safer and more comfortable with no vehicle driver (7 participants). This was mentioned by all female participants. At the same time, however, four female participants also spoke about the contradictory feeling of riding alone (mentioned above in negative feelings). Three participants mentioned the freedom of being "alone." Finally, four participants commented positively on proficiency in parking. All participants tended to mention negative emotions early in the journey but then decreased in frequency.

Third, regarding curiosities, although all participants knew the concept of autonomous driving, it was their first time experiencing it. They were curious about the first-ever devices they have seen (formally LiDAR on the top and Radar on the front)

(6 participants). In addition, there were questions in various contexts. This is due to the fact that there is no information other than the basic concept of a vehicle driving itself.

Fourth, regarding concerns about specific situations, all participants imagined various anxious situations. Basically, all the participants had a doubt about, "Can the autonomous vehicle skillfully respond to various unexpected situations?". In addition, specific issues that were previously recognized in other technical fields, such as hacking, personal information leakage, limitations of digital devices, and incompleteness of AI and machine learning, were specifically mentioned.

Fifth, regarding expectations and other else, expectations for voice interaction were high (5 participants), but there was also an opinion that it was unlikely to use voice interaction because a display device was provided (1 participant). Two participants mentioned expectation for precise rider's location tracking. Some participants expected to improve their driving guidance.

PAIN. In discussing the thoughts and feelings, many different opinions of the participants were expressed, but mostly negative thoughts or feelings. All participants judged that the driving of the shared autonomous vehicles was immature and incomplete.

The pain point was selected from all the PARTICIPANTS' anxiety and constant curiosity. Many participants pointed out the feeling that the shared autonomous vehicles were too stable.

"The door opens too late and frustrating." (Participant 2)
"I'll arrive later than in a non-autonomous car." (Participant 5)
"For Elderly people, it will be difficult to maneuver." (Participant 8)

GAIN. Riding comfortably alone was selected as the main advantage (4 participants). Two participants answered that they could do - other tasks instead of driving in a shared autonomous vehicle. The comfort factor for the inexperienced driver such as children and the elderly are also mentioned as an advantage.

Imaginary Technique Results. In the imaginary technique, as in the Empathy map, participants' opinions were collected on the requirements of Seoul's complex urban environment on a rainy night with heavy traffic jams. Participants recorded feelings centered on THINK & FEEL.

As a result of the analysis, all the participants had distrust about driving stability. In particular, there was a high level of concern about failing to board with the identical-looking cars waiting on the side of the road. All eight participants commented that seamless riding would be difficult. The distinction between vehicles was also perceived as difficult (7 participants).

There were also negative comments about the sophisticated operation of autonomous driving systems. All of the test participants expressed trust in the company Google, and even though they are engaged in the IT industry, they expressed anxiety about the sophistication and quality of driving.

Analysis of 1st Experiment at Touchpoints Aspect. Based on the SEE, SAY & DO, and HEAR results of the Empathy Map, the hindering factors of trust for each touchpoint were analyzed (Table 4).

Table 4. Touchpoints of Autonomous Vehicle that affect trust

Touchpoint	Number of participants who perceived	Contents hindering trust formation	Categorization
Mobile app	6	–	Mobile
Door	2	Button typed door	Unlocking, security
Display device	8	The sophistication of real-time modeling	Technology of autonomous vehicles
Handle	8	Moving by itself in the air	Design of Steering wheel
Camera	7	Whether it shoots me	Personal information, security
Speaker	2	Personalization and sentiment of voice messages	Voice Interaction
Digital blinker	7	Sound and design of blinker	Interior design
Emergency button	8	Existence of emergency buttons and its functions	Button in case of emergency

The Mobile Application. All the participants did not feel special emotions because it is similar to the experience of using the app of the existing shared vehicle service or the shared vehicle service.

Door. It is necessary to examine whether the button typed automatic door is a good method in terms of security and safety. Subsequent paragraphs, however, are indented.

Display Device. All participants focused on this device. However, real-time modeling of surrounding objects, traffic signals, and flickering features added with navigation looked too complicated to interfere with the participant's understanding.

Handle. Operating alone in the air gave most participants a sense of rejection. Research is necessary for terms of formative aspects and movements.

Camera. There was a negative opinion in terms of personal information leakage. Participants also expected emotional voice interaction instead of touch displays.

Digital Blinker. There were negative opinions about the operation principle, sound, and design. Further research is needed to optimize interaction.

Emergency Button. The emergency button was the touchpoint that passengers wanted to check first. Participants wanted to know the location and function of the emergency button to ensure their safety. Therefore, in order to resolve the user's anxiety, further discussions will be conducted in the 2nd workshop considering the improvement.

Analysis of 1st Experiment at User Context Aspect. Based on the THINK & FEEL, PAIN & GAIN results and IMAGINARY TECHNIQUE results, User contexts that undermine trust were derived (Table 5).

Table 5. User context of autonomous vehicles affecting trust

Type of user context	Number of participants who perceived
Doubts about sophisticated driving skills	8
Lack of information on how to handle an emergency situation	8
Whether a response scenario exists for a corner case	8
Unable to understand some parts of the system, curiosity	8
Driver's absence	5

Participants were all suspicious of the driving ability optimized for the context of shared autonomous vehicles. In addition, there was no detailed information on the basic driving ability (visual, reaction speed) of the vehicle, and thus participants feared an accident. In other words, since no data on actual performance was provided, trust factors related to performance could not be formed.

In terms of process, participants also questioned a lot of principles and situations that they faced for the first time. This made it impossible for the participants to fully understand the system and hindered the formation of predictability. Subsequent paragraphs, however, are indented.

In terms of purpose, participants were anxious because they could not know any information about the vehicle's ability to respond in the event of an accident.

Collectively, the above factors were hindering the building of trust in the service of shared autonomous vehicles (Table 6).

Table 6. Sub factors of trust which is not formed via 1st experiment

Performance	Process	Purpose
Competence	Predictability	Fiduciary responsibility
Ability	Understanding	Leap of faith
Functional/specific competence	Confidentiality	
Interpersonal competence		
Reliability		
Congeniality		
Trial and error experience		

4.2   The Result of the 2nd Experiment

Regarding contents derived as the main issues affecting user trust formation through the first experiment, people indirectly related fields such as autonomous driving, digital cockpit for the connected car, mobile, project owner, project manager, software engineer, hardware engineer were selected for the second process of this study.

The workshop was conducted in two groups with four people in one group. After discussing the contents that emerged as the main discussion points in the first experiment, future scenarios were prepared. The workshop for the first group took place from 18:00 to 20:00 on November 1, 2019, and the workshop for the second group took place from 18:00 to 20:00 on November 8, 2019. Information on the workshop participants is as follows (Table 7).

Table 7. Information of the workshop participants

	Group	Field of job	Position	Company
P1	A	Autonomous Driving Simulation S/W	Project owner	Automotive Artificial Intelligence GmbH, Germany
P2	A	Connected Car	UX Designer	Samsung Electronics, Korea
P3	A	Connected Car	UX Designer	Samsung Electronics, Korea
P4	A	Mobile	UX Designer	Samsung Electronics, Korea
P5	B	V2X (Vehicle to Everything) Communication	Project Manager	Samsung Electronics, Korea
P6	B	V2X (Vehicle to Everything) Communication	H/W Engineer	Samsung Electronics, Korea
P7	B	V2X (Vehicle to Everything) Communication	S/W Engineer	Samsung Electronics, Korea
P8	B	Connected Car	S/W Engineer	Samsung Electronics, Korea

Co-created Scenarios. Voting and discussion were conducted for participants in the 2nd Workshop on the factors that hinder trust formation in each touchpoint and contexts derived in Experiment 1. The discussions of the semi-professional groups who participated in the 2nd workshop on each element are as follows.

Touchpoint 1. Handle. Six out of eight participants agreed that if the handle is out, the user could feel the burden of driving and confusion about whether it could be driven. So they agreed to the internal mounting of the handle of shared autonomous vehicles.

Touchpoint 2. Emergency Button. All participants agreed that the emergency button is the focus of the users' attention and that the optimal location is important. In particular, the location of the button should be in consideration of the children or patients. 5 out of

8 agreed on the central position. All participants agreed that pressing the button should lead directly to troubleshooting.

Touchpoint 3. Display Device. Generation Z, the main customer of shared autonomous vehicles, is expected to be familiar with complex digital information, so most participants agreed to keep the current state of the display device.

Touchpoint 4. Door. The door lock should be opened after the identity verification is completed. This is a shared car, so it is impossible to apply the biometrics of each customer. Most of the participants agreed to be able to reliably complete the self-authentication with a mobile device owned by the individual.

Touchpoint 5. Camera. All workshop participants agreed to partial filming and limited streaming of users. In other words, if the guardian or the users ask to turn on streaming and turn on streaming in an emergency, but the users should be well informed.

Touchpoint 6. Voice Interaction. Since the voice guidance is awkward in the absence of the driver, most participants agreed to convey the friendly personality to the user through the personification of the service through voice guidance.

Touchpoint 7. Digital Blinker. The sound of the digital blinker is unnecessary but must be visible to the user and the outside.

User Context 1–3. Concerns. Participants in experiment 1 expressed doubt about the sophisticated driving capabilities of the shared autonomous vehicles. In addition, they were concerned about whether there was a corresponding scenario for the corner case. The workshop participant P1 said, "This anxiety was caused by the participants' lack of understanding of based technology of autonomous driving. "The response speed, judgment speed, and vision of the autonomous car are much better than those of humans."

Participant P8 said, "Even though they all trusted the company Google and all work in the IT field, everyone expressed anxiety because they had some knowledge of autonomous driving. Because it is a life-threatening task, they think the shared autonomous vehicle service should reflect the latest technology, and the complexity of the transportation system should be perfectly internalized.

User Context 4. Unable to Understand Some Parts of the System, Curiosity. Participants had at the same time comfort from the absence of the driver and anxiety that they were in a self-driving vehicle that was locked alone. Users should be familiar with the fact that they will eventually be provided with a complete response to emergency situations, which, like conventional aircraft, can be accessed such as guidance at the start of boarding.

User Context 5. Absence of a Driver. Participants felt comfortable in the absence of a driver but simultaneously were anxious to be alone in a locked autonomous car. Users should be aware that they can cope with emergencies perfectly. The solution to this problem can be approached by referring to the announcement before boarding the plane.

5   Conclusions

5.1   Research Summary and Implications

This paper focused on the process of user's trust formation in the design of autonomous vehicles and aimed to derive critical points of trust formation that enables users to accept and continue to use autonomous vehicles. In detail, focusing on shared autonomous vehicles that are currently highly feasible in terms of commercialization, a whole process of booking, boarding and getting off of shared autonomous vehicle is defined and trust formation points in each process were derived. Through this process, UI design scenarios were suggested that improved the factors that hinder the formation of trust in the user's autonomous vehicles experience.

When people think of fully autonomous cars, they are still hesitant and not completely comfortable. This is a critical point on how important it is to gain user trust in developing and simulating autonomous vehicles. In this respect, this study has academic implications in that it has found that the formation of a user's trust is an important factor for users to accept new technology and showed that the factors that form trust in autonomous vehicles must be identified from various angles.

5.2   Limitations and Suggestion for Future Research

This study has limitations in that it does not directly use the completed actual shared autonomous vehicles service and indirectly conducted the experiment through video in the first-person view. Future work needs to experiment with user experience in technologically advanced environments.

Previous studies mention that virtual experience testing has often been used as a practical approach to measuring confidence levels in new technologies. However, simulation tests alone cannot answer all consumer questions, such as software failures, bugs, and abnormal behavior, and inevitably face gaps in the actual driving environment. For example, in a simulated test, the driver can not be sleepy, but during long drives, the human driver can be sleepy [9].

In addition, there is a limitation in that it is not possible to derive trust formation according to various user environments in that the passenger condition is not classified in detail. The user has various differences in the degree of involvement in driving, driving ability and understanding of the traffic system. In future studies, various conditions of passengers can be subdivided into the study.

For citations of references, we prefer the use of square brackets and consecutive numbers. Citations using labels or the author/year convention are also acceptable. The following bibliography provides a sample reference list with entries for journal articles [1], an LNCS chapter [2], a book [3], proceedings without editors [4], as well as a URL [5].

References

1. SAE J3016 automated-driving graphic. (n.d.). Accessed 24 Feb 2020. https://www.sae.org/news/2019/01/sae-updates-j3016-automated-driving-graphic
2. Choi, J.K., Ji, Y.G.: Investigating the importance of trust on adopting an autonomous vehicle. Int. J. Hum.-Comput. Interact. **31**(10), 692–702 (2015)
3. Lee, J.D., See, K.A.: Trust in automation: designing for appropriate reliance. Hum. Factors: J. Hum. Factors Ergon. Soc. **46**(1), 50–80 (2004)
4. Lee, J., Moray, N.: Trust, control strategies and allocation of function in human-machine systems. Ergonomics **35**(10), 1243–1270 (1992)
5. Ferreira, B., Silva, W., Oliveira Jr., E.A., Conte, T.: Designing personas with empathy map. In: SEKE, vol. 152 (2015)
6. Osterwalder, A., Pigneur, Y.: Business Model Generation. Alta Books, Rio de Janeiro (2013)
7. 정예경. (2017). Co-design 관점의 장소브랜딩 연구: 폐철도부지 도심공원을 중심으로. 이화여자대학교 대학원, 석사학위논문.
8. Sanders, E.B.N., Stappers, P.J.: Co-creation and the new landscapes of design. Co-design **4**(1), 5–18 (2008)
9. Hussain, R., Zeadally, S.: Autonomous cars: research results, issues, and future challenges. IEEE Commun. Surv. Tutor. **21**(2), 1275–1313 (2018)
10. Etherington, D.: Waymo focuses on user experience, considers next step (2017). https://techcrunch.com/2017/10/31/waymo-self-driving-ux/

Employees' Vulnerability – The Challenge When Introducing New Technologies in Local Authorities

Ann-Marie Nienaber[1,3], Sebastian Spundflasch[2(✉)],
Andre Soares[1,4(✉)], and Andree Woodcock[1(✉)]

[1] Coventry University, Coventry, UK
{ann-marie.nienaber,andre.soares,
andree.woodcock}@coventry.ac.uk
[2] Technische Universität Ilmenau, Ilmenau, Germany
sebastian.spundflasch@tu-ilmenau.de
[3] University of Münster, Münster, Germany
[4] Nicolaus Copernicus University, Torun, Poland

Abstract. While it is well-known that the implementation of new technologies requires appropriate technical capabilities, research has for a long time almost neglected the behavioural capabilities of organisation's employees to adopt innovative technologies. Employees have to trust new technologies and thus, to be willing to become vulnerable when they adopting it as they have to cope with something they are not familiar with. This paper highlights the challenge for local authorities to cope with employees' unwillingness to become vulnerable when it comes to implementing new technologies in local authorities. Based on semi-structured interviews that have been conducted under the umbrella of the European project SUITS, we were able to identify two indicators for the unwillingness of employees to adopt new technologies - attribution of negative motives and incongruence of values. Furthermore, we show best practise examples how to overcome the negative consequences of the unwillingness to become vulnerable and to be able to implement new technologies successfully in the long-run. Our practical implications in the end are derived by the experiences when introducing new technologies in the partner cities of the SUITS project.

Keywords: Distrust · Local authorities · New technology adoption · Trust ·
Vulnerability · Mobility planning

1 Introduction

Since years local authorities are faced with a variety of political and societal requirements and restrictions in relation to future sustainable mobility planning. On the one hand, the mobility field has undergone significant changes in recent years and is becoming increasingly complex, for example numerous innovative forms of mobility and service providers entered the market. On the other hand, citizens' mobility requirements changed enormously. Due to recent trends such as 'Fridays for Future', citizens require sustainable thinking and prefer resource efficient ways of travelling.

© Springer Nature Switzerland AG 2020
H. Krömker (Ed.): HCII 2020, LNCS 12213, pp. 297–307, 2020.
https://doi.org/10.1007/978-3-030-50537-0_22

Many of these changes and challenges are associated with the use of new technologies for local authorities. In recent years, the term Smart City has gained high popularity and the mobility sector benefits greatly from new technologies [e.g. 1, 2]. Local authorities started for example to collect status data by innovative sensor technology to allow for better service organization (e.g. public transport, parking management) to meet the citizens' requirement on the one hand but also to be able to use the data for analysis, evaluation and further development of mobility services. However, the implementation of these new technologies has become a major challenge for local authorities, in particular for small and medium sized cities, which do not have the same amount of time, men/women power and budget like bigger cities. Challenges associated with the use of new technologies threaten public sector employees in a number of substantive ways and make them feel more vulnerable. The willingness to become vulnerable is urgently required when people have to cope with technology they are not familiar with as they have to have the positive expectations that this new technology will not harm them in the long-run [e.g. 3]. These two aspects, having positive expectations towards a new technology and the willingness to become vulnerable when adopting and actual using new technologies, are the key elements of trust [4, 5]. Employees in local authorities have to trust new technologies when local authorities want to implement them successfully [2].

When employees are not willing to become vulnerable and may not share positive expectations with a new technology, it has been shown that it is very likely that distrust may occur. Following Bijlsma-Frankema and colleagues [6] distrust is a psychological state, comprising the unwillingness to accept vulnerability, based on negative perceptions and expectations. Thus, distrust is in our context the unwillingness by individuals to become vulnerable and the expectation that a technology may harm.

In the literature distrust has been connected to a variety of negative consequences such as a lack of cooperation [7], the avoidance of interaction [8], or the unwillingness to share knowledge or information [6]. Distrust can therefore be conceptualised as an unrecognised and neglected hazard that derives from feelings of vulnerability, which consequently stifles knowledge-exchange and relationship building between parties. Thus, tackling distrust is crucial for forecasting people's attitudes and behaviours and therefore for accepting and adopting new technologies in local authorities [e.g. 9].

This paper highlights the relevance of employees' willingness to be vulnerable when local authorities want to implement new technologies successfully. Thus, we present a unique approach to hazard and resilience when introducing new technologies for future mobility planning in local authorities by using research from psychological and business-related scholars to re-imagine them as issues of distrust, understood via attention towards lived experiences of vulnerability. We adopt a sociotechnical approach to change in our paper. This approach is based on the sociotechnical systems theory that recognises the importance of behavioural change when implementing technological innovation [e.g. 10, 11, 12]. We want to outline how distrust as key obstacle was overcome by different public authorities to cope with the required behavioural change in the frame of the SUITS project.

Based on semi-structured interviews with different local authorities in Europe we will present best practise examples how to enable local authorities to reduce employees' vulnerability, enhance their employees' resilience and foster organisational innovativeness. Herewith, we contribute decisively to research but also to management in the field of public administration. The structure of the paper is the following. We will start to outline the theoretical foundation of vulnerability when trying to introduce new technologies in an organisation. Afterwards we will describe the negative consequences of the unwillingness to become vulnerable by local authorities employees and the development of distrust before we show examples of how four European local authorities managed to overcome the unwillingness to become vulnerable and to avoid or minimize distrust inside the organisation successfully. Finally, we want to describe clear best practises along different case studies from the field that show the outcomes and the learnings for other local authorities in Europe and worldwide.

2 Theoretical Perspective

2.1 Internal Challenges for Local Authorities

The implementation of new technology is constantly bringing new challenges for local authorities, in particular in relation to future sustainable mobility planning. Local authorities have to become more effective and resilient to new technologies and/or simply new ways of working. However, most change programmes that focus solely on technological or/and technical change and thus, highlighting primarily the importance of trainings and seminars to enhance employees' abilities to cope with these innovations, are still ignoring the importance of social and behavioural aspects, and end up by failing [e.g. 3, 13].

One of the key reasons for failing might be the individual's vulnerability when it comes to the implementation of new technologies. Indeed, employees' vulnerability has been overseen by researchers and practitioners for many years [5]. So far vulnerability has typically been mentioned in relation to trust in the management and psychological literature since scholars in these fields see the willingness to be vulnerable as one core aspect when defining trust, but not in relation to the introduction of new technologies in organisations. Rousseau et al. [14] define trust as the individual willingness to be vulnerable based on positive expectations that another party will not take advantage of this vulnerability. In relation to the implementation of new technologies, trust has to be defined as the individual willingness to be vulnerable based on positive expectations towards the new technology and its benefit for the individual. While the first key element of the trust definition "the willingness to be vulnerable" has been identified as the rather affective side of trust, the second key element "positive expectations" is called the cognitive side of trust [5].

To show the development of the concept of vulnerability, we firstly have to refer to the United Nations [15] which describes how general categories of factors determine a community's level of vulnerability. Beside this rather macroeconomic perspective on vulnerability, we align with the findings of Nienaber and colleagues [5] that one of the most dominant streams of research on vulnerability can be found in medical sciences. Here, vulnerability describes an individual's inability to protect and maintain her/his interests [16]. In context of sociological factors, Chambers [17] explained vulnerability by two indicators: external threats and a lack of internal coping mechanisms. While the external threat can be described as the implementation of new technologies in organisations, the complexity of potential service providers (stakeholder) and technological solution for becoming a 'smart city', the latter one is of key interest as this lack of internal coping mechanisms may be driven by the unwillingness to become vulnerable and missing positive expectations towards the new technology.

To address the affective side of trust in terms of willingness to be vulnerable, we further refer to sense-making theory that suggests that risky experiences such as unfamiliar situations are characterized by negative feelings in the form of disorientation or foreignness. Following Weick and colleagues [18], the key question related to sense-making is 'same or different?' As long as something seems to be similar to something well known already, individuals perceive it as less risky and thus, rather related to positive feelings. Meaning, when a new technology has to be implemented in a local authority and it is perceived as rather familiar by the local authorities' employees as it can be connected to something well known already, the implementation and the actual use of the new technologies is very likely. Whereas, something that is not known and absolutely new, might be very likely perceived as threatening and thus, employees do not want to become vulnerable and thus, are not going to actual use it with the consequence that the implementation will fail. While the majority of research in this area focuses on how awareness of a specific situation is formed and categorized, and how these processes influence individual's actions, little attention has been paid on the introduction of new technologies in an organisation in particular.

2.2 Employees' Unwillingness to Become Vulnerable and It's Negative Consequences for the Introduction of New Technologies

In line with trust research and recent definitions on distrust, the unwillingness to become vulnerable is one key element of distrust next to the expectations that – here – new technology may be harmful. Distrust research has gained enormous attention in psychology in recent years [e.g. 19, 20]. Research shows since years impressively the negative consequences of distrust that will harm organisations when trying to implement new technologies. Scholars highlighted for example: diminished cooperation [21, 22] or the avoidance of interaction [6, 8, 23]. The rationale for the reduced willingness or unwillingness to cooperate as an effect of distrust is that it results from an accumulating and reciprocal diminished willingness to act cooperatively. Avoidance is another documented effect of distrust and refers to attempts to reduce or prevent future harm. Furthermore, scholars were able to identify less knowledge sharing in organisations, enhanced levels of knowledge hiding and increasing amounts of conflicts inside an organisation [24]. All these consequences will become real obstacles when trying to introduce new technologies in local authorities. The unwillingness to interact with the new technology and the unwillingness to act cooperatively with the top management of a local authority will hinder a successful implementation of new technologies as the employees will not adopt and use the new technology in the end. Even worse, employees may perceive the new technology as harmful and thus, distrust may flourish inside the organisation and lead to a distrustful climate in the whole local authority and herewith affect all levels inside the organisation but also maybe spread to the wider stakeholder relationships (pervasiveness of distrust [6]).

3 Empirical Analysis

3.1 Method and Sample

Our data was gathered via in-depth semi-structured interviews each lasting around one to two hours. Our sample consists of four local authority partners in Europe, comprising the local authorities and their wider stakeholder network in Kalamaria (Greece), Valencia (Spain), Alba Iulia (Romania), Rome and Turin (Italy) and West Midlands (UK). These local authorities have been chosen because they are partners in the SUITS project and therefore are currently on a change journey that involves the adoption of new technology and working systems. The SUITS project is a four-year research and innovation action, intending to increase the capacity building of local authorities and transport stakeholders and to transfer learning to smaller sized cities, making them more effective and resilient to change in the judicious implementation of sustainable transport measures. In total we were able to collect information of 12 different individuals – all connected to the local authority partners we worked with over the period of 3 years. All of these interviews were conducted either in participants' workplaces or during video conferences, audio recorded and transcribed in full.

3.2 Data Analysis

The transcribed interviews were coded according to a priori codebook, developed from a rigorous literature review on employees' vulnerability when introducing new technologies in local authorities. In a first step we checked for the existence of actual unwillingness to become vulnerable, before we run in a second step the analysis to identify indicators and consequences of the existent unwillingness to be vulnerable. After an initial scoping exercise for the fit of the identified indicators for vulnerability when it comes to the implementation of new technologies, we run a thematic analysis to identify the key indicators for the unwillingness to become vulnerable and the major negative consequences from it. Thus we included codes for distrust and trust as belief and as behaviour [6, 20], and "trustworthiness" as well as 'distrustworthiness' (e.g. incompetent, self-interested, exploitative, volatile, opportunistic).

We coded the data at the explicit, rather than implicit, level, and organised our results thematically, based on the patterns which emerged from the discourse [25]. In this way we progressed from deductive 'first-order codes' to inductive 'second-order themes', guided as appropriate by useful coding (such as that listed above) and thematic terminology found in similar studies [6]. Our findings section constitutes the most frequently found themes in relation to our research aim and theoretical framework.

4 Results

The first section of the results show how we investigated whether employees' vulnerability is actually present in the local authorities or not (step 1 of the analysis). The second section will highlight the key themes that emerged out of the data in terms of indicators that fostered the unwillingness to become vulnerable and its negative consequences in the local authorities when it comes to the implementation of new technologies to foster sustainable mobility.

4.1   Existence of Employees' Unwillingness to Become Vulnerable

The interview partners made it very clear that one of the key obstacles for the successful implementation of new technologies is the unwillingness to become vulnerable by the majority of employees. Most employees felt uncomfortable using and adopting new technologies they are not familiar with. Taking the idea of sense-making theory into account, we can assume that those situations are perceived as "risky" by the employees and thus, lead to a rather negative feeling of not willing to become vulnerable. Even more, some interview passages seem to indicate a real unwillingness to become vulnerable as employees really distrusted new systems of working with each other such as open data systems, information systems for traffic, parking, or air quality. Here, we could spot a tendency of not willing to cooperate with stakeholders that required data for their forecasting analyses for example as they distrusted the technology whether they benefit from it in the end or not. One quote for example was "I am not familiar with that technology and I do not understand the benefit of it except that I have to invest time and effort and giving the data away – and how do I know what you are going to do with the data in the long run?" [local authority 4, representative from transport department]. This behaviour can be seen as indicator for the existence of the unwillingness to become vulnerable and negative expectations and thus, show tendencies towards distrust in the technology[1].

Afterwards we undertook the thematic analysis (step 2). Two indicators could be spotted based on the interview data that was conducted during SUITS and two major negative consequences will be presented in the next two sections.

4.2   Key Indicators for Employees' Feeling Actual Vulnerable in Local Authorities

Attribution of Negative Motives. Attribution is the process through which people try to explain their own and others' behaviors [28–30]. The proposed relation between the unwillingness to become vulnerable and motivational attributions is built on the notion that individuals feel the urge to interpret behaviors of others that are salient to self, such as harmful behaviors. Our data indicated several aspects that may be summarized as attributions of negative motives. Negative experiences with new technologies in the past foster such attributions. A once failed new technology will thus increase the attributions of negative motives and foster distrust towards innovations. Several interview partners referred to such negative experiences in the past. Either the technology was not well developed or the technology was very weak introduced by the top management in the

---

[1] In the transcripts, we made sure that the participants referred to the new technology they distrust and not the company or stakeholder behind the technology. This is important as research shows that we have to differentiate between levels and targets of trust respective distrust (Fulmer & Gelfand [35]).

local authority. Another example was a rather poor introduction of the new technology in the organisation. As long as the employees of the local authorities did not understand their benefit of the implementation of the new technology or at least the benefit for the citizens', they developed an unwillingness to become vulnerable as they expected rather negative motives in relation to the technology implementation. "I do not really understand how this should help us and how we can benefit from it. Thus, what are the motives behind the introduction of this technology in our department? I suspect anything positive." [local authority 2, representative of the mobility department]. Sometimes the local authority failed to implement the new technology and the organization moved back to the old ways of working. Such negative experiences even strengthen the attributions of negative motives in the future and increase the likelihood that the unwillingness to become vulnerable emerges in relation to new technologies.

Perceived Value Incongruence. Perceived value incongruence has been defined as "the belief that others adhere to values that are perceived as incompatible with the actor's core values" [6]. Thus, in that moment when an individual identifies that its own values are not compatible with the values of someone or something else, the unwillingness to become vulnerable emerges. Perceived value incongruence has been proposed as a determinant of negative perceptions and expectations of the trustee's motives and behaviors in studies of professionals [26] which we could for example observe while working with the local authorities. One typical example in the following: the majority of local authorities failed explaining a potential supplier how the particular technological requirements should look like in detail due to a lack of specific technological expertise. The supplier (stakeholder) perceived such a behaviour very likely as unwillingness to become vulnerable as the supplier would assume the local authority does not want to provide the detailed information that would be needed to make a sufficient offer to the local authority. The consequence is that the supplier has to offer a technology that may be perceived as rather a generic technological solution by the local authority due to the missing detailed information. The local authority in turn may recognize the fact that a rather generic product solution has been offered. This perception of the local authority may very likely feed the unwillingness to become vulnerable and foster negative expectations such as "the supplier does not really care about us. They just want to sell their product." [local authority 3, representative of the mobility department].

The next example also shows the negative circle of perceived negative behaviours that lead to perceived value incongruence between different stakeholder and local authorities when new technologies are requested by the local authority to cope with sustainable mobility. Two stakeholder during SUITS tried to convince the local authorities to share data with them for the purpose of testing a new system for handling big data. However, the unwillingness of the local authorities to share the data was driven by different underlying value systems of the involved parties. While the local authorities could not understand the reasons and benefits of the new technology for which they had to provide data, they suspected the motives of the two stakeholder companies. They raised concerns regarding the confidentiality of data and the long-term use of the data when they provide the data to the private companies. One statement was for example "We expect that we have little in common with the other and that the other intentions are different to ours in the long run. Maybe that can harm us sometime." [local authority 2, representative of the mobility department]. Distrust arises as others come to be characterized as unpredictable and threatening, thus fostering a sense of uncertainty and vulnerability [27].

4.3 Negative Consequences of the Unwillingness to Become Vulnerable in Local Authorities

Avoidance of Interaction. One of the key outcomes of the interviews has been the fact that the unwillingness to become vulnerable lead into the avoidance of interaction – in particular local authorities tried to avoid to interact with the new technology at all. This finding is in line with several findings in the field of trust research [e.g. 8, 23]. The interview partners referred to several examples that fostered their unwillingness to become vulnerable and thus, lead to distrust towards the new technology. For example, when employees have been disappointed in the past as their positive expectations towards new technologies have been not proven right, they became skeptical next time and tried to avoid to interact with that new technology and the whole implementation process. "We tried to not to be involved with the top management that wanted us to test the new technology. You know last time we put so much effort in it and I still do not see the benefit for us." [local authority 1, representative of the mobility department]. A new technology cannot be implemented by the local authority's top management alone as then the implementation would fail. The top-down approach only works if it gets a bottom-up support by the wider local authority or the wider group of stakeholder. For example, Alba Iulia was able to implement new guidelines for procurement, but without working trustful with their stakeholder, this new process would have not been successfully implemented in the long run. As the local authority and the private providers worked close together and were willing to become vulnerable towards each other, they were empowered to implement these guidelines successfully.

Lack of Knowledge Sharing. Another negative consequence of the unwillingness to become vulnerable is an identified lack of knowledge sharing which was mentioned several times by the local authorities in line with SUITS. West Midlands Combined Authority put this point on the top of their agenda as it is very decisive for the organisational success, in particular organizational performance. Organisations spent for example almost a trillion dollars annually to analyze, store, and retrieve knowledge [31]. The willingness of employees to share knowledge depends on the level that these employees trust the organisation, colleagues or the other stakeholders [e.g. 32, 24]. While technology is able to store explicit knowledge, tactic knowledge cannot be stored in technologies as it resides only in the minds of people and its availability depends upon their decisions and behaviours [24]. Thus, when employees start to distrust a new technology they are not willing anymore to share their experiences with that technology, or their learnings [33, 34]. These aspects are very important for the future developments of new technologies. Even worse is the negative culture that may evolve from the unwillingness to share information and knowledge due to the fact employees may distrust a new technology. Employees may likely feel more and more isolated and becoming less motivated which leads to lower levels of organizational performance in the long-run [3].

4.4 Guidelines to Avoid the Emergence of Unwillingness to Become Vulnerable

Be Transparent and Honest. One learning from our work during SUITS has been the fact that local authorities have to become more transparent when searching for the best technological solution. Local authorities should communicate honestly and maybe even show evidence that they are not able to provide detailed information as they do not have the technological expertise. On the other hand, suppliers must put more energy into understanding exactly what requirements, expectations and uncertainties exist on the side of the local authority.

Communicate Face-to-Face. Try to meet face-to-face with potential suppliers as this allows trust to emerge between the negotiating parties (individual level) which may allow for trust transfer to the technology (human-technology level). It was said in the interview that face-to-face meetings are the best way to develop a trustful relationship with each other that will affect future decision-making.

Get a Third Party on Board. Sometimes it may help to ask as a local authority for external support by former suppliers or experts when decision have to be made regarding the implementation of new technologies. While the education system usually has to be independent and less cost intensive as consultancy companies, the recommendation is to work closer with the academic expertise that is needed. This may be the technological side of a new product or service or the human behavioural side when it comes to the employees in local authorities that have to adopt and use the new technology.

Foster Knowledge Exchange with Other Local Authorities. As local authorities often times do not have the budget to hire expensive consultancy companies, an alternative could be a learning group. During SUITS such a learning group was set up in which Valencia asked the West Midlands Combined Authority to support them in their recent developments. In addition, Coventry City and Coventry University joint the team to allow for a trustful and fruitful knowledge exchange in the future.

Create a Guiding Coalition that Serves as Project Management. Development of a clear vision and definition of goals, how the new technologies and data should contribute to making processes more efficient. Most important as well to identify one key person that serves as role model for others and is able to motivate and convince colleagues to support the implementation of the new technology.

Understanding Political Interests and Affecting Political Decisions. A clear understanding of what the technologies are needed for and what they are supposed to deliver makes it easier to influence decision makers and obtain the necessary financing.

5 Conclusion

This paper highlights the relevance of the employees' willingness to become vulnerable when new technologies are implemented in local authorities. As long as employees are not willing to become vulnerable when adopting a new technology and expect the new technology to be harmful, the introduction of new technologies in local authorities will fail. Based on comprehensive transcript material and observations during the project SUITS, we are first, able to show two key indicators that foster the unwillingness to become vulnerable and thus, may be the reason for a failing introduction of a new technologies in local authorities. Second we demonstrate two major negative consequences, a lack of knowledge sharing and the avoidance of interaction, due to the unwillingness of the local authorities' employees to become vulnerable. Finally, we are able to provide practical guidelines to avoid the emergence of an unwillingness to become vulnerable which pave the way to a successful introduction of a new technology in a local authority.

References

1. Kitchin, R.: The real-time city? Big data and smart urbanism. GeoJournal 79(1), 1–14 (2014)
2. Lakshmanaprabu, S.K., et al.: An effect of big data technology with ant colony optimization based routing in vehicular ad hoc networks: towards smart cities. J. Cleaner Prod. 217, 584–593 (2019)
3. Nienaber, A.M., Romeike, P.D., Searle, R., Schewe, G.: A qualitative meta-analysis of trust in supervisor-subordinate relationships. J. Manage. Psychol. 30(5), 507–534 (2014)
4. Mayer, R.C., Davis, J.H., Schoorman, F.D.: An integrative model of organizational trust. Acad. Manage. Rev. 20(3), 709–734 (1995)
5. Nienaber, A.M., Hofeditz, M., Romeike, P.D.: Vulnerability and trust in leader-follower relationships. Person. Rev. 44(4), 567–591 (2015)
6. Bijlsma-Frankema, K., Sitkin, S.B., Weibel, A.: Distrust in the balance: the emergence and development of intergroup distrust in a court of law. Organ. Sci. 26(4), 1018–1039 (2015)
7. Cho, J.: The mechanism of trust and distrust formation and their relational outcomes. J. Retail. 82(1), 25–35 (2006)
8. Bies, R.J., Tripp, T.M.: Beyond distrust. "Getting even" and the need for revenge. In: Kramer, R.M., Tyler, T.R. (eds.) Trust in Organizations, pp. 246–260. Sage, Thousand Oaks (1996)
9. Van De Walle, S., Six, F.: Trust and distrust as distinct concepts: why studying distrust in institutions is important. J. Comp. Pol. Anal.: Res. Pract. 16(2), 158–174 (2014)
10. Cherns, A.B.: The principles of sociotechnical design. Hum. Relat. 29, 783–792 (1976)
11. Cherns, A.B.: Principles of sociotechnical design revisited. Hum. Relat. 40, 153–162 (1987)
12. Clegg, C.W.: Sociotechnical principles for system design. Appl. Ergon. 31, 463–477 (2000)
13. Nienaber, A., Spundflasch, S., Soares, A.: Sustainable Urban Mobility in Europe: Implementation needs behavioural change. SUITS Policy brief X. SUITS funded from the European Union's Horizon 2020 research and innovation programme under grant agreement no 690650. Mobility and Transport Research Centre, Coventry University (2019)
14. Rousseau, D.M., Sitkin, S.B., Burt, R.S., Camerer, C.: Not so different after all: a cross-discipline view of trust. Acad. Manage. Rev. 23(3), 393–404 (1998)

15. ISDR: Living with the risk: A global review of disaster reduction initiatives: Preliminary version, United Nations, Geneva, CH (2002)
16. CIOMS: International ethical guidelines for biomedical research involving human subjects, Council for International Organizations of Medical Sciences (CIOMS), Geneva, CH (2002)
17. Chambers, R.: Vulnerability, coping and policy. Ids Bull.-Inst. Dev. Stud. **37**(4), 33–40 (2006). Reprinted from IDS Bulletin, 20
18. Weick, K.E., Sutcliffe, K.M., Obstfeld, D.: Organizing and the process of sensemaking. Organ. Sci. **16**(4), 409–421 (2005)
19. Sitkin, S.B., Bijlsma-Frankema, K.M.: Distrust. In: The Routledge Companion to Trust, pp. 50–61. Routledge (2018)
20. Guo, S.L., Lumineau, F., Lewicki, R.J.: Revisiting the foundations of organizational distrust. Found. Trends® Manage. **1**(1), 1–88 (2017)
21. Deutsch, M.: Trust and suspicion. J. Conflict Resolut. **2**(4), 265–279 (1958)
22. Fox, A.: Beyond Contract: Work, Power and Trust Relations. Faber and Faber, London (1974)
23. March, J.G., Olsen, J.P.: The uncertainty of the past: organizational learning under ambiguity. Eur. J. Polit. Res. **3**(2), 147–171 (1975)
24. Schewe, G., Nienaber, A.M.: Explikation von implizitem Wissen: Stand der Forschung zu Barrieren und Lösungsansätzen. J. für Betriebswirtschaft **61**(1), 37–84 (2011)
25. Deacon, D., Pickering, M., Golding, P., Murdock, G.: Researching Communications: A Practical Guide to Methods in Media and Cultural Analysis. Bloomsbury Academic, London (2007)
26. Sorensen, J.E., Sorensen, T.L.: The conflict of professionals in bureaucratic organizations. Admin. Sci. Q. **19**(1), 98–106 (1974)
27. Sitkin, S.B., Roth, N.L.: Explaining the limited effectiveness of legalistic "remedies" for trust/distrust. Organ. Sci. **4**(3), 367–392 (1993)
28. Abramson, L.Y., Seligman, M.E., Teasdale, J.D.: Learned helplessness in humans: critique and reformulation. J. Abnorm. Psychol. **87**(1), 49 (1978)
29. Heider, F.: The Psychology of Interpersonal Relations. Wiley, New York (1958)
30. Kelley, H.H.: Attribution theory in social psychology. In: Levine, D. (ed.) Nebraska Symposium on Motivation. University of Nebraska Press, Lincoln (1967)
31. McAllister, D.J.: Affect-and cognition-based trust as foundations for interpersonal cooperation in organizations. Acad. Manage. J. **38**(1), 24–59 (1995)
32. Nienaber, A.M., Schewe, G.: Enhancing trust or reducing perceived risk, what matters more when launching a new product? Int. J. Innov. Manage. **18**(01), 1–24 (2014)
33. Lohr, S.: Gazing into 2003: economy intrudes on dreams of new services, The New York Times 30 December, 3 (2002)
34. Lucas, L.: The impact of trust and reputation on the transfer of best practices. J. Knowl. Manage. **9**(4), 87–101 (2005)
35. Fulmer, C.A., Gelfand, M.J.: At what level (and in whom) we trust: trust across multiple organizational levels. J. Manage. **38**(4), 1167–1230 (2012)

PRONTOMovel – A Way of Transporting Creativity and Technology

Regiane Pupo^(✉) (iD)

Federal University of Santa Catarina, Florianopolis, SC, Brazil
regipupo@gmail.com

Abstract. The implementation of digital manufacturing laboratories today in Brazil has become very common in universities, federal institutes and schools. The so-called FabLab's, Maker Spaces, Hacker Spaces, or any environment in which technology is considered an issue and where the experience is applied, can serve a variety of fields. However, due to a diversity of economic and social problems, sometimes it is impossible for students and teachers to attend or even visit universities or those spaces. The purpose of this paper is to show the benefits and realities of a mobile digital manufacturing laboratory that takes along experiences and solutions to communities, schools and events where technology is not yet reached. It is believed that children, teachers, parents or anyone who has contact with the lab and the real manufacturing possibilities may open their minds and realize that anything may be feasible and constructed to enhance education, life quality and open job opportunities. The intention of the research is also to identify, describe and exemplify the results of the mobile digital fabrication lab in different communities, schools and economic realities. Each place that the lab visits has different attributes and characteristics conducted by people that compose the place. The relevance of overpassing university boundaries and carrying out projects in other scenarios, provides knowledge and experience not only to students involved but also to the community.

Keywords: Digital fabrication labs · Mobile lab · PRONTOMovel

1 Introduction

The type of labs explored in this research aims the learning of technology as a practical and playful way where "hands-on" methodologies follow interactions in order to have committed, conscious and creative activities facing the challenges. Thereby, it is almost impossible to show how to take advantage of technology without touching it, seeing it and living it. The term "technology" is much broader than just computers and much more than a learning medium [1]. All the thinking, designing and manufacturing become a single cell in which the professional, student or researcher remain absolutely committed throughout the design process. This allows the creation and production of unique artifacts or components that can be distinguished by digital controlled variation [2].

© Springer Nature Switzerland AG 2020
H. Krömker (Ed.): HCII 2020, LNCS 12213, pp. 308–317, 2020.
https://doi.org/10.1007/978-3-030-50537-0_23

Hands-on learning provides knowledge in practice, in making, helping the learner to acquire knowledge and skills outside the classroom. The learning process can occur through work, entertainment or any life experience. That is the reason why practical learning plays an important role in pedagogy, where design-build practices provide students with the ability of designing their own ideas and build their projects with their own hands, under the supervision of master builders, instructors, designers or architects [3]. Any project, when touching is involved, "cherish the understanding of scale, proportion, detailing, constructive techniques, textures, materials and countless sensations" [4]. Hands help to "understand the deepest essence of matter", aiding in the ability to imagine, freeing itself from the limits of matter, place and time [4].

PRONTO3D is a digital fabrication laboratory, located at Federal University of Santa Catarina, in Florianópolis, south of Brazil, that consists of a space for teaching, research and extension activities in the area of form materialization through automated techniques. Technologies such as laser cutting, 3D printing, milling on CNC machines and vacuum forming are available in the lab. The lab has been established since 2013 and is part of a Network composed of four Digital Manufacturing Laboratories, PRONTO3D NETWORK, throughout the state of Santa Catarina. All the network labs use digital fabrication practices as part of design processes, creating collaborative and dynamic spaces, where there is a great interaction of students and researchers from different majors.

The branch in the city of Florianópolis, which is the capital of Santa Catrina State, has been certified as a FabLab from FanFoundation since 2016, and seeks, among its many functions, to offer support and materials to meet the needs of local productions not only from academia, but also from the community in general. According to [6], in addition to environments that propel innovation, FabLabs help the community with small productions, thus representing a new way of sharing knowledge. The space has its strategy based on five pillars that support its basic characteristics: efficiency, collaboration, versatility, creativity and ludic.

Besides the university academic community, the laboratory has a project entitled PRONTO Kids, which includes workshops for elementary school children (ages between 6 and 12 years old). The activities are developed within the laboratory with the purpose of clarify the use of automated technologies for materialization of form. Generally, the topics covered are those brought up by the teachers themselves, in which they normally work in the classroom and can be produced with the help of FabLab equipment.

However, sometimes, due to economic and social reasons, it is not always possible for schools to send their students to attend the workshops. There are so many issues that can hamper study visits, such as inappropriate transport, uninterested teachers, distance, and so on. For this reason, PRONTOMovel, a Mobile digital fabrication Laboratory is prepared for visits in schools, exhibitions or conferences, taking PRONTOKids project outside university boundaries.

2 The Mobile Digital Fabrication Lab (PRONTOMovel)

A mobile lab is a van equipped with a series of digital fabrication machines including a 3D printer, a CNC milling machine, a laser cutter and a vacuum forming device where it is possible to build "almost" anything. The use of a mobile lab focuses the spread of form materialization processes using digital fabrication equipment as a main design tool.

Since its conception, the target audience has been schools, universities, communities and technological events, where digital fabrication actions may introduce a new approach on ordinary activities using creativity and local reality. In many schools, "the current educational regime is based on a certain view about what kind of knowledge is important: 'knowing that', as opposed to 'knowing how' [5]. The analogy corresponds to universal versus personal knowledge, where the first can be transmitted to and from anyone, through speech or writing because it is generic. On the other hand, 'knowing how', is always tied to one's own experience. It cannot be transferred; it can only be lived [5]. This routine practice is experienced in a Digital Manufacturing laboratory. Their actions, behavior, styles, and standards vary from person to person, from job to job, however, they all point to practice and experimentation as systematic learning.

2.1 Making of

Mobile laboratories have become a reality in several countries, including Brazil. Their main objective is to spread creative and practical activities to schools, thus being able to awaken students and teachers to a new way of learning and teaching, and, at the same time, to establish a "hands-on" culture. Those actions might convert students in protagonists in the process of creating products of interest [9].

Aiming the expansion of practice using a maker space for non-academic community, in several areas of activity, PRONTOMovel (Fig. 1) was planned. The vehicle, which was internally remodeled by students of the Design course and laboratory fellows, aims to be used outside the walls of the university, disseminating and demystifying the use of digital manufacturing technologies in areas geographically distant from the university, in this case, the Federal University of Santa Catarina, in Florianópolis.

Fig. 1. PRONTOMovel

The vehicle is a 1996 Volkswagen model that had its mechanical parts replaced and its interior turned into a work space. It has counters, seats, drawers and a perfect illumination set to assure the quality of work and equipment demonstration. Sliding and back doors turn into tables so visitors have easy access to the work that is being shown. Besides, students planned special supports in order to assure that small machines and other gadgets do not fall off during transportation. Each equipment has its own space to ensure safety.

2.2 Technologies

The configuration of PRONTOMovel is composed of three automated production techniques: 1) Additive – FDM 3D printing; 2) Subtractive - Laser cutting and/or CNC milling and 3) Formative - Vacuum forming. One of the intentions of PRONTO3D Lab is to use, as much as possible, national products. For instance, four of its 3D Printers and two Laser cutters are made-in-Brazil equipment and in the case of the vacuum forming, the machine was built in the laboratory, as a result of a research project in 2017.

Furthermore, mockups and prototypes developed in researches and class works are always taken on display, so that it is possible for visitors to observe and handle finished objects, providing them knowledge of the production possibilities of the machines presented.

Oddly enough, the country holds a great deal of communities, schools or institutions where it is still a lack of understanding regarding procedures and standards for digital fabrication tools and techniques. The experiences stated in this paper, and described below, represent an innovation and a challenge not yet explored. It is the role of the university to spread and inform new possibilities of research that serve the community, conducted by committed and attentive teachers.

3 Experiences

Among all the experiences already carried out by PRONTOMovel, there are three that have involved other research programs and a diversified group of majors. Although the three situations have different interests and approaches, the use of digital fabrication technologies enhanced the events that took place in the city of Florianópolis. The three locations are: 1) a Surf Festival, 2) a Public Elementary School and 3) Projeto TAMAR, a Brazilian non-profit organization that aims protection to sea turtles from extinction in the Brazilian coastline and they are all described in detail below.

3.1 Surf Festival

At first, what does a surf festival have to do with digital fabrication? A lot of things. This experience has the partnership of UGA-BUGA Surf Sports, a local non-profit organization that stablish surf festivals, and PRONTO3D Labs. In the event, PRONTOMovel had the chance to create and fabricate the trophies for a surf festival and the hand surfboards.

The idea behind the activity is to reuse, create and surf. To do so, discarded skateboard shapes are collected and restored using a CNC machine to shape the pieces of wood so then become hand surfboards. After that, each piece is graphic designed, and a finish surface is created (Fig. 2). A workshop that includes how to collect, cut and finish the pieces was conducted by PRONTOMovel at the festival, and made it easier to attendees to use their own work to surf. The presence of PRONTOMovel at the festival, by the beach, was important to clarify the activity as much easier as it is. The use of this kind of initiative has been increased in the past years with the need of reusing as much as possible material that has no longer its own use.

Fig. 2. Hand surf boards manufacturing

Figure 3 shows some hand boards already painted and finished by attendees and the trophies they created using FDM 3D printing. At the end of the activity the attendees had learned what is the difference between additive and subtractive technologies and also how to turn any idea in real thing.

Fig. 3. Hand surf boards and festival trophies

3.2   Public Elementary School

On this activity, PRONTOMovel visited an Elementary School, located in a neighborhood called Ribeirão da Ilha, about 30 km south of Florianópolis (Fig. 4). The visit was attended by students, teachers and school staff, who were able to have contact, for the first time, with digital fabrication tools. After a brief explanation by the members of PRONTO3D about the technologies involved in the digital manufacturing processes used in the laboratory, the group had free access to learn about the ways to design and fabricate any desirable artifact. The students had the support of three interns from the laboratory who directed the replicability of the content involved and answered any questions and comments about digital prototyping technologies.

Fig. 4. Workshop in Ribeirão da Ilha

The visit created the opportunity for many students and teachers to learn about technologies with which they were not familiar, especially because they could be part of the design processes of a series of Math board games they were creating. In fact, the visit was feasible due to a previous activity student were conducting in partnership with PRONTO3D, on creating Math games and fabricating them [8]. Therefore, this was the perfect time to consolidate their ideas that had been created previously.

3.3 Projeto TAMAR

Projeto TAMAR (Portuguese for TAMAR Project, with TAMAR being an abbreviation of Tartarugas Marinhas, the Sea Turtles) is a Brazilian non-profit organization owned by Chico Mendes Institute for Biodiversity Conservation [10]. The main objective of the project is to protect sea turtles from extinction in the Brazilian coastline. According to [10] there are currently 22 bases of the project, spread all over the country coastline, covering a range of more than 1000 km, and one of them is in Florianopolis.

An opportunity to show the lab activities as well as to enhance the importance of animal preservation was the "Sunday at TAMAR" event, organized by Projeto TAMAR in Florianópolis, which runs every third Sunday of the month. The action brings together several projects about the preservation, not only of turtles, but of several species thus threatened (Fig. 5).

PRONTOMovel participation in the event aims at educational activities in the dissemination of materialization technology in conjunction with some actions in a

R. Pupo

Fig. 5. Sunday at TAMAR

playful and practical way. The fabrication of turtle models is presented to visitors together with information cards that describe each threatened species. The models are fabricated in FDM 3D printing and Laser Cutting and technologies.

In order to increase the interest in technology and animal preservation, a special activity was prepared and includes a "hands on" strategy that mixes the use of two different digital fabrication tools (laser cut and 3D printing) and silicon mold, to produce replicas of animals threatened with extinction (Fig. 6). The action is called "FABRICATING TURTLES TO UNDERSTAND TURTLES" and is a 2019 Chevron STEM Education Award Winner [11]. The activity is published at SCOPES-DF [12] which is the first project of its kind to specifically develop effective pathways and resources for using digital fabrication in STEM education. According to [13], SCOPES-DF offers educators new models and methods for teaching. It offers students relevant, engaging, applied learning opportunities.

Fig. 6. Molds for turtle replicas

The connection among different abilities and skills, flagged by gained experience, can be exemplified in activities that apply STEM education as a premise. The term is an acronym for Science, Technology, Engineering, Mathematics, developed by the

National Science Foundation (NSF) in the early 2000s. More than an abbreviation that identifies the four areas, STEM supports the incentive of learning in order to interconnect subjects, aiming at their practical application. In STEM education, students learn to solve real problems by making, establishing hypotheses and proposing ideas [7].

Studies about turtles from the Brazilian coast is essential for the preservation threatened species. For this purpose, this project aims at information and awareness of preservation, through automated materialization of five different species. Additive, subtractive and formative technologies are the major contributors to the success of this research, providing educational activities in the dissemination of materialization technology. One of the main objectives of this extension action, is to clarify the possibilities and alternatives of "fabricate to understand". It is believed that from the moment there are replicas identical to the samples in question, the importance of their care and preservation are understood.

Children, especially in this event, always showed an intense interest in technologies and their possible actions in different areas of knowledge (Fig. 7). Generally, regardless of their age-specific interests, financial conditions and tastes, this audience assimilates new technologies very quickly, directing their doubts and curiosity to new applications.

Fig. 7. Children interest in technology

4   Final Considerations

The importance on expanding automated manufacturing technologies opens a world of possibilities for different areas of society. Being able to reconcile teaching with playful activities using technologies available in a FabLab has been consolidated. In addition, it is possible to observe the relevance of going beyond the university limits and carrying out projects in other scenarios. The activity itself provides knowledge and experiences not only to students involved but also to the community.

Among the visits and activities listed in this paper by PRONTOMovel, around 1600 people were achieved in many ways (Table 1). The popularization of technologies such as laser cutting and 3D printing, today, is fast and awakens collective interest and seeks for specific assistance to the visitors of the events. Promptly, the laboratory fellows, in addition to 3 professors, were always able to demonstrate the use of technologies in their various forms of performance.

Table 1. Number of people achieved by PRONTOMovel in the last 3 years

Date	Activity	People achieved
2017	Surf festival	500
2018	Elementary school	150
2019	Projeto TAMAR	1000

It is worth to mention that during PRONTOMovel visits, it was visible the satisfaction of people in perceiving the manufacturing facilities with accessible technology. It is well known that an explosion of creativity and contentment leads the student, regardless of social position, age group or school performance, to create, to innovate and to realize that technology helps a lot when used with awareness and responsibility.

Acknowledgments. The author would like to thank Escola Básica Municipal Batista Pereira and TAMAR Project for allowing and supporting the projects carried out within their institutions, FAPESC/FINEP and DUE Technology for financing the equipment, Federal University of Santa Catarina and PRONTO3D internship students for the effort and support.

References

1. Bybee, R.: The Case for STEM Education: Challenges and Opportunities. NSTA Press, Arlington (2013)
2. Kolaveric, B.: Digital Morphogenesis, Architecture in the Digital Age: Designing and Manufacturing. Spon Press, London (2003)
3. Abdullah, F.: Getting their hands dirty: qualitative study on hands-on learning for architectural students in design-build course. J. Des. Built Environ. **8**, 55–84 (2011)
4. Pallasma, J.: The Embodied Image: Imagination and Imagery in Architecture, 2nd edn. John Wiley, London (2011). ISBN: 0470711906
5. Crowford, M.: The Case for Working with Your Hands or Why Office Work is Bad for Us and Fixing Things Feels Good. Penguin Group, London (2009)
6. Maslyk, J.: STEM Makers: Fostering Creativity and Innovation in the Elementary Classroom. Corwin, Thousand Oaks (2016)
7. Martini, S., Chiarella, M.: Didactica Maker. Estrategias colaborativas de aprendizaje STEM en Diseño Industrial. In: XXI SIGRADI - Congresso Internacional da Sociedade Iberoamericana de Gráfica Digital, pp. 158–164. Blucher, Chile (2017). ISSN 2318-6968
8. Pereira, J., Pupo, R.: Learning math and digital prototyping with mobile digital fabrication lab. In: 22th Conference of the Iberoamerican Society of Digital Graphics (SIGRADI), pp. 1078–1083. Blucher, São Paulo (2018). ISSN 2318-6968, https://doi.org/10.5151/sigradi2018-1659

9. Santana, A., Raabe, A., Santana, L., Vieira, M., Ramos, G., Santos, A.: Lite Maker: Um Fab Lab Móvel para Aplicação de Atividades Mão na Massa com Estudantes do Ensino Básico. UFUB, Uberlândia (2016)
10. https://en.wikipedia.org/wiki/Projeto_TAMAR
11. http://www.tamar.org.br/
12. https://www.scopesdf.org/scopesdf_lesson/fabricating-turtles-to-understand-turtles/
13. https://www.scopesdf.org/about-scopes-df-project/

Multimodal Mobility Packages – Concepts and Methodological Design Approaches

Ulrike Stopka[✉]

Technische Universität Dresden, Dresden, Germany
ulrike.stopka@tu-dresden.de

Abstract. With the increased need to pursue the climate protection goals much more intensively efforts in many countries are directed towards a perceptible change in the mobility behavior of citizens. The main priority here is to reduce the private car ownership significantly making transport as a whole more environmentally friendly. A great variety of mobility providers are entering the market and complete the network around public transport with a wide range of services such as car sharing, bike sharing, e-scooter sharing, ride hailing, ride pooling, car rental, taxi, on demand shuttle or bus services, booked and paid via electronic platforms and smartphone apps. The comprehensive approach to combine these different mobility services into bundled subscription packages with special pricing schemes is one of the most important features of the MaaS concept.

First of all, the article considers the definition and general concept of MaaS. Afterwards, a step by step conceptual approach for designing mobility flat rates and subscription packages to support multimodal transport behavior will be presented and its possible effects will be discussed.

Keywords: Urban mobility · Multimodal mobility · Mobility-as-a-Service · Product bundling · Mobility packages · Subscription behavior · Conceptual design · Product pricing · Willingness of pay · Flat rates · User requirements

1 Background and Motivation

According to the United Nations World Urbanization Prospects 55% of the world's population today live in urban areas, a proportion that is expected to increase to 68% by 2020 [1]. This drives transport demand because more people require more mobility associated with considerable burdens on urban regions (see Fig. 1).

Actually, 1,32 billion cars (multi-track vehicles such as cars, buses and trucks) are registered worldwide, and their number will more than double to 2,7 billion by 2050 [3]. Even taking into account that 9,8 million cars are being taken out of service each year, these significant growth rates of road vehicles mean a tremendous increase of road traffic worldwide with the well-known environmentally harmful consequences. Private cars and taxis are currently used for nearly 75% of urban passenger transport in OECD countries and over 60% in non-OECD countries [4, p. 28]. Private vehicles will remain the preferred mode of personal travel worldwide. But whereas the private urban

© Springer Nature Switzerland AG 2020
H. Krömker (Ed.): HCII 2020, LNCS 12213, pp. 318–339, 2020.
https://doi.org/10.1007/978-3-030-50537-0_24

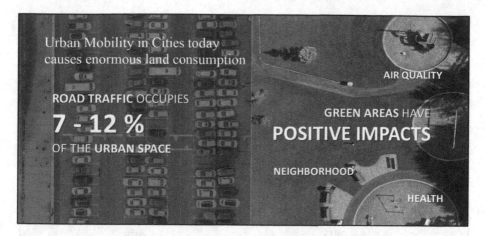

Fig. 1. Land consumption in cities by road traffic [2]

passenger transport will be slightly reduced in the OECD countries till 2050, it is expected to grow in the Non-OECD countries by 67% (Fig. 2).

Therefore, it is a primary concern of many megacities worldwide to reduce the number and usage of private cars significantly. Extensive measures are being taken in the capitals of many European countries to move toward this goal.

Madrid will be banning non-residents cars and heavy vehicles within a 500 acre perimeter in its core area. They are also repurposing 24 main arteries exclusively for bike and pedestrian use.

Copenhagen has pledged to become carbon neutral by 2025. The city currently has over 200 miles of bike lanes and 39 bike "Superhighways". Already half of the population bikes to work.

In Oslo all cars will be banished from the city center by the end of 2020. A countrywide ban of cars of city centers is planned for 2025.

Paris intends to eliminate all Diesel vehicles by 2024 and Petrol vehicles by 2030. By 2020, they will double the amount of bike lanes and make multiple streets exclusive to electric and no-emission vehicles.

In mid-2019, central London implemented zones for vehicles, including zero-emission zones in core business and tourist areas. Additionally, many protected bike lanes and pedestrian-only areas will be installed [5].

Helsinki wants to eliminate the need for private car ownership by 2025 based on plans to create an app-based, multi-modal, on-demand transport network throughout the city.

In Germany it is legally permitted for cities to ban diesel vehicles as a measure for air pollution control if NO_2 limits are significantly exceeded. Furthermore, it is the declared aim of the Federal Environment Agency to reduce the number of cars in cities from 450 to less than 150 per 1000 inhabitants.

All these measures are expected to shift people's travel behavior in cities towards public transport and shared mobility modes. In the current transport outlook scenario of the OECD International Transport Forum, the share of urban passenger kilometers

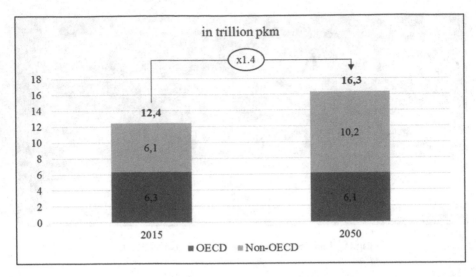

Fig. 2. Increase of urban private passenger transport from 2015 to 2050 [4, p. 28]

travelled in private vehicles (including individual taxi service) is forecasted to decline from around 70% in 2015 to 40% in 2050 [4, p. 77]. The possibilities to redesign the cities into green, low-noise and low-emission environmental friendly areas are very diverse and complex, including a great number of technological and business developments with disruptive potential (see Fig. 3).

This can be supported by the implementation of IoT-platforms performing an Intelligent Operation Centre (IOC) delivering multidimensional insights into a city from the macro to the micro level. Based on IoT-data the distribution and operating status of the municipal infrastructure and facilities including traffic, parking, power and water supply, heating and gas lines, street lights etc. can be monitored in real time. This also includes the operation of a complete urban mobility suite for vehicle operation and guidance, fleet management, public transport and intermodal trip management, incident management etc. Also traffic avoidance by innovative measures in the world of work, e.g. homeworking, co-working spaces or shared offices belongs to this[1].

In the mentioned overall concept smart mobility services for citizens undoubtedly represent a very significant and promising contribution. Urban areas have the most potential for game-changing mobility and transport innovations. That means to create and implement an app-based, multimodal, on demand mobility network throughout urban areas that makes the need for private cars obsolete. The user enters a destination in the app, then is presented the best door-to-door routing encompassing all possible

---

[1] In ITF Transport Outlook 2019, pp. 87–91, the impacts of increased teleworking and telepresence activities on urban passenger transport demand were investigated [4]. In the scenarios that simulate a disruptive development pathway for transport, teleworking and telepresence could affect between 3% and 30% of urban trips by 2050 and lead to a decrease in urban passenger-kilometers and related CO_2 emissions of around 2%.

Fig. 3. Possibilities to reduce private car ownership and redesign the cities green

travel modes: buses, trams, underground, trains, shared bikes, ride pooling/sharing, e-scooter, on-demand shuttles, taxis and ferries (see Fig. 4).

The payment for all used transport modes is also handled by the app. That means at least: intermodal transport from A to B has to become as comfortable as the own car. Because the preferences of citizens changes only very slowly towards new mobility services, especially to so-called active mobility modes such as cycling, micro mobility or pedestrian traffic, the question is which measures, offerings, approaches, incentives etc. can lead to effective transitions.

In this context, the term "Mobility-as-a-Service" (MaaS) has recently been intensively discussed in scientific literature and gradually implemented in transport markets, primarily in urban regions. The study "Mobility-as-a-service and changes in travel preferences and travel behavior" [7] provides a very comprehensive summary of the literature on this subject. Based on different definitions and a typology for "MaaS schemes", the preconditions for MaaS potential to challenge travel behavior pattern, the requirements and influencing factors for MaaS offers, the usage design and value contribution, the importance of travelers' characteristics and prerequisites for adoption of MaaS, the costs and willingness to pay are examined and described in detail.

Similarly, a special report by L.E.K. Consulting in partnership with Tourism & Transport Forum Australia and UITP Australia and New Zealand [8] highlights MaaS as the next transport disruption. Mulley et al. [9] elaborate in particular to what extent the Maas concept is relevant for public transport demonstrated in a case study with five Australian Community Transport Operators.

Findings and evaluation results from the UbiGo MaaS project in Gothenburg, Sweden are discussed in Karlsson et al. [10].

In addition to these literature reviews, the study WHIMPACT [11], delivers practical insights into the generated impacts of the world's first MaaS system operated by MaaS Global in Helsinki and several other European metropolitan regions such as Birmingham (UK) and Antwerp (Belgium).

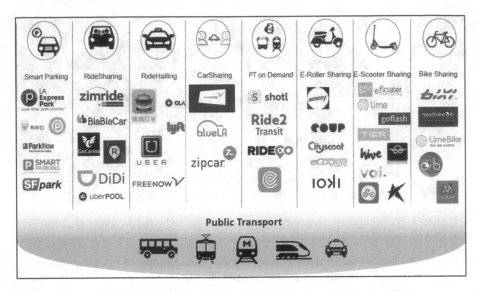

Fig. 4. Smart mobility services for citizens [6]

2 MaaS, Mobility Platforms and MaaS Ecosystem

Durand et al. [7] consider MaaS as "a new transport concept that integrates existing and new mobility services into one single digital platform, providing customized door-to-door transport and offering personalized trip planning and payment options" via a single app. The UITP [12] defines that MaaS is all about the integration of and access to different transport services in one digital mobility offer. This tailor-made service suggests to be the most suitable solution based on the traveler's needs, ensuring a door-to-door service and enabling life without having to own a car. Polydoropoulou et al. [13] see MaaS as a promising concept which aims at offering seamless mobility to the users and providing economic, societal, transport-related and environmental benefits to the cities of the future. A lot of similar definitions and explanations of the term MaaS can be found in [14–22]. The most important MaaS features are summarized in Fig. 5.

Moreover, MaaS requires a comprehensive ecosystem where multiple market players, transport companies, institutions, authorities and service industries act in collaboration leaving behind their traditional boundaries. That needs a high-performance mobility platform in the background, operated by a MaaS operator and MaaS integrator in the frame of a comprehensive MaaS ecosystem (see Fig. 6).

In the backend it is necessary to operate a central multimodal B2B platform that optimizes trip allocation, routing guidance, payment and tracking handling the data from the different transport companies and mobility service providers. The front-end application offers the customer interface with integrated functionalities (see Fig. 5) and additional services from other stakeholders.

In this context, we can distinguish between MaaS integrator and MaaS operator (see Fig. 6). The MaaS integrator is a company or organization that builds and manages the technical infrastructure shaping a comprehensive platform with standardized API's

- **One-stop access** to different public and private transport modes on demand based on consumer's preferences
- Information, registration, journey planning, booking and payment via a **single smartphone app**
- **Single contract with a Mobility operator/integrator**
  → registration, information, booking and payment via the appropriate operator's platform → **Single-Sign-On**
- **Single invoice** at the end of the month for all used mobility services, the operator handles the clearing between the mobility providers
- **Choice between different tariff options** (bundled mobility packages with a special price structure or "pay-as-you-go" scheme)

Fig. 5. MaaS features

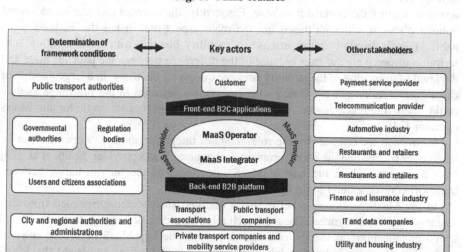

Fig. 6. Relevant stakeholders in the MaaS ecosystem [based on 35, p. 16]

for access and data exchange between different public and private transport companies and other mobility service providers offering on demand services, shared or value added services (see Fig. 4). On the one hand, the Maas Integrator must have access to all data concerning the vehicle availability or real time traffic data, and on the other hand, it must also be able to communicate the customers' request to the executing transport companies. This includes, for example, the booking of on-demand offers or the check of the allowance to open and use a vehicle in case of car sharing.

In summary, the Maas Integrator is responsible for the integration of data from multiple transport operators and infrastructure data and the MaaS operator builds the solution on the top of the integrated layer [12, p. 4].

The task of the Maas operator is to offer the customer different transport services from a single source. This includes, among other things, a user-friendly platform, usually in the form of a mobile application, for entering the customer's request and setting the best route suggestions, booking and paying. The app allows the customer to use various mobility services but it does not mandatorily include multimodal mobility packages when each trip is billed individually. With regard to MaaS the supply of mobility packages in a subscription scheme will be in the responsibility of the MaaS operator. In order to make the business model profitable for the MaaS operator, the revenues must exceed the costs. The revenue per customer is calculated from the prices for the multimodal mobility packages and other revenues for additional services. On the cost side, in addition to overheads (e.g. IT, marketing, sales, administration), the main factor is the purchase of transport services from the transport companies. There are three ways for operators to purchase transport services: Purchase of transport services before demand, standard settlement when the transport service is used or lump sum payments to the transport companies without exact offsetting, that means purchase of fixed contingents without single accounting. The latter is a higher risk for the MaaS operator because the demand is volatile. Frequently, the operator receives an additional volume discount, although this is not the main aspect that makes the business profitable. This is primarily given in case of mobility bundles with fixed monthly subscription respectively flat rates because of the difference between the package price and the level of utilization of the mobility package quotas by the users. In a large number of cases customers do not use their packet budgets or flat rates in the full range as they could. This difference is one of the most important sources of profit for the MaaS operator [24].

In practice we can see that the role of MaaS integrator and MaaS operator can be taken over by one and the same company as a general MaaS provider. Smith et al. [23] have a deeper look at the role of MaaS integrators as the intermediaries between transport companies and Maas operators.

In the MaaS ecosystem other business oriented stakeholders originated from different sectors are integrated. These are technology-specific actors, e.g. from the IT and telecommunications industries, offering support to the key MaaS providers, internet access etc. But also industries delivering value added services to extend the MaaS offerings belong to this stakeholder category, such as the utility and housing industry, finance and insurance industry, the media, local restaurants and retail industry, delivery services and logistics (see Sect. 6).

Furthermore, in the MaaS ecosystem, regulators, policy makers and governmental authorities are responsible to set market rules and support level playing field. But also the institutional coordination between different authorities in charge of the mobility system in cities or regions is an important prerequisite for the breakthrough of MaaS.

Polydoropoulou et al. investigated different prototype models for MaaS business ecosystems in three study areas (Budapest, Greater Manchester and Luxembourg) based on the generic business model Canvas [13, p. 160]. The results indicated that especially the public transport operators along with local/regional public authorities appear to be the best positioned players to fulfil the role as MaaS integrator/operator. Nevertheless, in the study areas they were not able to take this responsibility because of structural and resource constraints. There is a strong need for policy interventions to support the successful MaaS deployment with respect to create a truthful environment between the involved market players, to establish standards for compatible data formats and standardized open APIs or to foster the cooperation of all stakeholders.

3 Mobility Packages

Product bundling is a well-known strategy of marketing products or services in particular combinations for which a special bundle price is offered. Service and price bundling plays an increasing role in many industries, e.g. banking, insurance, software, automotive or telecommunications. Bundles are very often composed of complementary products which should augment the benefit for customers by delivering value-added features or services.

A distinction is made here between pure and mixed bundling as well as a variety of other special forms. In case of a pure bundle, the user is offered a completed predefined bundle. In the case of mixed bundling the user can either purchase individual products outside the bundle/package or single components of the bundle separately.

The motivation for providers of bundled products is more or less to increase the profit by using a discount to induce customers to buy more than they otherwise would have.

In addition, the topics of cross-selling, the concealment of the real unit product prices, the price image and customer loyalty plays a significant role. The customers profit from the higher functional benefits by price discounts and for convenience reasons. Often, bundling is also referred to as a win-win situation, although this is not generally valid [25–27].

In case of bundled mobility packages, the motivation is somewhat different. The most essential moment is not profit maximation but to support the market diffusion of new or underutilized transport modes which are much more environmental friendly than to use the private car. When we look on a single trip, the different transport modes can have definitely a substitutive character. For example, the user can decide to use bike sharing when the weather is good, while car sharing is preferred when the weather is bad or when larger items need to be transported.

With regard to product bundling, Reinders et al. [28] came to the conclusion that the combination of well-known and from the customer already used products or services and lesser known new products in a bundle makes it easier for customers to choose a bundled offer instead of single products, as the risk of a wrong decision is considered to be lower. According to this, Matyas and Kamargianni stated that "in case of MaaS, even though shared services are gaining wider acceptance, their use could be accelerated by bundling them together with more popular modes such as public transport or taxi" [29, p. 1955].

As shown in Fig. 5, mobility bundles are an essential component of the MaaS concept. The term "mobility package" or "mobility bundle" used to describe an integrated offer of various mobility services which vary in terms of different attributes and characteristics, such as number of rides, travel time or distances. They can have a flat rate character but with allowances for special transport modes or be offered as a pay-as-you-go variant. The contractual relationship between user and provider of mobility packages is usually based on monthly subscription.

Caiati et al. understand by subscription and bundling that various transport modes and services are converted into service packages, stimulating customers to buy these packages at a discounted price against different access prices for every single transport mode [30, p. 126].

Different studies and investigations tried to find out whether people are willing to use and to subscribe to mobility packages, how the bundles have to be designed, what pricing schemes are preferred, whether monthly subscription plans could be a promoter to change the mobility behavior of people towards the usage of more innovative environmental friendly transport modes, especially shared ones and so on [29, 30].

Therefore, in the following chapter a general step by step approach how to design mobility packages based on the MaaS concept is discussed.

4   Conceptual Approach for Designing Mobility Packages

For the conception of the design and pricing of mobility packages, numerous aspects have to be considered such as the determination of target groups, their mobility behavior, demand and preferences, type and scope of the mobility services included in the bundle, willingness to pay, ease of use and transparency or social influences.

The methodological approach to determine mobility bundles comprises in general two domains: The design-oriented aspect that helps to identify which among a feasible set of transport modes and complementary services should be combined in the bundle and the pricing-oriented aspects typically assume a service portfolio and propose the prices of the offered bundles [31, p. 18].

Figure 7 gives a more detailed overview about the sequential steps to determine the not independently existing partial aspects in the designing process of mobility packages.

With regard to *target groups and type of packages*, the mobility operator has in general three options for designing its products: a uniform product, target group-specific product variations or the user can assemble itself an own product from different allowed components (Create-It-Yourself) [32]. In the past, all customers were provided with more or less the same product. This reduces complexity and purchase prices due to high standardization but it severely limits demand, because the mobility needs and willingness to pay of the consumers are different. With the help of user segmentation into target groups, it is possible to counteract this problem by differentiating the products according to the customer's demographic, socio-economic, physiological and, in particular, behavioral characteristics [33, 34], but also regional or local specifics. This allows it to offer the customer freedom of choice without having to design an individual product for each customer. Finally, it is also possible to provide packages according to the modular principle where the customer can assemble the bundle himself (Create-It-Yourself). Here, the customer selects the preferred means of transport, the desired attributes and levels, while in the background the total price for the product is calculated. Such a high level of customer centricity reflects the mobility behavior of each individual best but is accompanied by a high degree of complexity. Nevertheless, different MaaS studies have shown that these offers do not in any case have the desired effect. For example, Ho et al. [32] conducted a stated choice survey in Sydney with 252 respondents, in which the users had to choose between various preselected and self-assembled choice sets. While 36.2% of the participants opted for a predefined package

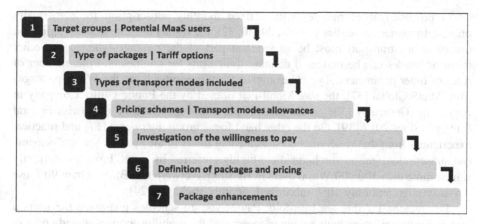

Fig. 7. Conceptual approach for the design and pricing of mobility packages

and 11.0% for a pay-as-you-use plan, only 32.1% chose a self-developed Choice Set. Ho et al. see reasons for this in particular in the fact that the respondents overestimated their mobility needs, leading to very high prices and that the predefined choice sets generally describe the requirements in a sufficient manner.

Conversely, MaaS providers that are already on the market for a longer period, such as MaaS Global with its Whim offer, see an urgent need in the near future to launch more target group-specific mobility packages better suited to the individual mobility behavior and the different requirements of users than it is the case today. Exemplarily MaaS Global communicates five different personas that can purchase different bundles with added service options according to their specific needs (e.g. family package, business world package, 15 min package) [36].

In terms of *package types and tariff options*, the question has to be answered whether multimodal transport services should be offered in bundles with a pure flat rate[2], i.e. without limits, as packages endowed with certain budgets for the included transport means or as mixed packages. Pure flat rate bundles allow the customer unlimited use of the integrated services [41]. Especially products with high fixed costs and marginal costs tending towards zero are predestined for flat rates.

In the mobility sector, which offers physical transport services, the circumstances in the provision of these services are completely different from those in telecommunications, for example, which offers virtual products in the form of the transmission of digitized data and is therefore more tending towards flat rate models. With regard to mobility bundles, the risk for pure flat rate tariffs is definitely too high for the MaaS provider because on demand taxi or car sharing services e.g. are not suitable for it due to considerable fuel and personnel costs. Therefore, mobility packages with budgets are preferred. There can be made a distinction between the possibility to divide the budget among several means of transport or to budget each mobility mode separately [35].

---

[2] An example for a pure flat rate is the classic monthly or yearly ticket for public transport in a city or limited region.

In practice, mixed bundles with a fixed monthly subscription fee are usually offered. In numerous studies [16–18, 29, 30, 32], the results indicate that the unlimited use of public transport must be an inherent part of the package whereas the other transport modes can be budgeted defining limitations for kilometers, time, number of rides or other parameters. Typical examples for this approach are the Whim packages from MaaS Global [37], the swa Mobil-Flat offered by the Public Utility Company in Augsburg (Germany) [38] or the SBB Green Class launched by the Swiss Federal Railways since 2017 [39]. On the other hand focus group interviews [17] and practical experiences from the MaaS markets indicate that offers in the "pay-as-you-go" scheme (no subscription fee, no surcharges) are highly preferred by users. From the currently signed more than 100.000 Whim users in Helsinki, Antwerp and Birmingham 90% use pay-as-you-go and only 10% subscribed monthly packages [24].

The reasons for this are manifold. Pay-as-you-go schemes guarantee the users a maximum flexibility without having to worry that the mobility services already paid in the subscribed package are not fully used. But they still have the comfort to invoice and pay all flexibly used services in a one-shop manner via their MaaS provider. Furthermore, the planned usage frequency of the transport modes included in the mobility package and the level of the user's monthly income play a significant role. While pay-as-you-go in particular is preferred by subjects without a fixed monthly income, subscription based packages have been more useful for the professionals.

The step of *product bundling* comprises selection of the various transport modes can be included in the multimodal mobility packages, such as bus, light rail (tram, underground etc.), regional train, bike sharing, car sharing, taxi, car rental, on demand transport services and micro mobility. This decision should be made primarily from the user's perspective to provide a real benefit for the customer.

Studies on this topic have been conducted, for example, in Leipzig (Germany) [42], London [29, 43, 44], Sydney [32], Helsinki [45], Amsterdam [46] and Switzerland [47]. The result of these studies is that in any case public transport is perceived as the backbone. For example, 96% of the respondents in Leipzig described the flat rate based public transport as very important or important within the mobility package that provides the greatest worth-part utility [42]. The studies conducted in London, Helsinki and Sydney have also shown that public transport has a significant positive influence on the decision whether the customer chooses MaaS or not [29, 32, 43, 44].

With regard to the evaluation of shared transport modes, such as car or bike sharing the findings diverge strongly. Whereas in Leipzig 52% and 41% of respondents consider respectively flexible or station-based car sharing as an important component of a bundled mobility tariff, it is not mentioned as an important component by the respondents in the other studies.

Furthermore, the results in [29] show "that even though respondents do not prefer shared modes in their MaaS plans, a significant number of them are willing to subscribe to plans that include these modes. Once they have subscribed, over 60% of them indicated that they would be willing to try transportation modes that they previously did not use if their MaaS plans included them" [29, p. 1951]. This can be seen as a strong promoter for changing the people's mobility behavior towards more traffic-reducing, private car abandoning transport modes and more sustainable ways of travelling.

Besides the decision what kind of transport services should be included in multi-modal mobility bundles it is necessary to determine the *pricing scheme*, that means in what way and to what extent these services can be used within a certain package price. Therefore appropriate attributes and their attribute levels or values have to be defined, as shown in Table 1 exemplarily.

Table 1. Examples for transport modes pricing schemes [17, p. 436]

	Attributes and attribute characteristics					
Transport modes	Free floating car sharing*	flat rate (unlimited) < 30 min	10 rides < 60 min	up to 15 h or up to 150 km	300 min	pay per ride with 50% discount
	Station-based car sharing	48 h with 500 free km**	10 × 3 h incl., 20 km/ride	3 × 3 h incl. 15 km/ride	1 × 3 h incl. 15 km/ride	30 h without limit of km
	Bike sharing***	12 h flat rate****	flat rate, <1 h	10 ×30 min	5 × 1 h	pay per ride with 50% discount per ride
	Taxi (within the city area)	flat rate from 9 pm–5 am	50% discount on every ride	5 rides	1 ride	taxi outside the city area 30 km included
	Public Transport	Subscription	50% discount on a single ticket	10 rides	4 rides	unlimited rides in one zone, outside the zone pay per ride

* every additional minute 0,30 €/min
** minimum lending time 1 h
*** every 30 additional minutes 1€
**** 0–6 am do not count to the 12 h

Additional attributes in case of station-based car sharing could be the vehicle type (petrol, diesel or e-vehicle) or vehicle category (mini, compact, luxury, van/minivan, SUV).

Caiati et al. included in their comprehensive investigations to get insight in customers MaaS subscription decision seven different transport modes with each four different attribute levels [cf. 30, p. 131]. As can be expected, the estimated utility of the bundles increases significantly for the subjects when public transport is offered in a flat rate manner with unlimited rides and it reduces in case of pay per ride with a certain discount on standard fare. Contrary, the characteristics for e-bike sharing were assessed completely differently. The subjects preferred to include this service in their bundle when e-bike sharing was offered with a two part-tariff, pay per ride with a 50% discount. Contrary again, the respondents decided to integrate the mobility service e-car sharing in their bundle when it offers a certain amount of included travel allowances (e.g. 120 min) instead of pay per ride or pay per ride with a discount, whereas if taxi is offered with a relative high amount of travel distance (e.g. 50 km) the utility for MaaS subscription tends to decrease [30, p. 138]. Regarding pricing schemes of car rental Caiati et al. found out that 4 days of rental included in the package has positive and significant effect on the probability of car rental to be chosen in a self-configured bundle. For the transport modes ride sharing and on demand bus the pricing schemes indicated that unlimited rides (e.g. flat rates) are most appropriate for the respondents.

With regard to *the willingness of pay* the consumers make their decision rationally, i.e. they weighs the benefits against the costs (homo economicus). Rising prices generally reduce the probability of the purchase decision. Many different studies in the MaaS environment (Ho et al. 2018 [32] and 2020 [40], Matyas and Kamargianni 2017 [44], 2018 [29], Ratilainen 2017 [45], Stopka et al. 2018 [17], Stopka 2020 [48], Caiati et al. 2020 [30]) have shown how strong the likelihood of a purchase at a certain price level is influenced by the disposable income.

As part of the "INTERmobil" research study for optimized pricing and product design of mobility packages conducted in 2018 by a research team of the Technical University of Dresden [42], approx. 300 subjects from the city and the district of Leipzig had to evaluate in a survey[3] various mobility bundles (S, M, L, XL) at monthly prices between 50 and 200 €. The calculation of the part-worth utilities (cf. Fig. 8) shows that the respondents were willing to accept a price up to a little more than 100 €. The strongest decline of the part-worth utilities for the attribute price, which means a decrease of the subjects' individual benefit, was observed in the case of a hypothetical price increase from 75 € to 100 €. For this reason, a subscription price of 75 € per month was favored for further studies and the intended market launch of the mobility package. In addition to an urban public transport flat rate, the package includes a bike-sharing flat rate for rides of less than one hour, 30 min of flexible and three hours of station-based car-sharing including 15 km, and one taxi ride per month in the city of Leipzig. 27.7% of the respondents would opt for such a multimodal mobility package [42, p. 18].

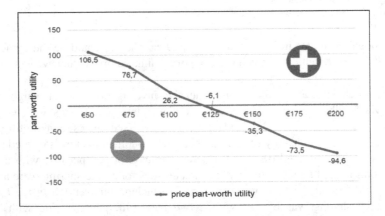

Fig. 8. Part-worth utilities of the monthly subscription price [42; p. 17, 35, p. 44]

---

[3] The investigation was designed as a stated preference survey and evaluated by a conjoint analysis.

In a study conducted in Helsinki in 2017 Ratilainen [45] compared the willingness to pay for mobility services included in bundles with the prices actually available on the market. The results of this comparison can be seen in Fig. 9.

means of transport	Quantity	Willingness to pay	Current price	Difference
Public transport	15 rides	56,50 €	32,70 €	+ 23,80 €
	Flat rate	117,14 €	52,40 €	+ 64,74 €
Bike-Sharing	6 rides	1,20 €	5,00 €	- 3,80 €
	Flat rate	6,65 €	4,20 €	+ 2,45 €
Taxi	3 x 10 km	11,57 €	63,00 €	- 51,43 €
	6 x 10 km	16,00 €	126,00 €	- 110,00 €
Car-Sharing	4 hours	-21,71 €	40,00 €	- 61,71 €
	8 hours	-24,93 €	80,00 €	- 104,93 €

Fig. 9. Comparison of willingness to pay for package included mobility services with actual prices [45, p. 59; 35, p. 45]

Remarkable are the differences in all involved transport modes. While the willingness to pay for six bike sharing rides per month is significantly lower than the prices actually available on the market (€1.20 vs. €5.00), the subjects are willing to pay a higher price for bike sharing when it is offered as flat rate (€6.65 vs. €4.20). This obviously shows the benefits the customers associate with a flat rate. The same can be observed for public transport. Contrary to bike sharing and public transport, the accepted reservation prices differed from the actual prices for taxis and car sharing significantly. Due to the high variable costs for these transport services, these offers are budgeted. For taxis, the willingness to pay varied depending on the number of included trips by a factor of five to eight from the current prices car sharing. Like some other studies revealed [29, 32, 43, 44] the benefit of car sharing within mobility packages is often evaluated negatively. Also in this case, the willingness to pay is far away from the actual prices on the market.

In summary, the studies have shown very clearly that the willingness to pay for each means of transport varies strongly. Only the diminishing benefit with increasing prices is generally valid.

In terms of *mobility package enhancements* a comprehensive range of additional features has been examined in various studies. Table 2 provides an overview about the analyzed features and their influences on the subjects' assessment of MaaS bundles.

Table 2. Impact of product enhancements on customers' purchasing decisions [cf. 35, p. 47f.]

Product expansion	Impact
Digital travel authorization (e-ticket) [49]	+
Compensation payments in case of malperformance (e.g. delay) [45]	+
Expansion to long-distance traffic [45]	+
Monthly changes/updates [45]	+
Discount for exclusion of rush hours (e.g. 9 o'clock ticket) [45]	+
Roaming [24, 50]	+
Guarantee of the pick-up time [24]	+
Telephone support/hotline [49]	+
Transferability (outside the household) [45]	+
Transferability of unused credits to the following month [45]	+
1st class transition [51]	+/o
Selection of different vehicle categories [45]	+/o
Combination with the residential rent [24]	+/o
Rewards for the use of environmentally friendly means of transport [24, 49]	+/o
Child seat included in the vehicle [45]	+/o
Transferability to household members [45, 52]	+/o
Real time alerts and notification about travel events (e.g. delay, disruption etc.) [30]	./.
App synchronization with personal agenda [30]	./.
Parking payment [30]	./.
Free non-renewable trial period [30]	./.
Subscription cancellation with no cancellation charge [30]	./.

+... positive, o... neutral, -... negative, ./. ... not available

It may be remarkable that all product enhancements have been evaluated positively or neutrally. An exception is the list of additional features that have been examined in the study by Caiati et al. [30, p. 132 f.]. Some of them are listed in Table 2. The total of 11 selected features with 30 different characteristics were offered to the respondents with extra costs of between 5–40% of the monthly basic subscription price. In the context of this study of Caiati et al. [30] based on a web-based survey conducted in the Netherlands (mainly in the Amsterdam and Eindhoven region) the respondents were asked to choose among two possible combinations of additional features supplementing to their chosen bundle leading to a price surcharge. The analysis and evaluation results of the collected data giving insights in the respondents' willingness to pay for their bundle enhancement are not yet available.

In general, an increased performance of mobility packages is appreciated by the customers. Some of the enhancements are associated with low or no additional costs for the MaaS provider and should be taken into account in the product design for an attractive MaaS bundle.

5 General Methodological Approaches to Determine the Demand for Multimodal Mobility Packages

In order to be able to launch multimodal MaaS packages on the market it is necessary to determine the demand for such offers, the product design and pricing. Therefore a distinction can be made between

- objective measurements of the customers real observed selection behavior
- surveys of realized decisions (revealed preference) and
- surveys of hypothetical preferences (stated preference) or decisions (stated choice) [40, 53, 54].

The first two survey methods can be deduced either from previous transport usage data without MaaS or from the results of pilot projects. MaaS concepts such as Whim, UbiGo, swa Mobil-Flat, SBB Green Class etc. [55–60] have been scientifically accompanied, so that first insights into the usage behavior as well as the willingness to pay are possible. But the number practical examples are not enough for generally valid conclusions and recommendations.

Many of the studies mentioned in the previous section of this paper used stated preference data collection methods to investigate individual requirements for different MaaS bundles and the subjects' configuration choice with regard of the included transport modes, various pricing schemes, willingness to pay or additional features and services. Matyas and Kamargianni [29] and Ho et al. [32, 40] conducted corresponding stated choice and stated preference studies in London, Tyneside (UK) and Australia. For the evaluation and interpretation of a stated choice or a stated preference survey design the conjoint analysis is used in the most cases.

The conjoint analysis is a multivariate method in which various alternatives (choice sets) are presented to the interviewee and the interviewee decides for one alternative (stated choice) or has to make an assessment (stated preference). Each choice set consists of a combination of attributes (e.g., means of transport included, travel allowances or budgets, packet price etc.). In each round, several choice sets are presented to the survey subjects. By repeating the rounds with different choice sets, the benefit of individual attributes for the subjects can be statistically determined and conclusions can be made for the design of the packages [35].

Caiati et al. applied a slightly different procedural method in the Netherlands study [30] to estimate customers' preferences for MaaS plans. They focused on a customized bundle approach where respondents configure their most suitable mobility package by picking the preferred set if items from any of the available transport modes, service attributes and characteristics, each presented with varying prices [30, p. 145]. It can be expected that this kind of menu-based bundling (Create-it-Yourself) will provide comprehensive support for the market penetration of multimodal mobility packages.

6 Further Development Needs and Outlook

MaaS and mobility service bundling is still in its infancy with a very low market penetration. The mentioned studies in different countries and regions indicate that the most transport users are not yet disposed to subscribe to this new service in a large number [30, p. 145]. To overcome this situation the following further development paths should be considered:

1. Extended range of customized MaaS packages

In future is needed a paradigm shift towards mobility bundling according to the modular design principle and calculating the total price for the packet in the background system as mentioned above. One of the early examples for this approach is the Green Class product of the SBB Swiss Federal Railways. It has a consistently modular structure with two fixed modules, named by SBB Public Transport and Electric Car Rental, which must be included in all cases. However, the customer can still select from a number of different attributes and characteristics of each module. Within the public transport module e.g., he can choose between a General Abonnement 1st or 2nd class includes the use of all classic public transport modes in Switzerland, a Half Fare Travelcard reduces travel costs by 50% or an individual public transport subscription. The second fixed module is the electric car rental. The customer can choose between three electric vehicle models including different mileages [59, 60].

In addition to the fixed components, the customer can optionally book further modules. These include the parking module, the Swisscharge 3000 module, which gives the customer an annual credit for charging his electric vehicle with a charging capacity of 3,000 km. The car sharing and bike sharing modules allow customers to use the vehicles at a reduced rate or to use PubliBike bicycles free of charge for the first 30 min of each ride. It can also be supplemented by bike sharing with e bicycles. The taxi module comprises 10 taxi vouchers worth CHF 25 per year [39]. The duration of the subscription package can be individually set by the customer between 12 and 48 months. Using an online configurator provided by SBB, interested customers can create their own desired modules and receive information on the respective prices. Due to the large number of different configuration options, the price varies greatly, but is at least CHF 699 per month and at most CHF 2,351 per month (as at September 2019) [39].

Such a high level of customer centricity reflects the mobility behavior and needs of each individual in the best way, but it also implies a hardly manageable complexity for both the MaaS providers and the customers. To reduce complexity, MaaS Global in the past used whim points for its MaaS packages depending on the size of the package. The customer could dispose of these points freely and allocate them to the most preferred transport modes. But nevertheless this approach was not very successful. According to Hietanen [24], the idea of whim points is being revived in order to actively influence the customer's travel choice. The user can earn Whim points by choosing environmentally friendly, sustainable means of transport and convert them into various rewards. This, in turn, closes the circle on the background and motivation of MaaS (see Sect. 1).

2. Roaming functionality

Roaming is a worldwide offered service mode at the telecommunications industry allowing the mobile subscribers the use of mobile phones outside the range of their home network. Such a functionality is also very attractive for users of mobility services if they could use the transport modes included in their bundles also in other cities or regions, offered by the own MaaS provider or other mobility service providers integrated into a cross-regional MaaS platform. The customer can be debited for roaming as follows: no additional charges, surcharge on the costs in the home region or purchase of special country packages, as common in the mobile phone market. MaaS Global already offers roaming without extra charging. This includes - as of the end of 2019 - the cities of Antwerp, Birmingham, Helsinki and Vienna [61]. The attractiveness for the customers grows exponentially with the number of the included regions.

3. Mobility package upgrading by industry-related services and products

Cross-selling in cooperation with companies from other industries offers numerous upgrading possibilities. These can lead to additional revenues for the MaaS providers or the third-party companies, but also to more convenience for the customer.

Utility Industry and Real Estate Sector
Due to the fact that municipal transport companies are usually part of the municipal public utilities (e.g. responsible for energy supply), there are certain synergies can be exploited in the context of MaaS. The integration of mobility packages into the rent and the elimination of mandatory parking spaces as a result of less private car ownership seem to be possible [35, p. 87]. For example, MaaS Global is testing such a pilot with the housing provider SATO in Helsinki, where residents can purchase mobility packages at reduced prices [24].

Media Industry
The demand to consume digital media on the move such as newspapers, videos and music can also be used in the context of MaaS by offering on-board entertainment in the vehicles of the participating transport companies. Regardless of whether this is provided to customers free of charge, limited or as a premium it represents an additional comfort for customers. The billing can be done by the MaaS operator across all means of transport or directly included in certain mobility packages [35, p. 89].

Local Gastronomy and retail
Vouchers, discounts and other marketing activities connected with the mobility packages can ensure a higher number of customers in local businesses such as shops, restaurants, cultural institutions tec. This leads to more traffic demand, which usually takes place outside peak hours, and generates additional revenue for the MaaS operator [35, p. 90].

The design of multimodal mobility packages based on MaaS concepts requires a consequent step by step customer-centric approach. Many different aspects have to be taken into account particularly customer behavior, regulatory, commercial and technical issues. Public transport in a flat rate scheme is seen as an indispensable component of any kind of mobility bundles. MaaS should be built around public transport. This is particularly noteworthy in the context that Luxembourg is the first country in the world offering all public transport (bus, tram and train except 1st class) free of charge since the end of February 2020, with the aim of making public transport as attractive and reliable as possible. For the Luxembourg state, this means additional expenditure of 41 million euros per year [62].

The creation and implementation of multimodal mobility packages on the market are a relatively new topic and we are in an early stage to understand what this special offer might mean for mobility demanders and suppliers. A great number of studies provide theoretical investigations to estimate demand, the main factors influencing purchasing behavior, customer preferences, business models, environmental impacts, etc. By presenting and linking experiences from these studies and practical examples, this paper tries to present promising approaches for the design of mobility bundles and to give recommendations for the implementation in practice.

References

1. United Nations: Revision of the World Urbanization Prospects (2018). https://www.un.org/development/desa/publications/2018-revision-of-world-urbanization-prospects.html. Accessed 17 Feb 2020
2. Deckwart, A.: Bewertung von Maßnahmen zur Reduktion des MIV. Master Thesis, Technische Universität Dresden (2019)
3. Live-Counter.Com: Weltbestand an Autos (2020). https://www.live-counter.com/autos/. Accessed 17 Feb 2020
4. ITF: ITF Transport Outlook 2019. OECD Publishing, Paris (2019)
5. ONO via Linkedin: Bans and measures to eliminate polluting and voluminous vehicles off of the roads (2019). https://www.linkedin.com/feed/update/urn:li:activity:6503614226509107200. Accessed 17 Feb 2020
6. Reinz-Zettler, J.: Conference on Smart Mobility Services. COSMOS (2019)
7. Durand, A., Harms, L., Hoogendoorn-Lanser, S., Zijlstra, T.: Mobility-as-a-Service and Changes in Travel Preferences and Travel Behaviour: A Literature Review. KiM Netherlands Institute for Transport Policy Analysis (2018)
8. Streeting, M., Edgar, E.: Mobility as a Service - The Next Transport Disruption. L.E.K. Consulting, Boston (2017)
9. Mulley, C., Nelson, J.D., Wright, S.: Community transport meets mobility as a service: on the road to a new a flexible future. Elsevier Res. Transp. Econ. **69**, 583–591 (2018)
10. Karlsson, M.A., Sochor, J., Strömberg, H.: Developing the 'Service' in Mobility as a Service: experiences from a field trial of an innovative travel brokerage. Elsevier Transp. Res. Proc. **14**, 3265–3273 (2016)
11. Hartikainen, A., et al.: WHIMPACT: Insights from the World's First Mobility-as-a-Service (MaaS) System. Ramboll (2019)
12. UITP: Ready for MaaS? Easier mobility for citizens and better data for cities, Policy Brief, May 2019
13. Polydoropoulou, A., Pagoni, I., Tsirimpa, A., Roumboutsos, A., Kamargianni, M., Tsouros, I.: Prototype business models for Mobility-as-a-Service. Transp. Res. Part A: Pol. Pract. **131**, 149–162 (2019)
14. Kamargianni, M., Li, W., Matyas, M.: A comprehensive review of "Mobility as a Service" systems. In: Proceedings of the 95th Annual Meeting of the Transportation Research Board, Washington, DC (2016)
15. Kamargianni, M., Li, W., Matyas, M., Schäfer, A.: A critical review of new mobility services for urban transport. Transp. Res. Proc. **14**, 3294–3303 (2016). https://doi.org/10.1016/j.trpro.2016.05.277
16. Matyas, M., Kamargianni, M.: Mobility as a Service plans: how much do we prefer flexibility? Transp. Res. Proc. **00**(September), 4–6 (2017)

17. Stopka, U., Pessier, R., Günther, C.: Mobility as a Service (MaaS) based on intermodal electronic platforms in public transport. In: Kurosu, M. (ed.) HCI 2018. LNCS, vol. 10902, pp. 419–439. Springer, Cham (2018). https://doi.org/10.1007/978-3-319-91244-8_34
18. Li, Y., Voege, T.: Mobility as a Service (MaaS): challenges of implementation and policy required. J. Transp. Technol. **7**, 95–106 (2017)
19. Jittrapirom, P., Caiati, V., Feneri, A.M., Ebrahimigharehbaghi, S., Alonso-González, M.J., Narayan, J.: Mobility as a Service: a critical review of definitions, assessments of schemes, and key challenges. Urban Plan. **2**(2), 13–25 (2017)
20. Lyons, G., Hammond, P., Mackay, K.: Reprint of: the importance of user perspective in the evolution of MaaS. Transp. Res. Part A **131**(2020), 20–34 (2020)
21. Schikofsky, J., Dannewald, T., Kowald, M.: Exploring motivational mechanisms behind the intention to adopt Mobility as a Service (MaaS): insights from Germany. Transp. Res. Part A **131**(2020), 296–312 (2019)
22. MaaS Alliance: What is MaaS? (2019). https://maas-alliance.eu/homepage/what-is-maas/. Accessed 21 Feb 2020
23. Smith, G., Sochor, J., Karlsson, M.: Intermediary MaaS integrators: a case study on hopes and fears. Transp. Res. Part A **131**, 163–177 (2020)
24. Hietanen, S.: Personal interview led by Nied, J. In: Conception, state of the art and further development of multimodal mobility flat rates & bundles. Master Thesis, Technische Universität Dresden, 2019, Appendix 2 (2019)
25. Alkas, H.: Preisbündelung auf regulierten Telekommunikationsmärkten. Köln, Universität zu Köln, Wirtschafts- und Sozialwissenschaftliche Fakultät, Dissertation (2007)
26. Roll, O., Pastuch, K., Buchwald, G.: Praxishandbuch Preismanagement: Strategien, Management, Lösungen. Wiley-VCH, Weinheim (2012)
27. Wübker, G.: Preisbündelung: Formen, Theorie, Messung und Umsetzung, Gabler, Wiesbaden (1998)
28. Reinders, M.J., Frambach, R.T., Schoormans, J.: Using product bundling to facilitate the adoption process of radical innovations. J. Prod. Innov. Manag. **27**(7), 1127–1140 (2010)
29. Matyas, M., Kamargianni, M.: The potential of Mobility as a Service bundles as a mobility management tool. Transportation **2019**, 1951–1968 (2018). https://doi.org/10.1007/s11116-018-9913-4
30. Caiati, V., Rasouli, S., Timmermans, H.: Bundling, pricing schemes and extra features preferences for Mobility as a Service: sequential portfolio choice experiment. Transp. Res. Part A **131**(2020), 123–148 (2019)
31. Venkatesh, R., Mahajan, V.: Design and pricing of product bundles: a review of normative guidelines and practical approaches. In: Rao, V.R. (ed.) Handbook of Pricing Research in Marketing, pp. 232–257. Edward Elgar Publishing Company, Northampton (2009)
32. Ho, C., Hensher, D., Mulley, C., Wong, Y.: Potential uptake and willingness-to-pay for Mobility as a Service (MaaS): a stated choice study. Transp. Res. Part A: Pol. Pract. **117**, 302–318 (2018)
33. Claudy, N.: Zielgruppe. In: Lewinski-Reuter, V., Lüddemann, S. (eds.) Glossar Kulturmanagement. VS Verlag für Sozialwissenschaften, Wiesbaden (2011)
34. Freter, H., Naskrent, J.: Markt- und Kundensegmentierung: kundenorientierte Markterfassung und -bearbeitung, 2nd edn. W. Kohlhammer, Stuttgart (2008)
35. Nied, J.: Conception, state of the art and further development of multimodal mobility flat rates & bundles. Master Thesis, Technische Universität Dresden (2019)
36. Hietanen, S.: Whim – Mobility as a Service: The End of Car Ownership? (2020). https://www.maas-market.com/sites/default/files/SAMPO%20HIETANEN.pdf. Accessed 15 Feb 2020

37. MaaS Global Oy: Find your plan (2020). https://whimapp.com/plans/. Accessed 15 Feb 2020
38. Stadtwerke Augsburg Holding GmbH: Mobilität (2020). https://www.sw-augsburg.de/mobilitaet/swa-bus-tram/. Accessed 26 Feb 2020
39. Schweizerische Bundesbahnen SBB: SBB Green Class – das individuelle Mobilitäts-Abo (2020). https://www.sbb.ch/de/abos-billette/abonnemente/greenclass.html. Accessed 26 Feb 2020
40. Ho, C., Mulley, C., Hensher, D.: Public preferences for mobility as a service: insights from stated preference surveys. Transp. Res. Part A **131**, 70–90 (2020)
41. Robbert, T., Roth, S.: The Fascination of Flat-Rates - How Tariffs Influence Consumption Behaviour, Conference Paper (2010)
42. Günther, C.: Mobilität als Paket - Präferenz und Zahlungsbereitschaft für multimodale Produkte: Results of an empirical study in the INTERmobil research project, Technische Universität Dresden (2019)
43. Kamargianni, M., Matyas, M., Li, W., Muscat, J.: Londoners' attitudes to-wards car-ownership and Mobility-as-a-Service: Impact assessment and opportunities that lie ahead (2018). https://docs.wixstatic.com/ugd/a2135d_33f08862a08148389c89de1e908ac8a0.pdf. Accessed 28 Aug 2019
44. Matyas, M., Kamargianni, M.: Stated Preference Design for Exploring Demand for "Mobility as a Service" Plans (2017b). https://pdfs.semanticscholar.org/4816/f687ffc0f8c3a64938b9ad58934b9d0fe93a.pdf. Accessed 15 Aug 2019
45. Ratilainen, H.: Mobility-as-a-Service: Exploring Consumer Preferences for MaaS Subscription Packages Using a Stated Choice Experiment. Delft, Delft University of Technology, Civil Engineering and Geosciences, Master Thesis (2017)
46. Alonso-González, M., van Oort, N., Cats, O., Hoogendoorn, S.: Urban Demand Responsive Transport in the Mobility as a Service ecosystem: its role and potential market share, Thredbo 15: Competition and Ownership in Land Passenger Transport, Vol. 60 (2017)
47. Axhausen, K., Simma, A., Golob, T.: Pre-commitment and usage: Season tickets, cars and travel, Arbeitsberichte Verkehrs- und Raumplanung, vol. 24 (2000)
48. Stopka, U.: Mobility-as-a-Service (MaaS) – intermodal mobility bundling and packages in urban areas. In: Böhm, S., Suntrayuth, S. (eds.) Proceedings of the International Workshop on Entrepreneurship in Electronic and Mobile Business (IWEMB), Vestfold (2019)
49. Sochor, J., Strömberg, H., Karlsson, M., 2014. Travelers' motives for adopting a new, innovative travel service: Insights from the UbiGo field operational test in Gothenburg, Conference Paper
50. Hietanen, S., Sahala, S.: Mobility as a Service: can it be even better than owning a car? (2016). https://www.itscanada.ca/files/MaaS%20Canada%20by%20Sampo%20Hietanen%20and%20Sami%20Sahala.pdf. Accessed 15 Feb 2020
51. Gebhardt, S.: Personal interview led by Nied, J. In: Conception, state of the art and further development of multimodal mobility flat rates & bundles. Master Thesis, Technische Universität Dresden, 2019, Appendix 4 (2019)
52. Sochor, J., Strömberg, H., Karlsson, M.: An innovative mobility service to facilitate changes in travel behavior and mode choice, Conference Paper (2015)
53. Kroes, E., Sheldon, R.: Stated preference methods: an introduction. J. Transp. Econ. Pol. **22** (1), 11–25 (1988)
54. Treiber, M.: Verkehrsökonometrie für Bachelor-Studierende, lecture notes, Technische Universität Dresden (2019). http://www.mtreiber.de/Vkoek_Ba_Skript/Verkehrsoekonometrie_BA.pdf. Accessed 15 Feb 2020
55. MaaS Global: Find your plan (2019a). https://whimapp.com/plans/. Accessed 15 Feb 2020

56. Ubigo Innovation AB: Allt resande i en app (2019a). https://www.ubigo.se/. Accessed 15 Feb 2020
57. Ubigo Innovation AB: UbiGo (Version 1.0.6+109) [Mobile application software] (2019b). https://play.google.com/store/apps/details?id=com.fluidtime.android.fluidhub.ubigo&hl=de. Accessed 15 Feb 2020
58. Stadtwerke Augsburg: Alles fahren zum fixen Preis: Die Mobil-Flat der Stadtwerke Augsburg - Zwei Preis-Pakete für 79 und 109 Euro monatlich (2019). https://www.sw-augsburg.de/ueber-uns/presse/detail/alles-fahrenzum-fixen-preis-die-mobil-flat-der-stadtwerke-augsburg-zwei-preis-paketefuer-79-und/. Accessed 15 Feb 2020
59. Schweizerische Bundesbahnen SBB: SBB Green Class lanciert neues Mobilitätskombi (2018). https://company.sbb.ch/de/medien/medienstelle/medienmitteilungen/detail.html/2018/4/1304-1. Accessed 15 Feb 2020
60. Schweizerische Bundesbahnen SBB: Abo-Module für SBB Green Class (2019). https://www.sbb.ch/de/abos-billette/abonnemente/greenclass/mobilitaets-abo/abo-module.html. Accessed 15 Feb 2020
61. MaaS Global: MaaS Global, the company behind the Whim app (2019b). https://whimapp.com/about-us/. Accessed 15 Feb 2020
62. https://www.swr.de/swraktuell/rheinland-pfalz/trier/Feier-zur-Verkehrswende-Bus-und-Bahn-werden-in-Luxemburg-gratis,bus-luxemburg-100.html

A Multi-device Evaluation Approach of Passenger Information Systems in Smart Public Transport

Waldemar Titov[1]([⊠]), Hoa Tran[2], Christine Keller[1],
and Thomas Schlegel[1]

[1] Institute of Ubiquitous Mobility Systems, Karlsruhe University
of Applied Sciences, Moltkestr. 30, 76133 Karlsruhe, Germany
{waldemar.titov,iums}@hs-karlsruhe.de
[2] FZI - Forschungszentrum Informatik, Haid-und-Neu-Str. 10-14,
76131 Karlsruhe, Germany
tran@fzi.de

Abstract. Adaptive passenger information for an enhanced mobility experience may be the next step towards a smart public transport. In our research project, we have developed a multi-device evaluation approach for adaptive passenger information systems of mobile public displays. An adaptive passenger information system needs to be aware of the passenger's context. In order to fulfill this requirement, we use the passenger's personal devices like smartphones or smart watches as context sources. In this paper, we describe our approach of a multi-device passenger information system evaluation focusing on privacy aspects. We present three different methods of pseudonyms that were used to visually link the personal information on passenger's private devices with the displayed information on the public display. In addition, we report on our evaluation results from a user study evaluating the acceptance and the intelligibility of the used visual pseudonyms.

Keywords: Smart public transport · Passenger information · Data privacy

1 Introduction

Public displays are often found in public places like railway stations, airports and touristic places. This technology is typically used to provide general information about places, flights and departing trains [1]. However, public interactive displays (PIDs) are used in commercial and research places like shopping centers or university buildings [2, 3]. The purpose of the PID technology is to provide specific and individual information such as wayfinding or route planning [4, 5]. In the context of public transport, most public displays are used as a source of information about upcoming trains and connections. However, nowadays most PIDs are used in a stationary context. They are built in existing or new infrastructure and, obviously, do not change places. In public transport vehicles, PIDs become mobile, which affects the information they present. While moving, their context constantly changes. Most PIDs in vehicles therefore show contextually adapted information, referring to the next stop for example.

H. Krömker (Ed.): HCII 2020, LNCS 12213, pp. 340–358, 2020.
https://doi.org/10.1007/978-3-030-50537-0_25

However, situated content that might be relevant for some passengers, may not necessarily be relevant to any individual. This becomes obvious when considering the varying information needs of different passenger having different destinations, tickets and different familiarity with the public transport network. The technology of PID has the potential to go beyond simple interactive information terminals and become truly smart technology providing situated, personalized and thereby relevant content for individual passengers.

Therefore, in our research project SmartMMI we are developing an adaptive passenger information system for a mobile public display focusing on improving passenger information along their travel chain. In this project, we research model and context based mobility information on smart public displays and mobile devices in public transport. We want to improve the information provision for passengers in every situation. Depending on the situation but also depending on the passenger, the need for information changes. This can happen in the event of a disruption, plan changes, the discovery of tourist destinations or of services available along the route. Our goal is to inform passengers appropriately in their individual situation. To this purpose, we combine a variety of data sources to form a smart public transport data platform that integrates real time public transport information, information on points of interests, but also information on bike or car sharing services along the route. The integrated data is then adapted to the passenger's context, and presented on semi-transparent public displays in public transport vehicles. These semi-transparent, multi-touch-enabled public interactive displays are built-in as windows in public transport vehicles and called "SmartWindows". Beyond the design of the interactive SmartWindow, we also research the interplay of mobile application and SmartWindow for situational passenger information.

The SmartWindow serves as a passenger information system and provides real-time traffic information, information about traffic disturbances as well as passenger's connections. Developing an adaptive passenger information, the passenger information system needs to be aware of the passenger's context. In order to fulfill this requirement, we use the passenger's personal devices like smartphones or smart watches as the context sources. In this paper, we present the results of our research evaluating a multi-device passenger information system focusing on privacy aspects. We present three different methods of pseudonyms that were used to visually link the personal information on the passenger's smartphone with the displayed information on the public display. Finally, we report on our evaluation results from the acceptance study and the user study evaluating the intelligibility of the used visual pseudonyms.

In the next section of our paper, we will look at approaches towards passenger information in public transport and at public displays and respectively their evaluation. We will then briefly describe the scope and intended configuration of our SmartWindow. The next section introduces the conducted user study and presents their result. In the last section of this paper, we will discuss the challenges and findings of our development and evaluation process and will give an outlook on our future work, pointing out evaluation steps we are planning in the upcoming months.

2 Related Work

Interactive digital displays are becoming a ubiquitous part of urban environments [6]. The use of public interactive displays varies greatly across different situations, yet the effectiveness of all public displays relies on the assumption that they will be noticed and used [7]. However, since public displays are used very contextually in various public spaces, lab studies can often not paint the whole picture and it is difficult to determine the real effectiveness of a public interactive display [5]. A user-centered design of public displays can ensure to keep the user's requirements in the specific application area in focus during the development process. In public transport, the user's requirements are hard to grasp, since the user group is as wide as it can get – almost everybody uses public transport at some point.

Personalizing the content of public interactive displays in order to make it more relevant to passers-by, many researchers have focused on user profiles. The personal profiles usually require to be set up by the user through a smartphone app, which can potentially compromise privacy and may create barriers for interaction as implementations. Parker et al. describe a deviating approach. The authors explore an implicit personalization approach for public interactive displays as an alternative to user profiles. Two evaluation studies of public interactive displays, implicitly adapting the display based on user's goals and characteristics, are presented [3]. Based on the carried-out studies, a definition of implicit personalization as well as the adaptation of the user interface of a PID is derived.

To understand the basic requirements of passengers in public transport, it helps to design personas [8]. Personas represent archetypical users and can facilitate the understanding of the user's behavior, needs, motivations and limitations. Hörold et al. describe personas and their interaction preferences in German public transport [8]. The personas we developed in our research project were based on this work. For the UK, Oliveira et al. described the development of personas in a collaborative research project of academic institutions and industry partners in the UK [9, 10]. Based on their personas, they found that newly designed technology could improve passengers' experience in public transport and argue for a solid understanding of the users and their needs when conceiving innovations in public transport. Oliveira et al. collected data by semi-structured face-to-face interviews and additionally used questionnaires. Questionnaires are a suitable method to have a closer look on which information passengers need in which situations. We also included a questionnaire in our design process.

Public interactive displays introduce new possibilities for transportation companies, like using less paper-based information at different stages of the journey. Hörold et al. suggest a user-centered design process and four different evaluation methods to identify where and how to apply public displays in public transport: expert workshops, comparative usability evaluations, two lab-based usability evaluations and an expert evaluation [5]. These multiple evaluation methods combine the knowledge of experts and the expectations of passengers as well as knowledge from transport companies. In our user centered design process, we also used several different evaluation methods. Public transport experts were involved in the development of our personas and requirements as well as regular passengers in lab-based evaluations. We also had the chance to evaluate our design with some media communications experts and reported on our finding in Keller et al. [26].

We argue to extend the range of evaluation methods by studies that involve public transport context, such as real public transport data up to field studies in real public transport settings, which is planned for our prototype at the beginning of next year. Ardito et al. also argue for field tests and report a certain tendency for more field tests in their survey on interaction with large displays [11].

3 Public Displays in Public Transport

Stationary public displays can be found in many places of mobility with a high density of people, such as airports, train and bus stations. Nowadays, many buses and trams are equipped with so-called mobile public displays through which the passenger information service is offered. Mobile public displays have attracted increasing attention in recent years. However, many of them do not offer any interaction possibilities but display advertising, upcoming departures or information about the current location. In future, the technology of semi-transparent public interactive displays implemented in public transport promises to improve the passenger information provision. First concepts of implementing interactive semi-transparent displays instead of usual windows and using them as passenger information systems already exist [12]. The concept of a SmartWindow as shown in Fig. 1 is interactive so that every passenger can work with it and get a variety of information, such as weather, next stops, detours and disturbances adapted to the passenger's route and passenger's preferences. However, challenges of visually linking the personal data with the data on the PIDs as well as data security arise.

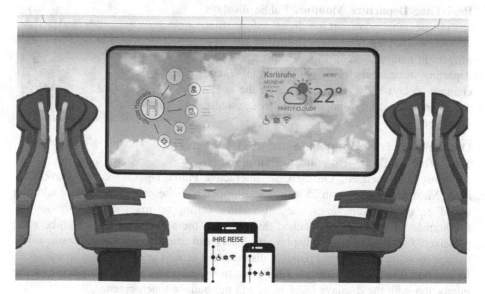

Fig. 1. Graphic design study of a mobile public interactive display called SmartWindow.

3.1 Public Displays at Stops

More and more public transport stops are being equipped with displays. The main purpose of these displays is to inform passengers about the departure times of buses and trains. Mostly, simple LED displays are used. A more sophisticated and modern technology is using LCD or plasma screens, which thanks to their significantly higher resolution are able to display information that is more complex. Three concrete examples are presented below.

TransitScreen. TransitScreen is a software solution from Multimodal Logic for displaying various transport information [13]. It enables the departure times of a large number of US transport companies (e.g. bus and train) to be displayed, but also bike and car sharing providers that offer the TransitScreen. The software shown in Fig. 2 can be used on almost all displays. This is to always ensure that only suitable information be shown for the location. As a site of operations, the manufacturer not only provides stops but also integrates them into shops or in the entrance area of companies, so that customers or employees can approach the stop according to the departure times. TransitScreen is designed as a pure display and does not allow any interaction by the user. Only if the lines and means of transport suitable for the user are shown on the display, the user can use the display.

Fig. 2. Exemplar of the TransitScreen [13].

Real-Time Departure Monitor. Public displays at stops of public transport are digital information systems displaying relevant information for passengers in realtime. As in Fig. 3 information like departing time and track of individual lines, changes, delays and in rare cases even the occupation rate of a transport system are usually displayed.

Fig. 3. Point of information [14].

On the Go! Travel Station. Since 2011, The New York MTA (Metropolitan Transportation Authority) has operated "On the Go!" Interactive Public Displays called Travel Stations. Meanwhile, more than 25 stops are equipped with these displays [15]. In addition to calling up departure times, they also allow calling up maps of the surrounding. Furthermore, routes can be planned directly, and the appropriate lines members are presented/displayed. Via the display, the users can also access current news, information about the weather and possible disturbances as well as other means of transport. In addition, the displays are used for advertising purposes. As shown in Fig. 4 the interaction with the displays takes place via the built-in touchscreen.

Fig. 4. On the Go! Travel Station in New York [15].

3.2 Public Displays in Public Transport Vehicles

Many public transport operators equip their buses and trams with screens through which a passenger information service is offered. These mobile public displays often display the current time and date, the upcoming stops and transfer options at the next stop as sown in Fig. 5. In addition to traffic relevant information, the mobile public displays are used for advertising and entertainment purposes. However, no passenger adaptation and interaction are possible.

Fig. 5. Displays in a tram presenting passenger information and entertainment.

4 Privacy in Public Transport

The visualization concepts being currently developed for the SmartWindow follow the general privacy rules. This means that personal passenger information like origin, destination and interchanges will remain in the private device while the omnipresent and big screen of a SmartWindow seems to be the proper way to visualize context-aware passenger information like a reminder to interchange. Since a public transport vehicle usually carries several passengers and many passengers can interact with the

SmartWindow at the same time, the information can potentially be seen by many passengers. As shown in Fig. 6, the developed visualization concept of a SmartWindow contains publicly accessible information. Consequentially, a graphical concept that visually links personal information with the information displayed on the PID needs to be developed. Additionally, passengers have a right of data protection, which means that context-aware passenger information displayed on the SmartWindow must be anonymous and cannot be affiliated with real passengers. This reveals the requirement of developing SmartWindow visualizations that must follow the privacy regulations.

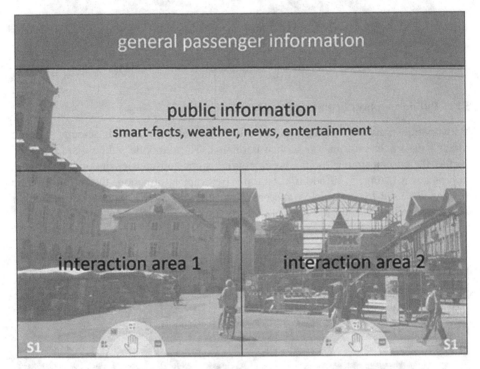

Fig. 6. Arrangements for interaction and information areas on the SmartWindow.

4.1 Legal Basis

The data transfer from a private device to the SmartWindow and the subsequent visualization of the data on the large window are subject to different legal hurdles, especially data protection regulations. Data protection refers to the protection of the citizen against the impairments of passengers' privacy by processing data, which relate to themselves. According to established case law, as an expression of human dignity as a self-determined individual, everyone must be able to decide which personal data should be accessible to whom, when and how [16]. In this specific case of using the SmartWindow, the passenger has to be able to decide independently, which data should be transferred to the service provider and which data could be visible on the SmartWindow. Like in all legal areas, these data protection principles consist of the interaction of a large number of standards and regulations.

In the international context, Article 8 of the European Convention on Human Rights (ECHR) plays a significant role. The ECHR is an international treaty signed by 47 countries [17]. The Convention binds all contracting states and, as a fundamental rights document from 1950, does not contain a particular basic data protection right. Instead, the basis for data protection is the right of respect for one's "private and family life, his home and his correspondence" guaranteed in Art. 8 Section 1 ECHR [18]. The scope of protection of Art. 8 ECHR is broadly interpreted and the protection of personal data and telecommunications secrecy is also understood as an expression of this right to privacy [19]. As a result of this interpretation, any collection, storage, transfer or other processing of personal data or human communication constitutes an encroachment in Art. 8 ECHR and must be justified.

All 27 member states of the European Union (EU) are also the contracting states of the ECHR. However, the EU itself as an organization has not joined the ECHR yet [20]. For this reason, the EU Primary Law, which results from the Charter of Fundamental Rights of the European Union (CFR), applies within the EU and between the member states. Art. 7 CFR standardizes the right of respect for private and family life, home and communications and largely corresponds to Art. 8 ECHR in terms of wording and warranty content. The European Convention, which has formulated the text of the Charter, states that the rights of Art. 7 CFR correspond to the rights guaranteed by Art. 8 ECHR [21]. Beyond this general principle, the CFR specifies data protection even further in Art. 8. This article is a lex specialis for the processing of personal data. This shows that the CFR was formulated more recently.

These principles of the ECHR and the CFR result in further, more specific EU data protection regulations, among which the European General Data Protection Regulation (GDPR) plays a central role. The GDPR has been directly applicable within the EU since May 25, 2018. On the one hand, the GDPR is supposed to unify members across the EU and create a solid and enforceable legal framework in the area of data protection [22]. On the other hand, it should offer uniform guidelines for equal economic conditions in the union and thus strengthen the internal market [23]. This is why the GDPR is the starting point for any data protection review. As part of this work, the basic structure and basic terms contained in the regulation are presented below, which are applicable to the topic of public transport. Since there are several different legal bases for data protection law in addition to the GDPR, the role of the GDPR that is applicable in our project is shown in Fig. 7.

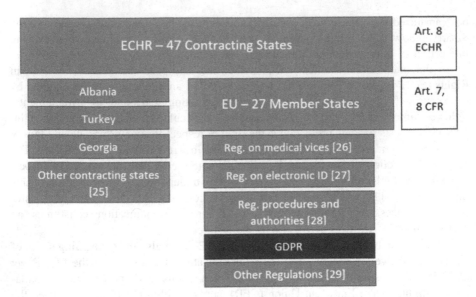

Fig. 7. Arrangement of the GDPR in the international context [24–27].

4.2   GDPR

As mentioned above, the user of the SmartMMI system should be able to decide which data, when, to whom and how he wants to make accessible. However, this does not affect all types of data, but only certain data that is protected in the GDPR. According to Art. 2 Para. 1 GDPR, the regulation applies to the wholly or partially automated processing of personal data [...]. Thus, only personal data are protected. According to Art. 4 No. a GDPR, personal data means any information relating to an identified or identifiable natural person. The identifiability of a person is particularly to be assumed, if it can be determined by association with an identifier such as a name, a location or one or more special characteristics, the expression of their physical, physiological, genetic, psychological, economic, cultural or social identity [28]. Therefore, personal information such as name, address, date of birth, age, origin, gender, education, marital status, eye color, fingerprints, X-rays, photos and video recordings, preference or attitude is to be regarded as personal data [29]. However, special personal data or sensitive data which according to Art. 9 Para. 1 GDPR enjoy higher protection requirements are defined more precisely. The reason for the increased need for protection is an increased risk of discrimination. Sensitive data includes information on:

- racial or ethical origin,
- political opinions,
- religious or philosophical beliefs,
- trade union membership,
- genetic & biometric data for the purpose of uniquely identifying of a natural person,
- health data and
- data on the sex life or sexual orientation of a natural person.

The data processed by the SmartMMI system can be classified into the following categories based on the above definitions:

Table 1. Categories of the data used in the SmartMMI-System.

Non-personal data	Personal data	
	Standard personal data	Sensitive data
Route	Name	
Start and end stop	Address	
Detour	Appointment calendar	
Weather	Tickets	
Alternative transportation - Carsharing - bikesharing	Alternatives transportation - Walkways to the destination	Health data (e.g. barrier-free selection could imply that the user is dependent on the wheelchair)
Degree of occupation of the train	Language	Language selection could imply ethical origin
Events in the area	Walking speed	
Language selection		
(1)	(2)	(3)

The data from (1) can be displayed on the SmartWindow in the area of public information, which is shown in Fig. 6, regardless of the user requirements. Data from (2) and (3) can only be processed with the consent of the user.

Independent of the consent regulation, the SmartWindow must not become a source of crimes. Therefore, measures must be taken that the sensitive data is not accidentally transmitted and/or displayed. For example, the walkways to the destination must not be displayed on the SmartWindow, in any case with or without consent, so that the possible address of the user can not be seen by other passengers. If there is a possibility to buy tickets via the system, then security measures must assure that the personal data of the customer, e.g. bank details, are not processed unprotected. One of the suitable technical and organizational measures for the purpose of data processing security, which is also mentioned in Art. 3 GDPR, is pseudonymization. The SmartMMI system also uses this method to ensure the required protection. The pseudonymization is explained in more detail in the following paragraph.

5 Pseudonymization of Passenger Data on Mobile Public Displays

Natural persons have the right of personal data protection. One way to ensure that the publicly displayed information cannot be assigned to a single person is the method of pseudonymization. Only the individual passenger knows which specific pseudonym represents the information on the SmartWindow that is intended for him or her. The method of pseudonymization applies the processing and alteration of personal data that cannot be traced back to the identity of a natural person without the inclusion of additional information and thus helps protecting a passenger's identity [3, 4]. Pseudonyms are widely used in programs that are used by several users simultaneously, such as GitHub, shown in or GoogleDocs.

Pseudonymization is mostly realized through different colors, symbols, avatars, signs or abstractions. The context awareness of the PID system is realized by the details-on-demand principle described by Shneiderman [30]. For this purpose, the passenger establishes a connection between the application installed on the mobile device and the PID displaying the passenger information service. This connection allows the system to track whether the previously planned trip and the planned connections are still feasible. In case of upcoming interchanges, delays or disruptions the system will inform the passenger on the omnipresent PID calling for action. Thereby, the pseudonymization will serve as an identification feature for the specific passenger. In the use case of SmartWindow, a pseudonym is assigned to a passenger when the personal smartphone and public SmartWindow are coupled to exchange those traffic relevant data. Since only the individual passenger knows which pseudonym is used by the coupled devices, the risk of publicly displayed data being traced back to the individual subject is minimized.

6 User Study

After analyzing the legal requirements, we conducted an experimental user study to evaluate the usefulness and intelligibility of the visual links between personal passenger data on passengers' smartphone and publicly visible data on the mobile PID called SmartWindow. Considering the multi-device evaluation approach, we designed paper mockups representing passengers' smartphone, as shown in Fig. 8.

Fig. 8. Paper prototypes of a smartphone.

For the SmartWindow we developed a high fidelity prototype based on HTML and JavaScript that implemented interfaces to our smart public transport data platform. The HTML prototype is shown in Fig. 9. In the design of this high fidelity prototype, we applied the results of all prior evaluation phases that were described in [31].

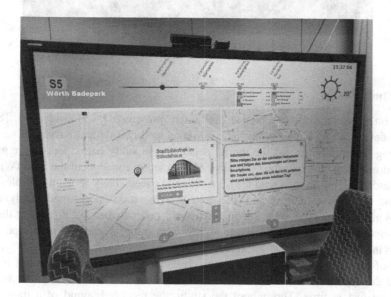

Fig. 9. HTML prototype of a SmartWindow.

The SmartWindow prototype runs on our 98-inch multi-touch display and can adapt its transparency as well as brightness. In our research project, we mainly research the configuration of a SmartWindow besides two rows of two seats facing each other. In order to reflect reality in our study, we arranged the seats correspondingly and adapted the size of the prototype to the exact size of vehicle windows in local trams. As shown in Fig. 10, test subjects will interact with the SmartWindow while seated.

Fig. 10. Study participant interacting with the SmartWindow.

6.1   Study Design

Based on public transport personae previously presented by Hörold et al. [8] and further developed by Keller et al. [32], we developed two typical passenger scenarios involving the commuter and the tourist persona. The study participants were given a short introduction covering the purposes and interaction concepts of the SmartWindow. Afterwards, the test persons read the scenarios and successively took on the roles of the personae of the commuter and the tourist. The tasks performed applying the SmartWindow prototype were set in rotation in order to be able to assess the conspicuousness of the individual pseudonyms. The subjects had the prototype of the SmartWindow on the display as well as a smartphone at their disposal. Applying the wizard-of-Oz method the smartphone was presented to the study participants in paper form at the appropriate time. Thus, the information needed to complete the tasks was passed on just in time. Throughout the study, the recorded sound of a tram ride is played in the background to make the scenario as realistic as possible. Subsequently, all test persons answered a questionnaire on general information, technological affinity, and knowledge of local public transport as well as rating the mockups and pseudonyms. It is ensured that no persons participate in the study who are familiar with the SmartWindow prototype, since prior knowledge may distort the results.

6.2 Study Procedure

The procedure of the study is organized in two parts, the study itself and the questionnaire at the end. In the first part, study participants are given paper prototypes of smartphone interfaces to obtain the necessary information. The smartphone contains personal data that legally may not be displayed on the SmartWindow such as origin or destination of a passenger. Applying the wizard-of-Oz method means that by designing mockups, a finished system can be simulated without being completely developed.

In addition to the wizard-of-Oz method, the thinking-aloud method was used to determine impressions and feelings of the participant during the user study. This enables the study director to make observations and record them. With the thoughts of the test persons, improvements or changes can be made after the study if necessary, and it becomes clear why the test persons act the way they act. This tool is often used in software development. It ensures that prototypes can be tested for user-friendliness and usability during the development phase helping to eliminate malfunctions [33].

During the study, each test subjects sit individually in a window seat in front of the display, as shown in Fig. 11. After a short introduction, they subjects are given tasks that need to be performed by interacting with the SmartWindow. These tasks include buying a suitable ticket or planning a route using the available devices SmartWindow and smartphone. The screens of the smartphone are available as paper prototypes and handed to the subjects as shown in Fig. 12.

Fig. 11. Study participant receives a scenario. **Fig. 12.** Participant receives a paper prototype.

The test subjects were not informed in advance about the fact that pseudonyms are used in the mockups in order to be able to assess the conspicuousness of these. All study participants, even those who did not notice any of the pseudonyms, were given the opportunity afterwards to familiarize themselves with the types of pseudonymization used on the SmartWindow as well as on the smartphone in order to be able to con-sciously answer the corresponding questions in the questionnaire. By utilizing the thinking-aloud method during the study, it became visible whether the method of pseudonymization was recognized by the test persons or whether they completed the tasks without the pseudonyms.

Once a study participant completed both scenarios, the questionnaire was answered online on a tablet computer. The questionnaire was divided into two parts. The first section contained general questions about the person, technology affinity and the knowledge of public transport. The second part asked questions about the study, such as the conspicuousness of the used pseudonyms as well as an assessment which of the pseudonyms the test subjects preferred for the SmartWindow in public transport.

Overall, 16 participants of different age groups, different experiences with public transport and affinity towards new technologies participated in our user study. The ages of the participants ranged from 23 to 34-years. 14 participants were students of different majors and two participants employed. Ten participants stated to use public transport regularly, four subjects stated one to three times a month and two respondent use public transport less frequently than once a month. All subjects stated to have a high technical affinity, were familiar with the operating technology, and could participate in the study without extensive technical introduction. The results give a good insight into the varying preferences of different passengers.

7 Results

All test subjects were able to empathize into the personae and the given scenarios. Some participants noted the typical background noise of a tram, which tried to make the scenario even more realistic. In general, the possibility of pseudonymization was stated as positive by all study participants in the context of being implemented in public transport on SmartWindow and smartphone in order to protect the data from unau-thorized access. Only two out of sixteen study participants did not see any of the three different types of pseudonyms used during the study. Those subjects still were able to successfully complete the tasks.

Twelve out of sixteen participants (75%) chose the symbols as shown in Fig. 13a as their favorite pseudonymization tool. The subjects described the symbol as most likeable, easy to understand and recognize, as well as most conspicuous. One partic-ipant rated the pseudonymization tool of abstractions as shown in Fig. 13b as appro-priate. The benefits were stated as easily recognizable to a person through the colors and applicable through prior knowledge of the type of pseudonymization. Others criticized the abstraction as requiring getting used to, strange and not easily recog-nizable. Equally, one participant rated the pseudonymization tool of numbers as shown in Fig. 13c as the best option. Other subjects mentioned the possibility of confusion with the numbers of lines seen on the departure monitor. However, the number as a pseudonym is described as most unobtrusive and thereby increasing the feeling of personal data security.

Fig. 13. Pseudonymization tools: a) symbols [left], b) abstractions [center], c) numbers [right].

The evaluation of the questionnaire shows that the respondents, who do not wish to display the sensitive data listed in Table 1, do not wish to do so on their own initiative. In addition, it becomes apparent that the subjects are reluctant to show sensitive, personal information in front of other persons. Subjects were more willing and likely to display data marked as insensitive on the SmartWindow than data marked as sensitive. One person stated to not hesitate in revealing the destination address. Another three subjects saw no problem revealing their ticket information.

Fig. 14. Study participant participating in our multi device evaluation study.

Ninety percent of the study participants would use the combination of SmartWindow and smartphone, shown in Fig. 14, in real life including pseudonymization and evaluated this concept extremely positively.

Summarizing the results, the study showed that most participants (83%) favored the pseudonymization method using symbols over the ones featuring abstractions and numbers, which both have been rated less noticeable on the big screen. Reasons for the preference of the symbols were their simplicity, readability, sympathy and abnormality.

The idea of pseudonyms on the big screen was unanimously seen as a good way of keeping personal information private as most participants mentioned a malaise to the idea of their sensitive data being visible for other passengers.

8   Discussion, Outlook and Limitations

Since the pseudonym can be found on both the smartphone and the SmartWindow, it is easy to find and recognize as a personal symbol. The disadvantage of this method is that only smartphone owners can use it. Potentially in future smartphone users may be able to determine what is displayed on the mobile PID themselves by setting the parameters in the application. Data actively released by self-determination is not covered by the EU data protection basic regulation. However, our study shows that participating subjects would not make sensitive data available for display on a SmartWindow. This shows that in the age of digitization, many people are more sensitive to how they handle their personal data and information.

Continuing this work, we are planning to carry out evaluations for other methods of pseudonymization, such as different color elements. Furthermore, we are planning on developing a real application and testing the pseudonymization with subjects who have prior knowledge of the pseudonyms. In this case, the information which pseudonym is dynamically assigned will be provided to the passenger while starting the application.

By using the thinking-aloud method, it was observed that the rotation of the tasks led to a recognition and understanding of the first pseudonym only when it was repeatedly displayed on the available devices. The corresponding question in the questionnaire revealed that some participants did not notice the use of pseudonyms at all. Since the symbol was the most conspicuous pseudonymization tool according to the majority of respondents, none of the other pseudonyms was recognized when the task began with the symbolization. In the other cases, when the pseudonyms number or abstraction were used first, all participants identified the pseudonyms.

Acknowledgements.   This work was conducted within the scope of the research project "SmartMMI - model- and context-based mobility information on smart public displays and mobile devices in public transport" and was funded by the German Federal Ministry of Transport and Digital Infrastructure as part of the mFund initiative (Funding ID: 19F2042A). We would like to thank Johannes Bauer and Sarah Eckert for their excellent contribution to this project.

References

1. Sahibzada, H., Hornecker, E., Echtler, F., Fischer, P.T.: Designing interactive advertisements for public displays. In: Proceedings of the 2017 CHI Conference on Human Factors in Computing Systems, New York, NY, USA (2017)
2. Storz, O., Friday, A., Davies, N., Finney, J., Sas, C., Sheridan, J.: Public ubiquitous computing systems: lessons from the e-campus display deployments. IEEE Pervasive Comput. **5**, 40–47 (2006)

3. Parker, C., Kay, J., Tomitsch, M.: Device-free: an implicit personalisation approach for public interactive displays. In: Proceedings of the Australasian Computer Science Week Multiconference, p. 8. ACM, January 2018

4. Kühn, R., Lemme, D., Schlegel, T.: An interaction concept for public displays and mobile devices in public transport. In: Kurosu, M. (ed.) HCI 2013. LNCS, vol. 8007, pp. 698–705. Springer, Heidelberg (2013). https://doi.org/10.1007/978-3-642-39330-3_75

5. Hörold, S., Mayas, C., Krömker, H.: Interactive displays in public transport – challenges and expectations. Proc. Manuf. **3**, 2808–2815 (2015)

6. Alt, F., Schneegaß, S., Schmidt, A., Müller, J., Memarovic, N.: How to evaluate public displays. In: Proceedings of the 2012 International Symposium on Pervasive Displays, New York, NY, USA, (2012)

7. Brignull, H., Rogers, Y.: Enticing people to interact with large public displays in public spaces. In: Human-Computer Interaction INTERACT 2003: IFIP TC13 International Conference on Human-Computer Interaction, Zurich, 1st–5th September 2003 (2003)

8. Hörold, S., Kühn, R., Mayas, C., Schlegel, T.: Interaktionspräferenzen für Personas im öffentlichen Personenverkehr. In: Mensch & Computer 2011: überMEDIEN| üBERmorgen, Chemnitz (2011)

9. Oliveira, L., Bradley, C., Birrell, S., Tinworth, N., Davies, A., Cain, R.: Using passenger personas to design technological innovation for the rail industry. In: Kováčiková, T., Buzna, Ľ., Pourhashem, G., Lugano, G., Cornet, Y., Lugano, N. (eds.) INTSYS 2017. LNICST, vol. 222, pp. 67–75. Springer, Cham (2018). https://doi.org/10.1007/978-3-319-93710-6_8

10. Oliveira, L., Bradley, C., Birrell, S., Davies, A., Tinworth, N., Cain, R.: Understanding passengers' experiences of train journeys to inform the design of technological innovations. In: Re: Research - the 2017 International Association of Societies of Design Research (IASDR) Conference, Cincinnati (2017)

11. Ardito, C., Buono, P., Costabile, M.F., Desolda, G.: Interaction with large displays: a survey. ACM Comput. Surv. **47**(3), 46:1–46:38 (2015)

12. Research Project SmartMMI: Model- and Context-based Mobility Information on Smart Public Displays and Mobile Devices in Public Transport. https://smartmmi.de/project/. Accessed 16 Dec 2019

13. TransitScreen. http://transitscreen.com. Accessed 15 Dec 2019

14. Fischer, K.P.: Digital Signage: Werbliche Kommunikation am Point of Sale auf Flachbild-schirmen. Theoretische Hintergründe, Aufgaben und Wirkungsmessungen. München, LMU München: Fakultät für Psychologie und Pädagogik, Diss., Januar 2011

15. MTA. Advertise with us - On The Go Travel Stations. http://web.mta.info/nyct/OntheGoAds/. Accessed 23 Dec 2019

16. Volkszählungsurteil (the census decision to German's Federal Constitutional Court), BVerfGE 65, 1, 41f.; Leeb/Liebhaber, JuS 2018, 534

17. Members of ECHR. https://www.coe.int/en/web/conventions/full-list/-/conventions/treaty/005/signatures?p_auth=pHWlcipN

18. Kühling, J., Seidel, C., Sivridis, A.: Datenschutzrecht (Start ins Rechtsgebiet), Recital 22

19. Leeb, C.M., Liebhaber, J.: Grundlagen des Datenschutzrechts. Jus: Juristische Schulung, (6), 534–537 (2018)

20. European Commission proposes negotiating directives for the Union's accession to the European Convention on Human Rights (ECHR) – FAQ. https://ec.europa.eu/commission/presscorner/detail/de/MEMO_10_84. Accessed 3 Jan 2020

21. Kühling, J., Seidel, C., Sivridis, A.: Datenschutzrecht (Start ins Rechtsgebiet), Recital 41

22. Recital 3 and 9 of the GDPR

23. Recital 5 and 10 of the GDPR

24. European Regulation on medical devices. https://eur-lex.europa.eu/legal-content/EN/TXT/?
uri=CELEX:02017R0745-20170505&qid=1576858417147
25. European Regulation on electronic identification and trust services for electronic transactions in
the internal market. https://eur-lex.europa.eu/legal-content/en/TXT/?qid=1576858555960&
uri=CELEX:02014R0910-20140917
26. European Regulation as regards the procedures and authorities involved for the authorization
of CCPs and requirements for the recognition of third country CCPs. https://eur-lex.europa.
eu/legal-content/DE/TXT/?qid=1576859018488&uri=CELEX:32019R2099
27. Database of European Union Law/Regulations. https://eur-lex.europa.eu/browse/directories/
legislation.html
28. Barlag in Europäische Datenschutz-Grundverordnung, Vorrang des Unionrechts – Anwend-
barkeit des nationalen Rechts, § 3 Recital 8
29. Paal/Pauly, Datenschutz-Grundverordnung, DS-GVO Art. 4 Recital 14
30. Shneiderman, B.: The eyes have it: a task by data type taxonomy for information
visualizations. In: Proceedings 1996 IEEE Symposium on Visual Languages, pp. 336–343.
IEEE, September 1996
31. Keller, C., Titov, W., Sawilla, S., Schlegel, T.: An evaluation approach for a smart public
display in public transport. In: Mensch und Computer 2019 – Workshopband, Hamburg,
Germany, 8–11 September 2019. Gesellschaft für Informatik e.V., S. 92-101 (2019)
32. Keller, C., Struwe, S., Titov, W., Schlegel, T.: Understanding the usefulness and acceptance
of adaptivity in smart public transport. In: Krömker, H. (ed.) HCII 2019. LNCS, vol. 11596,
pp. 307–326. Springer, Cham (2019). https://doi.org/10.1007/978-3-030-22666-4_23
33. Frommann, U.: Die Methode Lautes Denken (2005). http://www.eteaching.org/didaktik/
qualitaet/usability/Lautes%20Denken_eteaching_org.pdf. Accessed 29 Dec 2019

Investigating the Influencing Factors of User Experience in Car-Sharing Services: An Application of DEMATEL Method

Yufei Xie, Hanyue Xiao, Tianjia Shen, and Ting Han[✉]

School of Design, Shanghai Jiao Tong University, Shanghai 200240, China
xyfdesign@163.com, hanting@sjtu.edu.cn

Abstract. The user experience is a main determinant of the user satisfaction of car-sharing services. This research aims at investigating the users and their experience of using shared cars through interview and questionnaire survey. 14 influencing factors are extracted from literature research and user research. The depth interviews are deployed with Decision-Making Trial and Evaluation Laboratory (DEMATEL) questionnaires to evaluate the influencing directions and the degrees of the interactions among the 14 influencing factors of user experience in car sharing service. The findings show that three factors including "vehicle efficiency", "supporting service facilities" and "pricing model", are the three key factors influencing the satisfaction degree of the using experience. "Vehicle condition and performance", "complete supporting service facilities" and "vehicle service safety" have greater impact on other influencing factors. "Value identity", "brand and visual presentation" and "fun and novelty" are mainly affected by other factors. Based on the influence relationship and experience level of each factor, 14 factors are summarized into three levels: availability, easy-using and enjoyment. A user satisfaction optimization model of shared car travel service is proposed, aiming at providing a feasible way to enhance the user experience of car sharing service.

Keywords: Car sharing service · DEMATEL method · User experience · Influencing factor · Shared vehicle · User research

1 Introduction

As a new model of transportation, car sharing services are becoming an important mode of transportation. The shortage of traffic resources and road resources that cannot be changed in a short period of time determines that car sharing service has great development potential and strong vitality. As a new mode, car-sharing has aroused the interest of youth group while China's car sharing services are still in the primary stage of market operation. Many enterprises have explored the mode of car-sharing, but no mature solution has been developed for the user experience problems in operation. Issues of the user experience in car-sharing services still need to be resolved to enhance user's acceptance and satisfaction of shared vehicles.

© Springer Nature Switzerland AG 2020
H. Krömker (Ed.): HCII 2020, LNCS 12213, pp. 359–375, 2020.
https://doi.org/10.1007/978-3-030-50537-0_26

The objective, based upon interview and the Decision-Making Trial and Evaluation Laboratory (DEMATEL) analysis, is to identify critical factors influencing the user experience in car sharing services, as well as to assess the interrelationships among the main influencing factors, which will eventually serve as a novel tool model for future user experience improvement. This study aims to provide valuable references for the car sharing platforms to improve the user experience quality, clarify the design and development direction of related products and services, and advance the healthy development of the car sharing model.

2 Literature Review

The literature research is carried out through databases and search engines, primarily initiated through ScienceDirect, Web of Science, EBSCOhost, CNKI and Google Scholar but includes findings from papers reference sections as well. The research started from collecting studies from database, based on 6 keywords ("car sharing", "shared car", "shared vehicle", "user experience", "service quality" and "user satisfaction"), for the publication year range 2011 to 2019. The authors keep the relevant literatures according to the abstracts, and extract the relevant factors that affect the user experience of the car sharing service from the articles.

Lamberton et al. [1] proposed that the influence of price on transportation preferences equally applies to car sharing services. Joshua Paundra et al. [2] carried out that the three instrumental attributes, price, parking convenience and car type, influence people's intention to select a shared car. Stefano de Luca and Roberta Di Pace [3] found that the price increase of a car sharing service reduces the willingness to switch from private car to a car sharing service. Kaspi et al. [4] found that car sharing services are more attractive to potential users, if parking places are included in the service. Lower price, more convenience parking, and cars of the electric type linked to higher intention to select a shared car. Alexandre de Lorimier et al. [5] found that the combination of high availability and high vehicle usage ensures users satisfaction with the service and that vehicles are used. The number of vehicles parked at a station has the most effect on availability, and vehicle usage is affected by average vehicle age, and by member concentration in the vicinity of stations. According to the research of Cartenì et al. [6], users are sometimes attracted to carsharing service because of its positive environmental image. Stefan Illgen and Michael Höck [7] found that key success factors of a good car sharing service included advantageous relations between the market environment (e.g. electricity and fuel prices) and important characteristics of electric cars (e.g. price and range). Guido Perboli et al. [8] compared the business models of the four car sharing companies, and analyzed the advantages and disadvantages of each model. If car sharing and other shared services are to gain in popularity, there must be ample trust or reputation [9, 10]. Vehicle availability, brand reputation, price model, travel flexibility, charging convenience, station distribution, sharing model and web & app fluency are the influencing factors of the user experience that can be extracted from their research. Through depth interview with Sydney's car sharing users, Robyn Dowling et al. [11] explores the factors in car sharing, and divides them into three categories: Digital attributes (APP and web page, online payment, reservation system, etc.), material attributes (car cabin environment, parking space, car condition and performance, etc.) and social attributes (in car passengers, travel flexibility, entertainment, etc.). Xusen Cheng et al. [12] found factors that influence online and offline car sharing service quality, including structure assurance (safety, comfort, safe transaction environment), platform responsiveness (response speed and quality), information congruity (service conforms to the content displayed on the mobile platform) and reservation service. Min Qu et al. [13] developed an improved approach to evaluate car sharing options, and proposed 24 indicators from four dimensions: economic, environmental, social, and car sharing system performance. The influencing factors of user experience include mobility, accessibility, security, functional diversity, quality of service, accessibility for mobility impaired groups, etc. Jinhee Kim et al. [14] found that satisfaction significantly affects the car-sharing decision, and that car availability has a significant effect on the likelihood of joining a car-sharing organization. Francesco Ferrero et al. [15] mentioned that mode identifies the different ways in which a car-sharing service can be provided. Station-based mode [16] and the Free-floating mode [17] affects the flexibility of the shared car travel.

3 Method

3.1 DEMATEL Procedure

Decision-Making Trial and Evaluation Laboratory (DEMATEL) method is often used to analyze the complex relationship between management problems. It is a methodology put forward by Battelle Geneva Research Center in 1971 to solve the complex and difficult problems between science, technology and human beings in the real world and clarify the essence of the problems [18]. This approach has been widely used in many fields such as smart product service system [19], hospital performance management [20], sustainable online consumption [21] and logistics 4.0 [22]. As a type of structural modeling method, DEMATEL method can simplify the complex problems and prompt the logical relationship between the factors through determinant calculation [23, 24]. The method not only converts the interdependencies of factors into cause-and-effect relationships but also determines the critical factors of a system aided by impact relation diagrams [25].

The DEMATEL procedure in this study is carried out in the following steps: (1) identifying the main influencing factors that impact the satisfaction of user experience in the car sharing service; (2) collecting opinions of target users to estimate the influence of factors on each other using the DEMATEL questionnaire with one of four level values; (3) generating the direct relation matrix Z; (4) Calculating the λ value, the normalized direct relation matrix and the direct/indirect relation matrix T, and the corresponding Prominence value (D + R), and Relation value (D − R) of each influencing factor; (5) obtaining the influence relation map (IRM) with (D + R) value and (D − R) value of each influencing factor to help observe the interrelationship structure; (6) extracting the key influencing factors from the 14 factors to identify the relationship among all factors. Based on the previous research, this research explored the influence, causality and importance of the factors that affect the user experience in car sharing services.

3.2   Fourteen Main Influencing Factors

This study applies literature research and user research methods (questionnaire, observation and in-depth interview) to mine the influencing factors of user experience in car sharing service. A large number of target users were investigated to collect users' opinions toward the car sharing travel. The authors analyzed the information collected from related literatures and user research, to extract the factors that have impact on their perception of the satisfaction with car sharing service.

Users are divided into four groups according to the literature review and user research, including Loyal Sharing type, Restriction Alternative type, Business Travel type and Free-outing type. The Loyal Sharing group refers to the young people who have driver's license but have no car in the city, and use the sharing car to meet the rigid demand of travel. The Restriction Alternative group refers to the people whose private cars are restricted to drive on working days, and the shared cars are used to solve the restriction problem. The Business Travel group refers to the business travel group with high requirements for privacy, which needs to avoid the interference of noisy environment. The Free-outing group refers to young people/families who pay attention to the freedom of travel and people who try to use shared cars because of curiosity and freshness.

Different user groups have different emphasis on the evaluation of user experience. After analyzing the information collected by the above methods, the influencing factors categories of different user groups are sorted out. And as shown in Table 1, the following 21 preliminary influencing factors of the user experience in car sharing service are summarized and defined.

The 22 influencing factors are clustered by using card sorting method, and finally 14 main influencing factors are sorted out and given serial number Factor 1 to Factor 14 as follows: (1) Factor 1 - Vehicle Condition and Performance; (2) Factor 2 - Available Supporting Service Facilities; (3) Factor 3 - Safety of Vehicle Service; (4) Factor 4 - Pick Up & Return Mode; (5) Factor 5 - Price Model; (6) Factor 6 - Brand and Visual Image; (7) Factor 7 - Simple Learning & Using; (8) Factor 8 - Vehicle Efficiency; (9) Factor 9 - Traffic Transferring Convenience; (10) Factor 10 - Brand & Model Type Diversity; (11) Factor 11 - Environment Cleanliness; (12) Factor 12 - Real-time Feedback; (13) Factor 13 - Pleasure and Novelty; and (14) Factor 14 - Value Identification.

Table 1. 22 preliminary user experience influencing factors.

User group	Influencing category	Specific influencing factor
Loyal Sharing group	Vehicle usability Service system Economy Brand	Parking convenience Charging convenience App fluency Price model Brand reputation Payment security
Restriction Alternative group	Vehicle efficiency Dispatch efficiency	Reservation system efficiency Vehicle dispatch efficiency Vehicle identifiability Traffic transferring convenience
Business Travel group	Service quality Complete function	Vehicle diversity Driving comfort Cabin environment Digital Internet of vehicles Destination accessibility Pick up and return navigation
Free-outing group	Travel flexibility Simple operation Entertainment	Rental model Travel flexibility Psychological attribution Universal facilities Entertainment system

These 14 influencing factors form the items of DEMATEL questionnaire in this study, and the specific definitions of each influencing factor of car sharing service are shown in Table 2.

Table 2. Specific definitions of the 14 influencing factors of user experience.

Factors	Definitions
F1 - Vehicle Condition and Performance	Complete basic functions, brand-new vehicle, spacious space, sufficient battery power and good handling
F2 - Available Supporting Service Facilities	Parking spaces and charging facilities are widely distributed, easy to find and use
F3 - Safety of Vehicle Service	Vehicles and services that users can trust, including driving safety, payment safety and accident handling
F4 - Pick Up & Return Mode	Mode identifies the different ways in which a car-sharing service can be provided, including Two-way and One-way station-based mode, and the Free-floating mode
F5 - Price Model	The price and charging mode of shared car rental
F6 - Brand and Visual Image	Brand reputation and image, vehicle modeling, visual style and interface layout of digital products

(continued)

Table 2. (*continued*)

Factors	Definitions
F7 - Simple Learning & Using	The product provides the guidance of first use. The operation of vehicle and app is simple and smooth
F8 - Vehicle Efficiency	Users need less time to complete the whole process of sharing car rental and deal with problems
F9 - Traffic Transferring Convenience	It is convenient to transfer with other public transport modes (such as bicycle, bus, subway, etc.)
F10 - Brand & Model Type Diversity	The automobile brands and models used for sharing are various, meeting the personalized and diversified needs
F11 - Environment Cleanliness	The interior environment and body surface are clean
F12 - Real-time Feedback	The vehicle and platform use a variety of ways to timely provide the vehicle status, order progress information and other effective information to users
F13 - Pleasure and Novelty	The interesting interaction between vehicle use and user interface makes users feel happy, satisfied and fresh
F14 - Value Identification	Users' recognition of social values such as low-carbon and environmental protection of shared cars, and users' recognition of their own contribution value

3.3   DEMATEL Questionnaire

This study focuses on the general young users of car sharing service, and uses the 14 user experience factors summarized in the previous section to make the DEMATEL questionnaire. Expert interview and questionnaire were used in the process of investigation. The questionnaire was mainly distributed to the young people who have shared car use experience in the first and second tier cities.

The DEMATEL questionnaire (Table 3) is used for shared car users to estimate the direction of interaction and the degree of relative priority of each factor listed in first column to each factor listed in first row in Table 3. The priority degrees of factors in first column over factors in first row are defined in 0–4 five levels, i.e., value '0' means "no impact", value '1' means "low impact", value '2' means "distinct impact", value '3' means "big impact", and value '4' means "extreme impact".

Due to the poor readability of the form questionnaire, we transform the DEMATEL questionnaire into a normal questionnaire which is easy for users to understand and fill in for data collection. A total of 58 questionnaires were collected from May 29 to June 6, 2019, including 51 valid ones. And each questionnaire is filled in 15–40 min.

Table 3. The DEMATEL questionnaire.

	F 1	F 2	F 3	F 4	F 5	F 6	F 7	F 8	F 9	F 10	F 11	F 12	F 13	F 14
F 1	0													
F 2		0												
F 3			0											
F 4				0										
F 5					0									
F 6						0								
F 7							0							
F 8								0						
F 9									0					
F 10										0				
F 11											0			
F 12												0		
F 13													0	
F 14														0

4 Data Analysis and Results

4.1 Direct Influence Matrix Z

51 pieces of effective data were calculated based on the following equation.

$$Z = \frac{1}{n}\sum_{m=1}^{n}\left[z_{ij}^{m}\right], \quad i,j = 1,2,3,\ldots,k. \tag{1}$$

The elements in direct influence matrix **Z** of the interactions between 14 main factors are obtained as listed in Table 4.

Table 4. The direct influence matrix Z.

	F 1	F 2	F 3	F 4	F 5	F 6	F 7	F 8	F 9	F 10	F 11	F 12	F 13	F 14
F 1	0	2.55	3.16	2.39	2.1	2.68	2.77	2.74	2.32	2.26	2.35	2.29	2.48	2.55
F 2	2.35	0	2.71	3.06	2.87	2.55	2.71	3.06	2.74	2.16	2.13	2.65	2.42	2.42
F 3	2.94	2.58	0	2.52	2.68	2.71	2.45	2.65	2.32	2.16	2.65	2.23	2.61	2.52
F 4	1.90	2.58	2.03	0	2.32	2.29	2.42	3.06	2.77	2.03	2.00	2.45	2.16	2.35
F 5	2.13	2.26	2.16	2.32	0	2.61	2.19	2.55	2.16	2.48	2.19	2.26	2.13	2.48
F 6	1.74	1.81	1.84	1.87	2.45	0	2.16	2.03	1.77	2.61	1.94	1.84	2.65	2.65
F 7	2.29	2.06	2.19	2.35	2.06	2.39	0	2.97	2.58	2.13	1.94	2.29	2.58	2.45
F 8	2.29	2.32	2.10	2.68	2.39	2.19	2.61	0	2.61	2.23	1.94	2.58	2.52	2.42
F 9	1.97	2.52	1.87	2.77	2.61	2.32	2.29	2.81	0	2.29	2.03	2.39	2.39	2.48
F 10	2.16	2.19	2.06	2.10	2.42	2.68	2.13	2.10	1.90	0	1.94	1.97	2.58	2.58
F 11	2.06	1.94	1.94	1.74	2.68	2.90	1.81	1.90	1.84	1.97	0	1.97	2.39	2.55
F 12	2.35	2.45	2.55	2.65	2.52	2.23	2.52	2.84	2.42	2.03	1.87	0	2.48	2.35
F 13	1.81	1.61	1.74	1.77	2.13	2.77	2.42	2.23	1.84	2.35	1.77	1.81	0	2.45
F 14	1.68	1.81	1.68	1.45	2.19	2.45	1.77	1.58	1.68	2.23	2.16	1.81	2.45	0

Table 5. Matrix of total relations T.

	F 1	F 2	F 3	F 4	F 5	F 6	F 7	F 8	F 9	F 10	F 11	F 12	F 13	F 14
1	0.490	0.577	0.581	0.590	0.636	0.649	0.610	0.645	0.576	0.579	0.535	0.576	0.623	0.639
2	0.563	0.516	0.578	0.616	0.650	0.655	0.618	0.664	0.597	0.586	0.538	0.594	0.631	0.646
3	0.565	0.572	0.490	0.587	0.629	0.643	0.596	0.637	0.571	0.575	0.525	0.579	0.610	0.634
4	0.503	0.537	0.511	0.482	0.580	0.591	0.558	0.607	0.547	0.531	0.488	0.538	0.570	0.587
5	0.501	0.520	0.507	0.537	0.507	0.591	0.543	0.585	0.522	0.535	0.485	0.525	0.560	0.582
6	0.453	0.468	0.459	0.483	0.531	0.475	0.500	0.526	0.471	0.498	0.442	0.473	0.530	0.542
7	0.511	0.521	0.513	0.543	0.571	0.591	0.488	0.602	0.539	0.531	0.484	0.531	0.578	0.587
8	0.520	0.537	0.520	0.562	0.590	0.597	0.570	0.532	0.549	0.543	0.493	0.548	0.587	0.597
9	0.509	0.539	0.511	0.561	0.593	0.597	0.559	0.606	0.475	0.542	0.492	0.541	0.580	0.596
10	0.485	0.501	0.488	0.513	0.555	0.573	0.523	0.553	0.498	0.449	0.463	0.500	0.553	0.565
11	0.466	0.477	0.468	0.485	0.543	0.560	0.497	0.529	0.478	0.487	0.393	0.482	0.530	0.545
12	0.528	0.547	0.538	0.568	0.601	0.605	0.575	0.617	0.551	0.545	0.497	0.484	0.593	0.603
13	0.446	0.454	0.448	0.471	0.512	0.540	0.497	0.521	0.464	0.482	0.429	0.463	0.448	0.527
14	0.418	0.433	0.422	0.437	0.486	0.503	0.453	0.475	0.434	0.453	0.416	0.437	0.487	0.431

4.2   Matrix of Total Relations T

Users only estimated the direct influence of factors on each other. Hence, the total influence matrix T is obtained from matrix X by applying the transition theory and summing up all direct and indirect effects. The matrix of total relations T (Table 5) was derived, by using Eqs. (2)–(4). Matrix X is the normalized direct relation matrix, and I is the identity matrix.

$$X = \left[x_{ij}\right]_{k \times k} = sZ, \tag{2}$$

$$s = min\left(\frac{1}{\max\limits_{1 \le j \le k}\sum_{i=1}^{k} z_{ij}}, \frac{1}{\max\limits_{1 \le i \le k}\sum_{j=1}^{k} z_{ij}}\right), \tag{3}$$

$$T = \lim_{m \to \infty}\left(X + X^2 + \ldots + X^m\right) = X(I - X)^{-1}. \tag{4}$$

The upper quartile Q3 of all the elements in the total relations matrix T (keep 3 decimal places), i.e., 0.580, is taken as the threshold to measure the strength of interactions between factors. If all values in the row and the column that correspond to a factor in the matrix T are below the threshold value, this factor and the corresponding row and column will be removed.

Then, the upper third of Q3 (0.617) is taken as the boundary value of strong influence relation and general influence relation. Values higher than 0.617 are considered as strong influences, and values between 0.617 and 0.580 are considered as general influences. As shown in Table 6, the general influences are marked with underlines, and the strong influences are marked with bold type and underlines.

Table 6. Matrix of total relations T.

	1	2	3	4	5	6	7	8	9	10	11	12	13	14
1	0.490	0.577	0.581	0.590	**0.636**	**0.649**	0.610	**0.645**	0.576	0.579	0.535	0.576	**0.623**	**0.639**
2	0.563	0.516	0.578	0.616	**0.650**	**0.655**	**0.618**	**0.664**	0.597	0.586	0.538	0.594	**0.631**	**0.646**
3	0.565	0.572	0.490	0.587	**0.629**	**0.643**	0.596	**0.637**	0.571	0.575	0.525	0.579	0.610	**0.634**
4	0.503	0.537	0.511	0.482	0.580	0.591	0.558	0.607	0.547	0.531	0.488	0.538	0.570	0.587
5	0.501	0.520	0.507	0.537	0.507	0.591	0.543	0.585	0.522	0.535	0.485	0.525	0.560	0.582
6	0.453	0.468	0.459	0.483	0.531	0.475	0.500	0.526	0.471	0.498	0.442	0.473	0.530	0.542
7	0.511	0.521	0.513	0.543	0.571	0.591	0.488	0.602	0.539	0.531	0.484	0.531	0.578	0.587
8	0.520	0.537	0.520	0.562	0.590	0.597	0.570	0.532	0.549	0.543	0.493	0.548	0.587	0.597
9	0.509	0.539	0.511	0.561	0.593	0.597	0.559	0.606	0.475	0.542	0.492	0.541	0.580	0.596
10	0.485	0.501	0.488	0.513	0.555	0.573	0.523	0.553	0.498	0.449	0.463	0.500	0.553	0.565
11	0.466	0.477	0.468	0.485	0.543	0.560	0.497	0.529	0.478	0.487	0.393	0.482	0.530	0.545
12	0.528	0.547	0.538	0.568	0.601	0.605	0.575	**0.617**	0.551	0.545	0.497	0.484	0.593	0.603
13	0.446	0.454	0.448	0.471	0.512	0.540	0.497	0.521	0.464	0.482	0.429	0.463	0.448	0.527
14	0.418	0.433	0.422	0.437	0.486	0.503	0.453	0.475	0.434	0.453	0.416	0.437	0.487	0.431

4.3 Prominence and Net Effect

Value D and R are achieved by summing up the rows and columns of total relation matrix T, using Eqs. (5)–(6).

$$D_i = \sum_{j=1}^{n} t_{ij}, (i = 1, 2, 3, \ldots, n) \tag{5}$$

$$R_i = \sum_{j=1}^{n} t_{ij}, (i = 1, 2, 3, \ldots, n) \tag{6}$$

Value D is called the degree of influential impact, and value C is called the degree of influenced impact. D_i is the sum of the ith row of matrix **T** and represents all the direct and indirect effects which are dispatched from Factor i to other factors. And R_j is the sum of the jth column of matrix **T** and represents all the direct and indirect effects that Factor j receives from the other factors.

The prominence value (D + R) is obtained by adding the influence degree (**D**) and the influenced degree (R) of the factor.

The prominence value (importance) played by the factor in a system is represented by (D + R) value, where (D − R) value stands for the net effect (i.e. the degree of importance) of the factor contributed to the system. The factors are categorized into two groups of cause (driver) and effect (receiver) factors. If the (D − R) value of a factor is positive, this factor is grouped under the category of driver factors which has an influence on other factors; if the value of (D − R) is negative, this factor is grouped under the category of receiver factors which receives influence from other factors.

The value D, R, (D + R), (D − R) of Factor 1 to Factor 14 were listed in Table 7.

Table 7. Prominence (D + R) and Net effect (D − R) values.

D		R		Prominence (D + R)		Net effect (D − R)	
2	8.4516	6	8.1676	8	15.8424	1	1.3491
1	8.3071	8	8.0992	2	15.6487	2	1.2545
3	8.2120	14	8.0802	5	15.4839	3	1.1766
12	7.8510	5	7.9835	1	15.2650	12	0.5782
8	7.7433	13	7.8809	3	15.2474	9	0.4272
9	7.7000	7	7.5847	7	15.1762	11	0.2603
4	7.6298	4	7.4339	12	15.1239	4	0.1958
7	7.5915	10	7.3373	4	15.0637	7	0.0068
5	7.5004	12	7.2728	6	15.0188	10	−0.1186
10	7.2186	9	7.2728	9	14.9728	8	−0.3558
11	6.9384	2	7.1971	13	14.5835	5	−0.4830
6	6.8511	3	7.0354	10	14.5559	13	−1.1784
13	6.7026	1	6.9579	14	14.3642	6	−1.3165
14	6.2840	11	6.6781	11	13.6166	14	−1.7962

The average value of (D + R) of each factor is 14.997 (three decimal places are reserved). The (D + R) value of nine factors is higher than the average (D + R) value, so the importance of factors affecting the user experience in car sharing service is ranked as follows: Factor 8 - Vehicle Efficiency, Factor 2 - Available Supporting Service Facilities, Factor 5 - Price Model, Factor 1 - Vehicle Condition and Performance, Factor 3 - Safety of Vehicle Service, Factor 7 - Simple Learning & Using, Factor 12 - Real-time Feedback, Factor 4 - Pick Up & Return Mode, Factor 6 - Brand and Visual Image.

From the (D − R) ranking, we find that Factor 1 (Vehicle Condition and Performance), Factor 2 (Available Supporting Service Facilities) and Factor 3 (Safety of Vehicle Service) are the three important factors that directly influence other factors; and Factor 14 (Value Identification), Factor 6 (Brand and Visual Image) and Factor 13 (Pleasure and Novelty) are the three factors that are most influenced by other factors.

4.4   The Influential Relation Diagram

The influential relation diagram can be plotted in the $(D_i + R_i, D_i - R_i)$ layout, using $(D_i + R_i)$ as the horizontal axis and $(D_i - R_i)$ as the vertical axis. Fourteen user experience influencing factors of car sharing services are marked in the coordinate axis, and the influential relation diagram is made. According to the position of each factor in the diagram, the ones which have a great effect on other factors or have a complicated relationship with other criteria can be found.

The relationship with the influence strength lower than 0.580 will not be marked. The relationship with the influence strength between 0.580 and 0.617 will be regarded as general influence and marked with dotted line; the relationship with the influence strength greater than or equal to 0.617 will be regarded as strong influence and marked with solid line. The specific location of each factor is shown in Fig. 1.

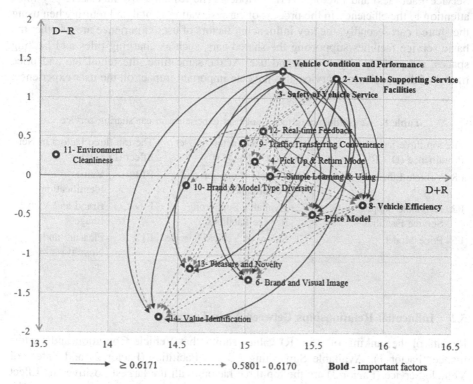

Fig. 1. The influential relation diagram of 14 influencing factors of user experience in car sharing service.

As can be seen from Fig. 1, Factor 1 (Vehicle Condition and Performance), Factor 2 (Available Supporting Service Facilities) and Factor 3 (Safety of Vehicle Service) are the most important influencing factors in the car sharing travel service. Factor 14 (Value Identification), Factor 6 (Brand and Visual Image) and Factor 13 (Pleasure and Novelty) are the three most directly/indirectly affected factors in the car sharing experience. Factor 8 (Vehicle Efficiency) and Factor 5 (Price Model) are two important influencing factors in the shared car service.

5   Discussion and Conclusion

5.1   Key Factors Influencing the User Experience in Car Sharing Service

The first three critical factors of car sharing user experience, the first three influencing factors and the influenced factors of the car sharing user experience are summarized in Table 8.

It is found that the first three key influencing factors of user experience in car sharing services are Factor 8 (Vehicle Efficiency), Factor 2 (Available Supporting Service Facilities) and Factor 5 (Price Model). The results show that users pay most attention to the efficiency in the process of car reservation, rental and return when using the shared car; secondly, the key influencing factors of user experience are whether the basic service facilities supporting the shared cars, such as charging piles and parking spaces, are perfect, easy to find and use. At the same time, the rental fee and price model of the car sharing service also have an important impact on the user experience.

Table 8. Key influencing factors of user experience in car sharing service.

The top three factors in Prominence (D + R)		The top three factors in Net effect (D − R) ≥ 0		The top three factors in Net Effect (D − R) ≤ 0	
F8	Vehicle Efficiency	F1	Vehicle Condition and Performance	F14	Value Identification
F2	Available Supporting Service Facilities	F2	Available Supporting Service Facilities	F 6	Brand and Visual Image
F5	Price Model	F3	Safety of Vehicle Service	F13	Pleasure and Novelty

5.2   Influential Relationships Between Main Factors

Results of the ranking of (D − R) values shows that Vehicle Condition and Performance (Factor 1), Available Supporting Service Facilities (Factor 2) and Safety of Vehicle Service (Factor 3) are the top three factors with the largest positive Net Effect values, implying that these three factors have the strong impact on other factors. The results demonstrated that the basic driving functions and maneuverability of the vehicle itself are the most important factors affecting the user experience of shared vehicle, which are the most basic factors for creating a good experience that directly affects the user's evaluation of the whole experience. This analysis found evidence that 1) users attach great importance to the vehicle efficiency, so good car sharing service needs to start from each node of the use process to improve efficiency; 2) the perfection and usability of charging piles, parking spaces and other supporting facilities are essential for user experience; and 3) the safety of vehicle driving and mobile payment also has an important impact on the car sharing experience.

According to the analysis of the affected factors, Value Identification (Factor 14), Brand and Visual Image (Factor 6) and Pleasure and Novelty (Factor 13) are the ones being influenced most greatly.

5.3   Analysis and Discussion on Specific Factors

The essence of Factor 1 (Vehicle Condition and Performance), Factor 2 (Available Supporting Service Facilities) and Factor 3 (Safety of Vehicle Service) is the basic needs of shared car users. In the influential relations diagram, we can see that these three factors are not affected by other factors, but mainly have strong impacts on other factors, and they are completely dominant influencing factors of user experience.

As the most intuitive experience content, the usage functions of vehicles are the basis of a good car sharing travel, which have the most direct impact on the evaluation of the whole experience. Therefore, when improving the user experience of shared cars, we should first ensure the improvement of basic functions and facilities. Only when these three factors are satisfied can the optimization of other factors be effective. On the premise of meeting these three factors, the optimization of other factors can achieve better results.

Factor 8 (Vehicle Efficiency) is located on the right-most side of the graph, which shows that it is strongly influenced by four factors (F1, F2, F3 and F12), and influenced by three factors (F4, F7 and F9) in general. Factor 8 also affects four factors (F5, F6, F13 and F14). The (D + R) value and (D − R) value of vehicle efficiency are both high, indicating that Factor 8 has strong influential relationships with other factors. Therefore, Factor 8 is a critical factor of user experience in car sharing service.

Vehicle efficiency is closely related to user experience. Most shared car users are urban youth, and most of the use scenarios are commuting, party and outing on weekdays, so there is a high demand for vehicle efficiency. Efficient car sharing service needs to be built from each node of the use process: for example, simplifying car rental procedures and operations, providing more renting entrances to enable users to reserve vehicles faster; increasing the number of outlets and providing beginning/ending navigation to enable users to get to the parking location faster; improving the features of shared vehicles, enabling users to find the reserved vehicles faster; optimizing the accident-handling process to reduce the time required to remove obstacles. It could improve the efficiency of car sharing service by shortening the time required by each node, which is conducive to improving users' satisfaction of the whole travel.

Factor 5 (Price Model) is strongly influenced by three basic factors (F1, F2 and F3), and generally influenced by four factors (F4, F8, F9 and F12). At the same time, it affects Factors 6 and Factor 14, and has a lot of mutual interrelationships with other factors, so it is also an important factor affecting user experience and their satisfaction.

Rental costs and price model of car sharing service will affect users' perception of a car sharing service. The results show that users will mainly evaluate the vehicle price charging mode according to the vehicle condition and performance, supporting service facilities and vehicle service safety; then consider the service quality of the platform and the rapidity and convenience of the service system, and judge whether the price is equivalent to the service. At the same time, the pricing of shared car service will affect the user's feeling of shared car brand image. Shared car users can be divided into price-sensitive users and quality-oriented users, pursuing high performance-price ratio and high quality respectively. Car sharing platforms should fully consider their brand positioning and target groups when pricing, to formulate a reasonable pricing model.

The interrelationship between Factors 10 (Brand & Model Type Diversity) and other factors is weak. It is only affected by Factors 2 (Available Supporting Service Facilities). Thus, it can be concluded that the contribution value of providing diversified brands and models for users is relatively low in improving the user experience of the whole service system. Therefore, the priority of brand and model diversity is lower than the previous factors when optimizing the car sharing service.

The four factors, Real-time Feedback (Factor 12), Traffic Transferring Convenience (Factor 9), Pick-up & Return Mode (Factor 4) and Simple Learning & Using (Factor 7), are concentrated in the middle of the relationship diagram. These factors do not have influence on each other, and their interrelationships with other factors are mostly general influences. On the premise of guaranteeing the availability factors in car sharing service, optimizing and upgrading the above factors (Performance Quality) will also enhance the user satisfaction accordingly. The only strong influence among the four factors is that Factor 12 (Real-time feedback) significantly affects Factor 8 (Vehicle Efficiency). It can be explained that the real-time feedback of mobile APP (such as vehicle status, location information, order progress information, etc.) can inform users the current status of service, interact with users and give users a sense of control, which plays an important role in improving user experience.

Value Identification (Factor 14) and Pleasure and Novelty (Factor 13) refer to the interest, emotion and value connection established between shared vehicles and users. From the perspective of user demand in Maslow's demand hierarchy theory, these factors belong to the upper level demand. Users agree with the social values of low-carbon and environmental protection of car sharing, and participate in sharing activities to gain a sense of self-achievement, to achieve self-worth recognition. In the process of using the shared car, through interesting user interface/interaction and the creative facilities, users will have the sense of pleasure, satisfaction and freshness. It makes users have the will to use continuously. In the process of the gradual improvement of the car sharing system, it should also start from establishing emotional relationship with users to enhance good experience and user viscosity.

Factor 11 (Environment Cleanliness) has no significant interrelationship with other factors. However, according to the interview results, the cleaning of the car interior and car body has an important impact on the user satisfaction. It is a visual and olfactory factor that can be directly observed. Therefore, although the interaction between environmental cleanliness and other factors is weak, it is still important to ensure the regular cleaning of car interior and car body in a good car sharing service.

6 Conclusion

This study uses DEMATEL method to analyze the impact of 14 influencing factors on the sharing car user experience. Vehicle condition and performance, available supporting service facilities and safety of vehicle service are the basis of the car sharing service system. As the most intuitive experience content, they have the most direct impact on the evaluation of the whole experience. Good car sharing service should first ensure the availability of vehicles.

Vehicle efficiency is an important key influencing factor of the user experience in the car sharing service. Therefore, all aspects of the process should be designed to improve the efficiency of the whole process. Price model is also one of the most important influencing factors. The price model should conform to the brand positioning, and need to develop reasonable rental mode according to different user groups.

Combining the influence relationships and experience level among the influencing factors in the influential relations diagram, the 14 factors affecting user experience can be divided into three layers (Able-to-use, Easy-to-use and Enjoyable-to-use) to set up the user experience optimization model of car sharing experience. The influence of factors on user experience decreases from bottom to top, and gradually transits from functional factors to emotional factors. Car sharing service system can be optimized by referring to this model: 1) Completing the usability construction of basic functions and service facilities; 2) Designing the usability aspects of the sharing system to simplify the user's operation and improve efficiency; 3) Establishing an emotional connection with users to make them enjoy the using process and even achieve self-worth.

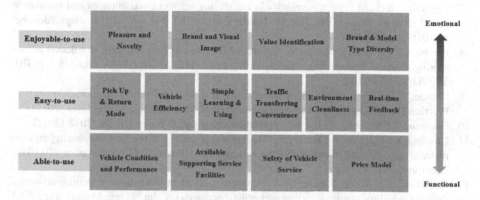

Fig. 2. The user experience optimization model of car sharing experience.

In conclusion, this research provides references and proposals for the user experience design of car sharing service from specific factors, and advances the healthy development of the car sharing model. However, there are limitations in this study. The sample size in this study is small, so in the future research, the research scope should be expanded to make the research more rigorous. In addition, due to the excessive number of questions, respondents may lose patience when filling in questionnaires or participating in the interviews, which may affect the authenticity of the results. In the future, researchers should explore a simpler form of questionnaire to make the results more accurate.

References

1. Lamberton, C.P., Rose, R.L.: When is ours better than mine? A framework for understanding and altering participation in commercial sharing systems. J. Mark. **76**(4), 109–125 (2012). https://doi.org/10.1509/jm.10.0368
2. Paundra, J., Rook, L., van Dalen, J., Ketter, W.: Preferences for car sharing services: effects of instrumental attributes and psychological ownership. J. Environ. Psychol. **53**, 121–130 (2017). https://doi.org/10.1016/j.jenvp.2017.07.003
3. de Luca, S., Di Pace, R.: Modelling users' behaviour in inter-urban carsharing program: a stated preference approach. Transp. Res. Part A: Policy Pract. **71**, 59–76 (2015). https://doi.org/10.1016/j.tra.2014.11.001

4. Kaspi, M., Raviv, T., Tzur, M.: Parking reservation policies in one-way vehicle sharing systems. Transp. Res. Part B: Methodol. **62**, 35–50 (2014). https://doi.org/10.1016/j.trb.2014.01.006

5. de Lorimier, A., El-Geneidy, A.M.: Understanding the factors affecting vehicle usage and availability in carsharing networks: a case study of communauto carsharing system from Montréal, Canada. Int. J. Sustain. Transp. **7**(1), 35–51 (2013). https://doi.org/10.1080/15568318.2012.660104

6. Cartenì, A., Cascetta, E., de Luca, S.: A random utility model for park & carsharing services and the pure preference for electric vehicles. Transp. Policy **48**, 49–59 (2016). https://doi.org/10.1016/j.tranpol.2016.02.012

7. Illgen, S., Höck, M.: Electric vehicles in car sharing networks – challenges and simulation model analysis. Transp. Res. Part D: Transp. Environ. **63**, 377–387 (2018). https://doi.org/10.1016/j.trd.2018.06.011

8. Perboli, G., Ferrero, F., Musso, S., Vesco, A.: Business models and tariff simulation in car-sharing services. Transp. Res. Part A: Policy Pract. **115**, 32–48 (2018). https://doi.org/10.1016/j.tra.2017.09.011

9. George, C.M.: User Adoption of Innovations in a Sustainability Transitions Context: A Practice Theory Analysis of Car Sharing and Urban Mobility Behavior in Oslo (2017)

10. Botsman, R.: Welcome to the new reputation economy. J. Wired UK **20**, 2012 (2012)

11. Dowling, R., Maalsen, S., Kent, J.L.: Sharing as sociomaterial practice: car sharing and the material reconstitution of automobility. Geoforum **88**, 10–16 (2018). https://doi.org/10.1016/j.geoforum.2017.11.004

12. Cheng, X., Fu, S., de Vreede, G.-J.: A mixed method investigation of sharing economy driven car-hailing services: online and offline perspectives. Int. J. Inf. Manag. **41**, 57–64 (2018). https://doi.org/10.1016/j.ijinfomgt.2018.03.005

13. Qu, M., Yu, S., Yu, M.: An improved approach to evaluate car sharing options. Ecol. Ind. **72**, 686–702 (2017). https://doi.org/10.1016/j.ecolind.2016.07.018

14. Kim, J., Rasouli, S., Timmermans, H.: Satisfaction and uncertainty in car-sharing decisions: an integration of hybrid choice and random regret-based models. Transp. Res. Part A: Policy Pract. **95**, 13–33 (2017). https://doi.org/10.1016/j.tra.2016.11.005

15. Ferrero, F., Perboli, G., Rosano, M., Vesco, A.: Car-sharing services: an annotated review. Sustain. Cities Soc. **37**, 501–518 (2018). https://doi.org/10.1016/j.scs.2017.09.020

16. Nourinejad, M., Roorda, M.J.: Carsharing operations policies: a comparison between one-way and two-way systems. Transportation **42**(3), 497–518 (2015). https://doi.org/10.1007/s11116-015-9604-3

17. Firnkorn, J., Müller, M.: What will be the environmental effects of new free-floating car-sharing systems? The case of car2go in Ulm. Ecol. Econ. **70**(8), 1519–1528 (2011). https://doi.org/10.1016/j.ecolecon.2011.03.014

18. Gabus, A., Fontela, E.: World problems, an invitation to further thought within the framework of DEMATEL (1972)

19. Chen, Z., Lu, M., Ming, X., Zhang, X., Zhou, T.: Explore and evaluate innovative value propositions for smart product service system: a novel graphics-based rough-fuzzy DEMATEL method. J. Clean. Prod. **243**, 118672 (2020). https://doi.org/10.1016/j.jclepro.2019.118672

20. Jiang, S., Shi, H., Lin, W., Liu, H.-C.: A large group linguistic Z-DEMATEL approach for identifying key performance indicators in hospital performance management. Appl. Soft Comput. **86**, 105900 (2020). https://doi.org/10.1016/j.asoc.2019.105900

21. Song, W., Zhu, Y., Zhao, Q.: Analyzing barriers for adopting sustainable online consumption: a rough hierarchical DEMATEL method. Comput. Ind. Eng. **140**, 106279 (2020). https://doi.org/10.1016/j.cie.2020.106279

22. Torbacki, W., Kijewska, K.: Identifying key performance indicators to be used in Logistics 4.0 and Industry 4.0 for the needs of sustainable municipal logistics by means of the DEMATEL method. Transp. Res. Procedia **39**, 534–543 (2019). https://doi.org/10.1016/j.trpro.2019.06.055

23. Kumar, A., Dash, M.K.: Using DEMATEL to construct influential network relation map of consumer decision-making in e-marketplace. Int. J. Bus. Inf. Syst. **21**(1), 48–72 (2016)

24. Wu, H.-H., Tsai, Y.-N.: An integrated approach of AHP and DEMATEL methods in evaluating the criteria of auto spare parts industry. Int. J. Syst. Sci. **43**(11), 2114–2124 (2012). https://doi.org/10.1080/00207721.2011.564674

25. Dalvi-Esfahani, M., Niknafs, A., Kuss, D.J., Nilashi, M., Afrough, S.: Social media addiction: applying the DEMATEL approach. Telematics Inform. **43**, 101250 (2019). https://doi.org/10.1016/j.tele.2019.101250

Assistive Systems for Special Needs in Mobility in the Smart City

Chuantao Yin[1], Bertrand David[2(✉)], René Chalon[2], and Hao Sheng[3,4]

[1] Sino-French Engineer School, Beihang University, Beijing 100191, China
chuantao.yin@buaa.edu.cn
[2] Université de Lyon, CNRS, Ecole Centrale de Lyon, LIRIS, UMR5205,
69134 Lyon, France
{Bertrand.David,Rene.Chalon}@ec-lyon.fr
[3] State Key Laboratory of Software Development Environment,
School of Computer Science and Engineering, Beihang University,
Beijing 100191, China
[4] Beijing Advanced Innovation Center for Big Data and Brain Computing,
Beihang University, Beijing 100191, China
shenghao@buaa.edu.cn

Abstract. The Smart City is a multifaceted objective aiming at increasing viability of the city for its inhabitants. Mobility in the city is an important dimension of the Smart City for humans and goods, which can either move without assistance or may need special assistance. The objective of this paper is to investigate the role of assistive systems for the mobility of humans and goods in order to take into account special needs. We discuss and classify different situations for assistance of humans and goods either by means of technology or humans (and technology) and present different assistive system behaviors already used or to be implemented in the future.

Keywords: Smart City · Mobility of humans and goods · Assistive systems for mobility · Special needs

1 Introduction

1.1 Mobility in the Smart City

The Smart City is a very hot and important application field of ICT (Information and Communication Technologies), able to receive a wide variety of contributions and using multiple declinations of ICT aspects (data management, data processing, Human-Computer Interaction, collaboration, information accessibility in mobility, automated or on-demand information access, IoT, AI contribution, etc.). This area is so large that it is not easy or possible to find a commonly accepted definition. However, to clarify our view we decided to take into account the following definition: "The Concept of "Smart Cities" describes how investments in human and social capital and modern Information and Communication Technologies (ICT) infrastructure and e-services fuel sustainable growth and quality of life, enabled by a wise management of natural resources and through participative government" [1].

© Springer Nature Switzerland AG 2020
H. Krömker (Ed.): HCII 2020, LNCS 12213, pp. 376–396, 2020.
https://doi.org/10.1007/978-3-030-50537-0_27

At present, the Smart City is still a concept undergoing evolution and experimentation. It aims at highlighting the role of ICT in a modern city, as well as integrating and optimizing the resources of a city, to make city life more efficient, energy-economic and intelligent. The Smart City is not only the application of new information technologies, but also concerns the participation of citizens in the various activities of the city with the intelligence of humans, combined with Artificial Intelligence in different forms. Different classifications try to clarify this huge domain. One of them proposes six dimensions: environment, economy, living, mobility, people and government. It is also possible to ask several questions and try to use the relevant answers to clarify this domain.

Smart City to whom: there are many answers, from a generic one "to all", to more specialized and precise ones (by age: kids, teenagers, adults, seniors), by city status, by implication in the city (citizen, neighbor, student, worker, administrator, governor); and also "to all" from the point of view of deficiencies.

Smart City by what: better common being and neighborhood, energy, transportation of goods and passengers, information dissemination on culture, sports, social services, etc.

The Smart City as an area of use for all ICT possibilities: data accessibility, data processing, information access exchange and manipulation in static and mobility situations, using wired or wireless networks, in appropriate applications taking into account contextual situations.

The context of the different contributions presented in this paper consists of transportation of passengers and freight in the city, as well as the associated services, which can be naturally added to this field. In these applications, we mainly base our contributions on the use of Internet, Internet of Things (IoT), Location-Based Services (LBS) and Big Data and Artificial Intelligence with Deep Learning.

1.2 State-of-the-Art

To carry out a complete state-of-the-art for the Smart City would appear a totally impossible goal as the amount and diversity of work are large and diversified. Merely to establish a common definition and classification is hard and has not yet been achieved, while to list all approaches and contributions seems totally impossible. We thus propose that the reader consults our paper [2]. In relation to transportation aspects a relevant reference is [3].

2 Mobility Characterization

Different kinds of mobility can be identified in the Smart City: (1) Human individual mobility, in which the human decides, organizes and does (executes) his or her mobility; (2) Good mobility based on standard shipping procedures. These two classical cases are out of the scope of our study. We will only mention these cases and indicate the assistive tools or systems used.

Table 1. Identified situations and characteristics taken into account.

Case #	Situations	Pedestrian	Private Transportation	Public Transportation	Humans: Children, Elderly, Disabled	Human-based	Technology-based	Goods	Synchronous Collab.	Asynchronous Collab.	Personal diary	Limited list	Large list	Security/safety	Rewards
		Nature of mobility				Assistance			Collab.		Selection in				
0a	Classical Human mobility	Or	Or	Or			X								
0b	Classical Goods delivery	Or	Or				X	X			Or	Or		Tr	
1	Indoor Navigation**	Or	Or	Or	Or		X		X						N
2	School Walk-bus	X			X	X			X			X		SS	N
3	Dynamic Lane Allocation**		&	&			X		X					Tr	N
4	Delivery Area Reservation**		X			Or	Or	X	Or	Or	Or	Or		Tr	N
5	Carbon-free goods transportation*			X		Or	Or	X	Or	Or			X	Tr	Y
6	Store client based distribution*	Or	Or	Or			X	X					X	Tr	Y
7	Goods transportation for colleagues*	Or	Or	Or		Or	Or	X		X	X			Tr	Y
8	Goods transportation for neighbors*	Or	Or	Or		Or	Or	X		X		X		Tr	Y
9	Escorted Children*	Or	Or	Or	X	Or	Or		X		Or	Or		SS	Y
10	Escorted Elderly persons*	Or	Or	Or	X				Or	Or	Or	Or	Or	Se	Y
11	Escorted Disabled*	Or	Or	Or	X				Or	Or	Or	Or	Or	Se	Y

*Human- & Technology-assisted

**Technology-based only

Or - Possible diversity	SS – Security/Safety
& – Mix between public & private transportation	Se – Security
X - Concerned	Sa – Safety
Y – Yes	Tr – Traceability
N – No	

We are concerned with assisted mobility of humans and goods characterized by specific needs. This mobility can be walk-based or use different kinds of vehicles (car, train, subway, tram, bicycle, scooter, etc.), and assistance can either be fully techno-logical or human- and technology-based. We try to elaborate a classification allowing us to identify main elements and associated solutions. The following characteristics are taken into account (Table 1):

- **Nature of mobility:** pedestrian, by personal vehicle, and by public transportation.
- **Who is moving: humans** (children, adults or elderly persons) possibly with dif-ferent kinds of disabilities,... or **goods** (parcels, letters, shopping (books, clothes, food), construction supplies)
- **Nature of assistance:** human- and/or technology-based
- **Nature of the escort:** from start to end, by segment (with or without continuity)
- **Nature of collaboration:** synchronous or asynchronous escorting
- **Selection process** of the escorting person (identified, pre-recorded, known as col-league or neighbor) in small, medium or large lists.
- **Level of security** to ensure (traceability, security, safety)
- **Kind of reward** ...

In Table 1 we indicate the cases studied, which are explained in this paper in relation with identified characteristics.

3   Assistive System Role and Functions

To provide appropriate services organized in an assistive system, it seems important to identify the main aspects of this system. Globally, ITS (Information & Communication Systems) systems are responsible for taking into account data collection, storage and management, appropriate manipulations in different working situations, providing communication over short and large distances by wireless and wired networks, in mobility, and interacting with the environment in which active and passive things are organized in IoT (Internet of Things).

Two important aspects are HCI (Human-Computer Interaction) and new services based on AI (Artificial Intelligence). The main HCI objectives are related to multi-modal and multi-channel communication with various devices and interaction styles as touch-based and speech-based interactions, as well as AR (Augmented Reality) approaches. These interactions can occur either on desktop computer or with mobile devices allowing individual and collaborative situations. The Mashup approach for implementation of appropriate services is proposed.

From the AI point of view, Big Data and Deep Learning are mainly used to procure appropriate information from the huge amount of data available and to syn-thetize observed behaviors in typical data. An interesting summarization is proposed in Fig. 1 [4].

To organize all these aspects, it seems appropriate to create an **Intermediation platform.** This is an assistive system, the objective of which is the Vitalization of data [5]. Its architecture is summarized in Fig. 2a.

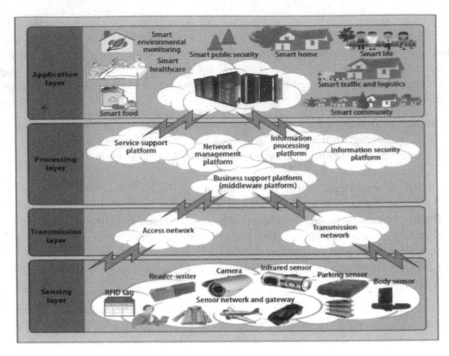

Fig. 1. Smart city architecture summarized by Liu and Peng [4].

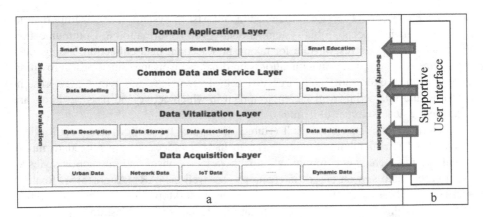

Fig. 2. a/ Architecture of an Intermediation platform with Data Vitalization [5], b/ Supportive User Interface [6].

3.1   Intermediation Platforms

Intermediation platforms connect people, services and even things in ways that have been unthinkable until now. Search engines provide relevant references for people searching for information. Social networks connect users in their environment. Carpooling systems link drivers and passengers whose goal is to take the same routes.

Intermediation platforms use big data to fuel the services they offer. All these services are evolving extremely quickly but are almost unnoticed.

All intermediation platforms essentially rely on the same structure. To begin with, they collect huge amounts of data, which can derive from the outside world (e.g., web pages for search engines) or are hosted by the platform (e.g., social networks). However, they are never produced by the platform itself but rather by the people, services or things around it. These primary data are then indexed and transformed to extract information that fuels the primary services offered.

Users' activities on the platform generate secondary data. These secondary data mainly consist of traces that the platform generally has exclusive rights to use, as well as to create secondary services. A key example of this is the precise profiling of users, which permits personalized and customized services: personal assistants trace users as they go about their day-to-day activities, not only on-line but also in the physical world through the use of geo-localization or quantified-self means.

3.2 Mashup Approach for Intermediation Platform Adaptation

A generic intermediation platform needs to be adapted to different working situations (as for our cases noted in Table 1 and described later). The objective is to provide appropriate collection of data, services and acquisition sensors and valuators, which are able to work together in a vitalization approach and create adapted User Interfaces allowing appropriate interactions. The Mashup approach is an interesting way to support these adaptations. Mashups are defined as "the perspective of software engineering. A Mashup is constructed by the assembly and combination of several existing functions integrated into a new application" [7].

The term "mashup" was defined initially in the field of music, where it consists in remixing two (or more) sounds in order to obtain a new one. Mashup is primarily and usually performed for the so-called "drag&drop" applications from different sources. The Mashup architecture is made up of three elements according to Merrill [8]: Data, Services and User Interface. Mashup aims at the composition of a three-tier application: (1) Data (data integration), (2) Application logic (process integration), and (3) User interface (presentation integration). Integration of heterogeneous data sources uses two main technologies: web services and Mashup. Integration implies that all relevant data for a particular bounded and closed set of business processes are processed in the same software application.

Moreover, updates in one application module or component are reflected throughout the business process logic, with no complex external interfacing. Data are stored once, and are instantaneously shared by different business processes enabled by the software application [9]. In the Mashup, every user can compose his/her own service with other services in order to create a new service. Mashup is a "Consumer Centric Application". It describes web 2.0 sites combining functions of one site with another site. Different pieces of UIs (User Interfaces) are integrated into a new web application. This approach requires composition and orchestration of web services.

Application mashuping in the context of intermediation platforms is a much appreciated approach for creating appropriate applications based on reuse of existing ones. It is based on a four-layer structure (Fig. 2a). To update or adapt data and services, a particular user interface called Supportive UI is used to introduce appropriate adaptations concerning data, services, IoT exchanges and HCI. This supportive user interface, which is also Mashup application creation- or evolution-oriented, proposes a programming interface for experienced developers and a visual programming-oriented approach for experienced users (Fig. 2b). A typical example of mashuping is visual selection of actors as we will see later in this paper.

By mashuping between OpenStreetMap with subway stations and potential actor locations, we create an application in which the choice of the accompanying escort can be manually selected directly on an interactive map. For example, Fig. 3 shows a journey from the Croix-Paquet to Masséna subway stations in Lyon to be completed between 4.30 and 6 pm. All nearby actors around the Croix-Paquet station are shown, and the user can select the right person by checking that her/his starting time and final destination are compatible with the journey.

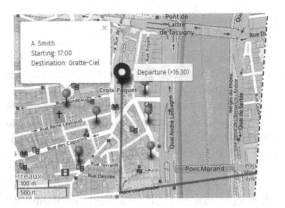

Fig. 3. Mashup between OpenStreetMap and actors' locations.

3.3 Focus on Two Important Services

The first service we wish to explain is the **Collaboration between humans participating in transportation of parcels or escorting situations:** In human- and technology-based assistive systems it is important to determine collaboration organization between participating humans. We identified two main collaboration situations: synchronous and asynchronous.

As we will explain later (Sects. 5 and 6), we studied with students parcel distribution and human escorting based on public city transportation in Beijing, in which we expected to use the subway. The main idea is to use subway users as transporters of parcels or as escorts of people who cannot travel alone. We identified two solutions. The first is based on **synchronous collaboration** (Fig. 4). At the starting station, the "sender" (person who needs to send a parcel or who has a person to escort in the subway) gives responsibility for the parcel or person to the "transporter" and, at the final station, the "transporter" hands over responsibility of the parcel or person to the "receiver". In this case, the process is fully human-oriented without need of techno-logical infrastructure related to public city transportation.

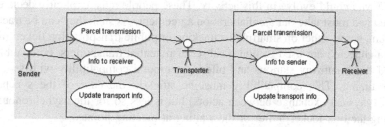

Fig. 4. UML Use-Case Diagram of the synchronous solution

The second solution is based on **asynchronous collaboration** (Fig. 5).

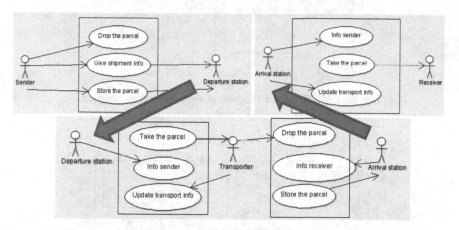

Fig. 5. UML Use-Case Diagram of the asynchronous solution

In this case, the "sender", "transporter" and "receiver" move independently. To coordinate their collaboration, a subway infrastructure in each subway station is pro-vided. For parcels, a box-locker is provided to act as a support of the asynchronous buffer between sender & transporter and transporter & receiver. For human escorting in each subway station, a location called a "meeting point" is provided. At this location, the escorted human can wait for the new accompanying escort, either transporter or receiver. Figure 5 shows the UML Case Diagram of the asynchronous solution based on box-lockers.

Figure 6 shows the Sequence Diagram of the synchronous solution with the parcel transportation and confirmation message exchanges, while Fig. 7 gives the UML Sequence Diagram of the asynchronous solution also in the parcel context. Transpo-sition to human escorting is relatively easy, consisting in replacing lockers with meeting points. Confirmation messages are the same but with semantic changes.

The second service to be explained relates to **Participant selection from a list of potential actors.** In the approach of parcel distribution or human escorting based on public transportation clients, it is mandatory to discover potential travelers ("transporters") who could evolve in this activity. These people must explicitly declare to be interested and must take out a collaboration agreement. As such they can be tracked by the system, based either on smartphone tracking or public transportation ticket tracking. The data collected are used to find a subset of potential transporters, compatible with start and finish subway stations and offering temporal compatibility with sender and receiver timing. This compatibility must be strict in the case of the synchronous solution (physical exchange between actors) but is less so for the asynchronous solution using the box-locker buffer or the waiting at meeting points.

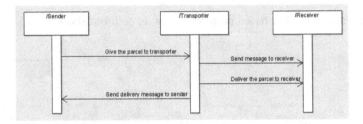

Fig. 6. UML Sequence Diagram of the synchronous solution

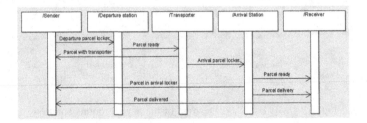

Fig. 7. UML Sequence Diagram of the asynchronous solution

This research of appropriate escorting/transporting persons varies in complexity according to the size of the list in which the selection occurs. Selection in a personal diary or a short list is easy, for example by scrolling. It can be proposed in a visual way (as mentioned earlier and shown in Fig. 3).

The selection process in a large list can become more complicated, and appropriate algorithms must be provided for this intermediation process.

Intermediation between individuals and services is also considered as a **recommendation problem,** that is to say, correct services or items are recommended as a ordered list to users by the intermediation platform. Many algorithms can be applied in intermediation platforms, of which Collaboration Filtering is the earliest and most famous algorithm. It can find users' preferences by mining their historical behavior data, divide users into groups based on different preferences, and recommend products with similar preferences. User-based collaborative filtering algorithms and item-based collaborative filtering algorithms are two principal categories of algorithms.

Both algorithms have the same main idea: calculate the similarities between users (or items), then recommend similar items to similar users. Calculation of similarities is often based on distance calculation or clustering methods. We studied in [10] several matching or clustering algorithms (GN, CNM, LCV, EACH, etc.) and proposed a general clustering algorithm, CDHC (Community Detection based on Hierarchical Clustering). The CDHC Algorithm first creates initial communities from global central vertices, and then expands the initial communities layer by layer according to the link strength of vertices and communities, before finally merging some very small communities into larger communities.

In more detail, in the intermediation process, the platform proposes to use the collaborative filtering method and the CDHC algorithm [10]. Once the public transportation users' data have been collected, we can try to build a complex network of transporters. The problem is then transformed into the detection of communities in which transporters have similar properties.

A city has one or more centers and is expanded layer by layer around the centers. The closer to the center the layer is, the denser the connections to the layer are. Inspired by the hierarchical structure of cities, a community of transporters should have one or more global central nodes, and be expanded layer by layer around the global central nodes. The nodes in layer n are mainly connected by nodes in layer n + 1 and layer n − 1. The connection number of one node with other nodes is known as its degree. Community detection consists in initializing communities, expanding communities, merging small communities and choosing the best result.

In the first step of initializing communities, we first sort all the nodes by degree in descending order, before choosing k nodes with maximal degree as global central nodes. The node with the maximal degree is then assigned to the first community and marked as the first community's central node. For each node of the remaining k − 1 global central nodes, its similarity with each initialized community's central node is calculated. If there is a similarity greater than a given threshold, the node is assigned to the community maximizing the similarity. Otherwise, a new community is initialized, assigned to the new community, and marked as the new community's central node.

After finishing the process of initializing communities, we need to expand these communities. The process of expanding communities consists in marking node level and calculating link strength. All the global central nodes are marked as first level. If a node is connected to the first level and is not yet marked, its level is marked as two. By repeating this we can mark all the node levels, if we assume that the network is connected. Starting from each level-two node, the link strength between the node and each community is calculated. The node is then assigned to the community with the maximal link strength, until the nodes of each level have been assigned to communities.

The communities, in which nodes have strong connections and good similarities, have now been detected. The final step is to merge some small communities into large ones. Different community detection solutions with similarity thresholds could be evaluated by extensive modularity, and the best solution found. Some experiments on peoples' network datasets were conducted, and the result showed that the algorithm is effective [10].

CDHC can help us find compatible transporter groups for trajectory similarity, based on their daily traffic tracked data. During the searching process, if the historical transporters cease to be available, similar transporters are recommended first. They will get the information before the others and have priority to accept the order.

4 Mobility With and Without Specific Needs

As we mentioned earlier, our objective was to study and characterize mobility of humans and goods in the city, distinguishing between classical and new ones. We examine in-depth the role played by assistive tools and systems in both situations, as well as for humans and goods.

4.1 Classical Mobility Situations (Cases #0a, #0b)

Human movement in the city (Case #0a) can use different kinds of transport such as walking, individual transportation such as bike, scooter **or** car and their e-versions. People can also use public transportation. In these situations, they can either move without assistance, in the case of well-known trajectories (everyday paths) or can use multiple tools to prepare, assist and consolidate the path. Tourist websites can be used to prepare the trip, GPS for guidance, mobile websites to find out and follow path trajectories, etc. Mono and multimodal transport trajectory guidance systems are also very interesting tools. Movement using taxi or Uber is also possible. For the latter, assistive tools are used by the driver.

Standard movement of goods (parcels) (Case #0b) is carried out by official transporters such as the Post Office, UPS, etc. services. The senders are exempted to use assistive tools unless for location of Post or UPS office. They can use tracking tools to know at any time where their goods are located during the delivery process. Guarantee of professional delivery can be also assisted by tools or systems, either standard as mentioned earlier (GPS) or more specific.

4.2 New Mobility Situations with Assistive Functions (Cases #1, #2, #3 and #4)

Navigation Assistance for the Blind in Indoor Situations (Case #1). Assistive technology for indoor navigation is very important for blind people, especially when they have to go to new or rarely visited places. Navigation technologies rely primarily on the precise location of the person in order to provide the right navigation instructions at the right time.

Outdoor localization is easy to obtain throughout the planet by using global navigation satellite systems (GNSS) like GPS or Galileo, although this may be imprecise.

However, for indoor localization, GNSS signals are too weak or absent and thus many alternatives are proposed. If the location has WiFi coverage, the signal emitted by the access points can be used for location, but generally accuracy is insufficient for navigation. Alternatively, RFID tags can be incorporated into the soil and a blind man's cane equipped with a RFID reader can get the unique identifier associated with the tag. A similar approach is to use BLE beacons such as iBeacon, which are placed on walls, and to use an application on a mobile phone equipped with a Bluetooth device to retrieve the identifiers. In all cases, it is necessary to establish a precise cartography of the place with the exact locations of the access points, tags or beacons and their identifiers, which will allow localization by the navigation application [11].

A concrete application, Wayfindr [12] proposed by the Royal London Society for Blind People (RLSB's), was successfully tested in the London Underground. When the smartphone application retrieves a signal from a specific beacon, it triggers the playback of an audio message through a bone conduction headphone, to give the user turn-by-turn directions.

Alternatively, Okeena developed a solution called Naviguo+Hifi [13] which is currently being deployed throughout the Parisian metro network. Audio beacons are installed at strategic locations in the subway corridors; these beacons broadcast audio messages when activated on-demand by remote control or smartphone app. The remote control uses the same standard already used by blind people to activate French pedestrian traffic lights.

Walk-bus – escorting groups of children to school (Case #2) is another assistive system application, aimed at creating journeys between a distant starting point and a school, the goal of which is to collect children at well identified stops allowing parents to limit their implication in their children escorting over short distances instead of full trajectories between their home and the school (and back). The walk-bus is managed by identified persons (mainly parents), who are in charge of respecting the journey schedule and collecting new children at identified stops and taking them to the school. The traceability procedure is clearly defined between parents and walk-bus "driver" in relation with organizers (usually school – parents associations). One of the possible solutions is use of a QRCode for each child, which is scanned when the child joins or leaves the Walk-bus. This enables child transportation to be recorded long-term, with precise information on conditions of use (boarding and unboarding locations, dates and times), which can be used for security reasons. These data can also be used for more general studies, such as journey schedules, or for modifying journey profiles.

Dynamic lane management (Case #3) on which we worked [14] aims at creating dynamically lanes reserved for buses in order to accelerate transportation time for public transportation in traffic jam situations. The objective is to create these lanes dynamically only when buses are present and to leave all lanes open to general traffic when buses are not present. As such, general traffic speed is managed. The main technologies used are: A Location-Based Service integrating bus detection sensors; an intermediation platform collecting sensor information and determining dynamic bus section activation and deactivation; in-the-field infrastructure and/or embedded vehicle interface receiving instantaneous information of selected situations (Fig. 8).

Fig. 8. Dynamic lane management [14].

Better and easier parcel delivery to inhabitants or stores (Case #4) is another supportive system on which we worked [15]. The main idea was to propose a reservation system of delivery areas in the city in order to shorten delivery time by better management of delivery stops based on delivery area reservation [15]. In this context, the main technologies used are: Address delivery based on a list of parcels to be distributed, geographical location of delivery areas to be reserved. A trip elaboration algorithm or HCI based tool creating a time integration schedule description with stop delivery locations and delivery addresses. A mobile interaction tablet-based tool allows the driver to modify the trip if a traffic jam problem or other problems occur (Fig. 9).

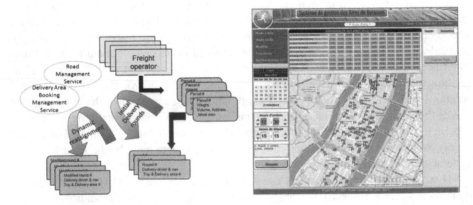

Fig. 9. Parcel distribution based on Delivery area reservation [15].

5 Case Studies for Parcel Distribution

Parcel distribution is a very important activity, time-consuming, and a source of pollution and traffic jams, mainly in large cities. In relation with the last application we continued to work in this field. Large distribution companies very often subcontract to small companies to perform this unprofitable activity. Several reasons account for this non-profitability: access to downtown is complicated by circulation constraints, traffic jam problems, lack of parking availability, as well as the limited presence of receivers at destinations. This problem is increased by the number of distributors with a relatively small number of parcels to distribute, thus generating increased traffic problems in the same district. A variety of solutions were tested in several cities: identification of distributors for each district, thus reducing the number of vehicles entering each district and increasing the volume of distributed parcels. From a collaborative economy point of view, a more original and effective approach, which is carbon-free, is based on the use of existing movements of persons and vehicles to take advantage of these existing trips to have them transport parcels. We propose to examine three of the different working situations that we have identified: the first is based on public city transportation users, the second relies on supermarket clients, while the third is based on a closed network of small craftsmen. The latter situation is based on pedestrian box-lockers to which pedestrians go to collect the goods they have ordered.

5.1 Public City Transportation Based Parcel Distribution (Carbon-Free Parcel Distribution Case [16]) – (Case #5)

In a large city like Paris, Lyon or Beijing, it is possible to imagine high-speed transportation of letters and small parcels by users of this public transportation (buses, trams and subway). The process is based on identification of segments defined by start and destination public transportation stations, identification of start date and time (availability of the parcel at the start station) and expected destination time (availability of the parcel at the destination station). The next step is to find a transporter, a city transportation user, who is "interested" in carrying out this transportation. This requires an available list of potential transporters with their history of movement on the city transportation network. An algorithm can determine potential transporters with appropriate profiles (segment used, in the proposed time interval and transportation conditions such as parcel weight and transportation remuneration). A preliminary contact with this identified transporter is established via the intermediation platform in order to validate this transaction.

In this case, two different solutions can be used for the exchange of parcels at the start and destination stations (see Sect. 3.3). The first (synchronous approach) is based on physical exchange between the persons involved in transportation, between sender and transporter first, then between transporter and recipient. The second solution is based on an asynchronous approach using technological support for parcel exchange based on box-lockers located at each city transportation station and electronic key access, which is shared, respectively, between sender and transporter and transporter and recipient. In this case, there may be a delay between deposit and withdrawal at each extremity of the segment. This solution is efficient and flexible (no need for synchronous presence of two actors at each segment point). This is also important from a security point of view, as it integrates delivery traceability. In the first case, scanning of the parcel by each actor must be introduced to allow its follow-up. In the second case, box-locker handling is charged with scanning the parcel in both cases (In & Out).

These two approaches (synchronous and asynchronous) can either be based on start to end transportation or on splitting the trip into several segments working separately. This is possible if no direct transportation is found.

The asynchronous approach box-locker solution, which may be considered to be expensive and space-consuming, can be replaced by a cheaper solution based on special bags with electronically controlled opening and handling, allowing the bag to be attached to an exchange post located in each station.

5.2   Market Client Based Parcel Distribution (Case #6)

Another possible approach for carbon-free distribution is based on market clients who are neighbors of the client waiting for his/her parcel at home. The starting point is the internet site of the shop. The goods are ordered and purchased by a distant person who is the client and receiver of the corresponding parcels. The internet purchase is connected to the delivery service in the market shop. The process of discovering a transporter who is a neighbor of the ordering client (receiver) can be based on different situations. The first is a totally open situation in which the goal is to find and create dynamically an association by checking the proximity of geographical locations between ordering persons and potential transporting persons. The presence of potential transporting persons can be discovered either by explicit declaration (signaling of presence in the shop or shopping mall) or by contextual location detection (by smartphone localization or other implicit identification). In this case, the intermediation algorithm must be able to take into account a huge amount of data.

In a more collaborative situation, parcel transportation can be an interesting functionality of a neighbor association collaborative system or an intergeneration collaborative system in which interpersonal help, information and cooperation are supported. Through this system, its members have access to a list of members and can either pre-establish potential relationships between the ordering person and the transporting person (one or more) or have access to less accurate information providing a list of potential transporting persons. These data can be supplied to the delivery service of the shop or mall. In this case, the intermediation algorithm works on a limited set of data.

Tracking of goods carriage is important in order to check that delivery is complete. Traceability is also required by scanning at departure and arrival (delivery to receiver), often by smartphone apps.

Assessment of all actors is also essential for managing their participation in subsequent actions. Naturally, unappreciated transporters will be excluded from future matching processes. This also applies to clients whose requirements are considered to be excessive (deadline, weight, etc.).

5.3   Supply Distribution Between Members of a Network of Small Craftsmen (Case #7)

Small craftsmen work on different sites. Their activities mainly consist in studying the work to be performed (discussion with the client), in choosing and purchasing the appropriate items (supplies) in DIY stores, and, when necessary, in transporting them to the work site, where the work can be performed. This intermediate activity of choosing, supplying and transporting is a time-consuming activity which could be reduced by a more appropriate organization. If the craftsman knows exactly what he needs, he can order by Internet and, when necessary, ask a colleague from the network of craftsmen, who is working near their current work site, to transport the objects to them.

In this case, it is also important to find a colleague who is in contact with the same DIY store and who is currently working near their work site.

This situation occurs in closed environments, in most cases a club or an association of local craftsmen. They are interested in a relationship allowing collaboration in large construction or reconstruction projects and mutual assistance in problematic situations. A collaborative system is then naturally used to support these activities. Its extension to shared transportation is a logical step.

In this case the set of potential transporters is more limited, and the potential intermediation algorithm may be less sophisticated and less efficient.

Of course, appropriate authorizations must be established between actors and the relevant DIY store so as to trace all purchases and deliveries.

In all cases, compensation must be decided: reciprocity in carrying, financial compensation (free subway tickets, etc.) or inter-generation assistance association action, etc.

5.4 Pedestrian Drive: Fresh Box-Locker Based Food Distribution [17] (Case #8)

We are currently studying another shared economy support system by transforming a well-known supermarket drive for pedestrians. The supermarket drive allows consumers to order their shopping on-line and then go to the supermarket only to collect the ordered items. The supermarket drive is usually located near the supermarket and is organized to minimize loading time.

The pedestrian drive is a variant of this concept, specially designed for pedestrians. Located in the city and accessible 24 h a day, it allows ordered goods to be collected at any time. It is able to store not only ambient temperature products but also fresh and frozen goods.

This fresh product box-locker can be either owned by a supermarket firm and totally integrated into the ordering, management and logistic process of the firm, or can be managed independently. This case is an interesting support system for shared economy, as the role of the pedestrian drive manager is to manage such use for multiple providers. His/her role is to ensure Internet access for all provider offers and to organize the global supply chain with consumer information on availability of ordered goods in the box-locker. In this way, circular economy as a short circuit of agricultural goods can be supported.

In this case, the intermediation platform is more open than in association-based situations (such as craftsmen associations and intergeneration associations), while not as large as the Airbnb or the BlaBlaCar breeding ground.

6   Case Studies for Mobility of Persons with Special Needs

6.1   Escorting Children in Public Transportation (Case #9)

The case of parcel and letter transportation by public transportation users (Case #5) can be adapted to human escorting of children, elderly and disabled persons.

In the case of escorting children (**Case #9**), the main characteristics are:

- Escorting the full trip (from starting station to final station)
- Synchronous escort, i.e. entire trip
- Choice of escort from a list of carefully selected and identified persons
- Security and safety of the child is mandatory

6.2   Escorting Elderly Persons in Public Transportation (Case #10)

In the case of escorting elderly persons (**Case #10**), the main characteristics are:

- Escorting can be either for the entire trip (from starting station to final station) or by section, i.e. with exchange of escorting person if it is not easy to find someone to escort the entire trip
- Synchronous or asynchronous escort, i.e. exchange of escorting persons in real time or with a waiting period when the elderly person is alone. This is provided, of course, that psychological conditions allow this
- Choice of escort from a list of selected and identified persons is mandatory
- Security of the elderly is mandatory

6.3   Escorting Disabled Persons in Public Transportation (Cases #11)

In the case of disabled persons (**Case #11**), the nature of the disability is taken into account, and the main characteristics are modified, namely:

- Escorting can be either for the entire trip (from starting station to final station) or by section, i.e. with exchange of accompanying person if it is not easy to find someone to escort the entire trip
- Synchronous or asynchronous escort, i.e. exchange of escorting persons in real time or with a waiting period when the disabled person is able to remain alone. This is provided, of course, that psychological conditions allow this.
- Choice of escort from a list of selected and identified persons is mandatory
- Security of the disabled is mandatory

We have not yet discussed **reward conditions**. It is not easy to enact general rules. In each situation, for goods and humans alike, it is important to define clearly what seems appropriate. Financial rewards can be possible, either rewards in kind such as transportation tickets or other compensation. Reciprocity can also be an appropriate solution for transportation of goods between colleagues or neighbors.

7 Application of Deep Learning to Mobility

In several cases it seems appropriate to work on accumulated data and use them for Deep Learning in order to obtain interesting solutions. We can take as examples Walk-bus (Case #2) and Carbon-free goods transportation (Case #5). This treatment can also apply to other cases.

For the Walk-bus, it seems interesting to study the evolution of trip trajectories, taking into account schoolchildren, including the schedule and home addresses of the transported children. The objective is to determine the appropriate trip trajectory and timetable.

For goods transportation by public transportation users, it is important to start with a case-by-case study, before going on to progressively capitalize these data in order to find main transportation trajectories and propose them more quickly.

Deep learning is hugely popular today. The past few decades have witnessed its tremendous success in many applications. Academia and industry alike have competed to apply deep learning to a wider range of applications due to its capability to solve many complex tasks while providing state-of-the-art results [18]. Recently, deep learning has also revolutionized intermediation architectures, providing more opportunities to improve matching performance. Recent advances in deep learning-based intermediation platforms have gained significant attention by overcoming obstacles of conventional models and achieving high recommendation quality. Deep learning catches the intricate relationships within actual data, from abundant accessible data sources such as contextual, textual and visual information.

In most cases, intermediation is deemed to be a two-way interaction between user preferences and service features. For example, matrix factorization decomposes the rating matrix into low-dimensional user/service latent factors. It is natural to construct a dual neural network to model the two-way interaction between users and services. Neural Network Matrix Factorization (NNMF) [19] and Neural Collaborative Filtering (NCF) [20] are two representative works.

The Deep Structured Semantic Model (DSSM) [21] is a deep neural network for learning semantic representations of entities in a common continuous semantic space and measuring their semantic similarities. It is widely used in information retrieval areas and is supremely suitable for service intermediation and recommendation. DSSM projects different entities into a common low-dimensional space, and computes their similarities with a cosine function.

Autoencoder is a very effective deep learning method. Autoencoder can be applied to intermediation platforms in two general ways: (1) using it to learn lower-dimensional feature representations at the bottleneck layer; or (2) filling the blanks of the interaction matrix directly in the reconstruction layer. Almost all the Autoencoder variants, such as denoising autoencoder, variational autoencoder, contractive autoencoder and marginalized autoencoder, can be applied to the recommendation task. AutoRec [22] takes user partial vectors r (u) or service partial vectors r (i) as input, and aims to reconstruct them in the output layer.

Convolution Neural Networks are very powerful processors of unstructured multimedia data with convolution and pool operations. Most CNN based intermediation models utilize CNNs for feature extraction. CNNs can be used for feature representation learning from multiple sources such as image, text, audio, video, etc.

CNN-based collaborative filtering may prove very useful in intermediation. Directly applying CNNs to vanilla collaborative filtering is also viable. For example, He et al. [23] proposed using CNNs to improve NCF, and presented ConvNCF. This uses the outer product instead of the dot product to model user-item interaction patterns. CNNs are applied over the result of the outer product and could capture high-order correlations among embedding dimensions.

Deep learning-based intermediation has proved very effective. However, the key problem for deep learning in different application scenarios is always data. Once data have been sufficiently collected and tagged, some deep learning-based solutions could be studied and applied.

8 Conclusion

In this paper we studied the mobility aspect of the Smart City, which is one of the main issues raised here. After a general introduction, we listed different approaches of mobility for humans and goods, and split them into groups: mobility with and without special needs. We limited our observation of non-assisted mobility to two significant examples for humans and goods, and focused our paper on assisted mobility with special needs. We identified eleven situations (summarized in Table 1), and characterized them by a series of characteristics. We also identified the main services needed, and organized them into an intermediation platform architecture.

Finally, we described the finality of eleven situations identified and gave potential applications of different selection algorithms from simple scrolling to more sophisticated approaches. This was based on graphical selection on transportation maps and Big Data selection in a huge collection of data taken from the general public transportation system. A proposal for the Deep Learning approach is also briefly described, taking into account evolution of walk-bus routes for picking up and returning schoolchildren from/to their homes and schools. We also show how to synthetize all used transportation paths and discover those most frequently used in order to propose them first and foremost.

This work is open to multiple extensions, not only for in-city mobility but also for intercity mobility, which has equally emerged as an interesting area of research.

Acknowledgements. This study is partially supported for Chinese part by the National Key R&D Program of China (No. 2019YFB2102200), the National Natural Science Foundation of China (No. 61977003, No. 61861166002, No. 61872025, No. 61635002), the Science and Technology Development Fund of Macau SAR (File no. 0001/2018/AFJ) Joint Scientific Research Project, the Macao Science and Technology Development Fund (No. 138/2016/A3), the Fundamental Research Funds for the Central Universities and the Open Fund of the State Key Laboratory of Software Development Environment (No. SKLSDE2019ZX-04). For French part main supports were by the French Ministry of Ecology (No. 09MTCV37 and PREDIT-G02 CHORUS 2100527197).

References

1. Caragliu, A., Del Bo, C., Nijkamp, P.: Smart cities in Europe. J. Urban Technol. **18**(2), 65–82 (2011)
2. Yin, C., Xiong, Z., Chen, H., Wang, J., Cooper, D., David, B.: A literature survey on smart cities. Sci. China Inf. Sci. **58**, 100102:1–100102:18 (2015). https://doi.org/10.1007/s11432-015-5397-4
3. Kolski, C.: Human-Computer Interactions in Transport. Wiley, Hoboken (2011)
4. Liu, P., Peng, Z.: China's smart city pilots: a progress report. Computer **47**(10), 72–81 (2014)
5. Xiong, Z.: Smart city and data vitalisation. In: The 5th Beihang Centrale Workshop, 7 January 2012
6. Atrouche, A., Idoughi, D., David, B.: A mashup-based application for the smart city problematic. In: Kurosu, M. (ed.) HCI 2015. LNCS, vol. 9170, pp. 683–694. Springer, Cham (2015). https://doi.org/10.1007/978-3-319-20916-6_63
7. David, B., Chalon, R.: Orchestration modeling of interactive systems. In: Jacko, J.A. (ed.) HCI 2009. LNCS, vol. 5610, pp. 796–805. Springer, Heidelberg (2009). https://doi.org/10.1007/978-3-642-02574-7_89
8. Merrill, D.: (2009). https://www.ibm.com/developerworks/library/x-mashups/index.html
9. Grammel, L., Storey, M.: An end user perspective on mashup makers. University of Victoria Technical Report DCS-324-IR (2008)
10. Yin, C., Zhu, S., Chen, H., Zhang, B., David, B.: A method for community detection of complex networks based on hierarchical clustering. Int. J. Distrib. Sens. Netw. **2015**, 9 (2015). https://doi.org/10.1155/2015/849140
11. Ahmetovic, D., Gleason, C., Kitani, K.M., Takagi, H., Asakawa, C.: NavCog: turn-by-turn smartphone navigation assistant for people with visual impairments or blindness. In: Proceedings of the 13th Web for All Conference, p. 9. ACM (2016)
12. Wayfindr. https://www.wayfindr.net/
13. http://www.okeenea.com/navigueo-hifi-audio-beacon
14. Wang, C., David, B., Chalon, R., Yin, C.: Dynamic road lane management study: a smart city application. J. Elsevier Transp. Res. Part E Logist. Transp. Rev. **89**, 272–287 (2015). https://doi.org/10.1016/j.tre.2015.06.003
15. Patier, D., David, B., Deslandres, V., Chalon, R.: A new concept for urban logistics: delivery area booking. In: Taniguchi, E., Thompson, R.G. (eds.) The Eighth International Conference on City Logistics, Bali, Indonesia (2014). Procedia Soc. Behav. Sci. **125**, 99–110. ISSN 1877-0428
16. David, B., Chalon, R., Yin, C.: Collaborative systems & shared economy (uberization): principles & case study. In: The 2016 International Conference on e-Learning, e-Business, Enterprise Information Systems, and e-Government, IEEE 2016, Las Vegas, Nevada (US), 28 July 2016, pp. 134–140. (2016). HAL: hal-01496630
17. David, B., Chalon, R.: Box/Lockers' contribution to collaborative economy in the smart city. In: 2018 IEEE 22nd International Conference on Computer Supported Cooperative Work in Design (CSCWD 2018), Nanjing (Chine), 11 mai 2018, pp. 802–807 (2018). https://doi.org/10.1109/CSCWD.2018.8465151
18. Paul, C., Jay, A., Emre, S.: Deep neural networks for YouTube recommendations. In: Recsys, pp. 191–198 (2016)
19. Dziugaite, G.K., Roy, D.M.: Neural network matrix factorization, arXiv preprint arXiv:1511.06443 (2015)

20. He, X., Liao, L., Zhang, H., Nie, L., Hu, X., Chua, T.-S.: Neural collaborative filtering. In: Proceedings of WWW 2017, pp. 173–182 (2017)
21. Guo, H., Tang, R., Ye, Y., Li, Z., He, X.: DeepFM: a factorization-machine based neural network for CTR prediction. In: Proceedings of IJCAI 2017, pp. 2782–2788 (2017)
22. Huang, P.-S., He, X., Gao, J., Deng, L., Acero, A., Heck, L.: Learning deep structured semantic models for web search using click through data. In: CIKM, pp. 2333–2338 (2013)
23. He, X., Du, X., Wang, X., Tian, F., Tang, J., Chua, T.-S.: Outer product-based neural collaborative filtering. In: Proceedings of IJCAI 2018, pp. 2227–2233 (2018)

Author Index

Printed in the United States
By Bookmasters